Li–S Batteries

The Challenges, Chemistry, Materials and Future Perspectives

Li–S Batteries

The Challenges, Chemistry, Materials and Future Perspectives

Editor

Rezan Demir-Cakan

Gebze Technical University, Turkey

World Scientific

NEW JERSEY • LONDON • SINGAPORE • BEIJING • SHANGHAI • HONG KONG • TAIPEI • CHENNAI • TOKYO

Published by

World Scientific Publishing Europe Ltd.
57 Shelton Street, Covent Garden, London WC2H 9HE
Head office: 5 Toh Tuck Link, Singapore 596224
USA office: 27 Warren Street, Suite 401-402, Hackensack, NJ 07601

Library of Congress Cataloging-in-Publication Data
Names: Demir-Cakan, Rezan, editor.
Title: Li-S batteries : the challenges, chemistry, materials and future perspectives /
 edited by Rezan Demir-Cakan (Gebze Technical University, Turkey).
Description: New Jersey : World Scientific, 2017.
Identifiers: LCCN 2016049015 | ISBN 9781786342492 (hc : alk. paper)
Subjects: LCSH: Lithium sulfur batteries. | Electric batteries. | Selenium cells. |
 Lithium cells. | Sodium compounds. | Electrochemistry.
Classification: LCC TK2945.L58 L5278 2017 | DDC 621.31/2424--dc23
LC record available at https://lccn.loc.gov/2016049015

British Library Cataloguing-in-Publication Data
A catalogue record for this book is available from the British Library.

Copyright © 2017 by World Scientific Publishing Europe Ltd.

All rights reserved. This book, or parts thereof, may not be reproduced in any form or by any means, electronic or mechanical, including photocopying, recording or any information storage and retrieval system now known or to be invented, without written permission from the Publisher.

For photocopying of material in this volume, please pay a copying fee through the Copyright Clearance Center, Inc., 222 Rosewood Drive, Danvers, MA 01923, USA. In this case permission to photocopy is not required from the publisher.

Desk Editors: Anthony Alexander/Mary Simpson

Typeset by Stallion Press
Email: enquiries@stallionpress.com

Printed in Singapore

Preface

Energy storage is one of the main challenges of our century. Rapid societal and industrial developments dictate energy consumption, which is foreseen to grow by 56% by 2040. However, most energy sources are scarce and cause an obvious environmental impact on climate, with global warming potentially resulting in a temperature increase of 4–6°C by 2100. New storage technologies are therefore needed if more renewable energy sources are to be employed on the electrical grid; likewise, the electrification of the transport industry also requires more economical and higher energy storage systems. The need for energy storage requires a range of solutions, including batteries which are always a choice between energy, power, cost and life cycle capacity.

Lithium-ion (Li-ion) batteries have played a crucial role in the development of such energy storage technologies since their first launch by Sony in 1991. Although today lithium-ion batteries are used in most portable electronic devices and electrical vehicles (EVs), they are capable of delivering a limited gravimetric energy density, 100–200 Wh/kg, which cannot fulfill the goal of replacing combustion engines, while 1 kg of gasoline provides around 44,000 kJ/kg (~12,000 Wh/kg). Considering the differences between the fuel tank size of a combustion engine (around 60 L) and the weight of a battery pack of an electrical vehicle (could reach up to 500 kg), the running time of a car powered by batteries alone is still between 5 and 10 times shorter than with gasoline. Thus, if we want to achieve or even approach the goal of longer driving ranges using battery powered vehicles, we need to explore new batteries which are different from existing Li-ion technology and which offer further innovation in energy storage.

One of the possibilities is lithium–sulfur (Li–S) battery technology, the principle of which has been known for almost five decades, but which until now has failed to be commercialized. Theoretically, Li–S batteries can meet all the requirements of the intelligent vehicle battery system since they possess a high volumetric and gravimetric energy density. The advantage of Li–S batteries in terms of capacity and price in comparison with Li–ion technology is clear and must be carefully investigated in order to provide evidence for market adoption.

With extensive research in the past decade, the field is evolving rapidly, with successes in the laboratory in terms of long cycling stability and high rate performances. However, the commercial application of Li–S batteries is still facing barriers of the complexity of manufacturing thick electrodes, utilization of high sulfur, minimizing electrolyte/sulfur ratio or use of Li metal anode.

This first book provides the reader with an excellent review and analysis of the current rechargeable Li–S battery research. Starting with a brief history of Li–S batteries and underlining the challenges, this book introduces the past and recent improvements for each cell compartment; namely the sulfur cathode, the Li metal anode and the electrolyte which are equally responsible for the limited performance of modern Li–S cells. Moreover, new designs of Li–S batteries with a Li metal-free anode, catholyte concept as well as characterization tools for fundamental understanding and modeling of Li–S batteries have been addressed.

The authors of this book are strongly engaged in research for the progress of Li–S batteries and concentrate on the fundamental understandings of the cells while attacking the challenges found therein. As the editor, I would like to acknowledge all of the authors for their outstanding contributions to this book.

Rezan Demir-Cakan,
Istanbul, August 2016

Contents

Preface		v
About the Editor		ix
Chapter 1	Introduction to Rechargeable Lithium–Sulfur Batteries *Rezan Demir-Cakan*	1
Chapter 2	Sulfur Cathode *Elton J. Cairns and Yoon Hwa*	31
Chapter 3	The Use of Lithium (Poly)sulfide Species in Li–S Batteries *Rezan Demir-Cakan, Mathieu Morcrette and Jean-Marie Tarascon*	105
Chapter 4	Lithium–Sulfur Battery Electrolytes *Patrik Johansson, Rezan Demir-Cakan, Akitoshi Hayashi and Masahiro Tatsumisago*	149
Chapter 5	The Lithium Electrode Revisited through the Prism of Li–S Batteries *Marine Cuisinier and Brian D. Adams*	195
Chapter 6	Analytical Techniques for Lithium–Sulfur Batteries *Manu U.M. Patel and Robert Dominko*	275

Chapter 7	Other Sulfur Related Rechargeable Batteries: Recent Progress in Li–Se and Na–Se Batteries *Rui Xu, Tianpin Wu, Jun Lu and Khalil Amine*	309
Chapter 8	Computational Modeling of Lithium–Sulfur Batteries: Myths, Facts, and Controversies *Alejandro A. Franco*	335
Chapter 9	Conclusion: Challenges and Future Directions *Elton J. Cairns*	351
Index		355

About the Editor

Rezan Demir-Cakan graduated from Yildiz Technical University, Chemical Engineering Department and obtained her master degree from the same university. After spending one year in Cambridge/UK, she moved to Berlin/Germany to start her PhD work. She received her PhD at the Max Planck Institute of Colloids and Interfaces in the group of Professor Markus Antonietti in 2009. Thereafter, she joined Laboratoire de Réactivitéet Chimie des Solides in Amiens/France as a postdoctoral researcher working mainly on lithium–sulfur batteries under the supervision of Professor Jean-Marie Tarascon and Dr. Mathieu Morcrette.

Most of the studies of Dr. Demir-Cakan were devoted to energy related subjects. She received several prestigious awards including "the Japan Carbon Award for PhD Student" offered by the Japan Carbon Society in 2008, "Young Investigator Award" in 2012 from IMLB in South Korea, Science Academy's "Young Scientist Award" in 2015, "the L'Oréal-UNESCO For Women in Science" in 2016.

Since August 2012, Rezan Demir-Cakan has been working as an Associate Professor at the Gebze Technical University in Turkey running her own battery research group in Turkey which incorporates the application of materials chemistry towards sustainable chemical and energy technologies.

demir-cakan@gtu.edu.tr

Chapter 1

Introduction to Rechargeable Lithium–Sulfur Batteries

Rezan Demir-Cakan

*Department of Chemical Engineering,
Gebze Technical University, 41400 Gebze, Turkey
demir-cakan@gtu.edu.tr*

1.1. Introduction and History of Lithium–Sulfur Batteries

1.1.1. Introduction

Rapid societal development and economic growth dictates the use of energy sources which are mostly based on fossil fuels. However, fossil fuels are neither continuous nor environmentally benign, leading to negative consequences in the environment caused by human energy needs. It is projected that the carbon dioxide (CO_2) emissions will cause global warming resulting in a 4–6°C temperature raise by 2100.[1] Likewise, fossil fuel depletion has been identified as a future challenge since it is foreseen that oil reserves will be terminated within the next 40 years while coal and natural gas may last at most for another 150 years.[2] Hence, researchers are responsible for realizing new ideas on how to exploit renewable energy resources such as wind, water or sun in the most efficient manner without causing any further ecological calamities. However, most renewable energy sources are intermittent; thus, energy storage is one of the essential components of the forthcoming energy supply system to make use of renewable energy sources with fluctuating power output.[3]

Technological demands for higher capacity and cost-effective energy storage options provide an incentive for exploration of alternative electrochemical energy storage systems. Technologies that can provide economic growth as well as CO_2 emission-free transportation by replacing the internal combustion engines with electric traction should be highlighted. Providing required flexible electricity generation and demand between daytime and night are highly important for grid operation.

New energy politics has led to the energy storage subject especially battery technologies becoming an important topic due to the development of mobile applications (i.e. electrical vehicles (EVs) or cellular phones). Over the past 25 years, Li-ion batteries (LIBs) played a crucial role in the development of such energy storage technologies. Today LIBs are used in 90% of rechargeable portable electronic devices. Although great improvements have been accomplished, and while active research for further developments continue, current Li-ion technologies provide a limited gravimetric energy density (100–150 Wh/kg for a full system) which cannot compete with either fulfilling the goal of replacing combustion engines or meeting large mass energy storage backup dictated by solar farms or wind turbine plants. For instance, EVs powered by LIBs result in an inadequate driving range (160–200 km) (Figure 1.1). Therefore, more drastic approaches are necessary to go beyond this limit and to accomplish "The Holy Grail" of a 500 km driving range at low cost, without the need for hybridization with conventional internal combustion engines.

Earth abundant sulfur is one alternative to reach such goals. Lithium–sulfur (Li–S) batteries offer around three fold increase in energy density compared with present Li-ion technologies. Li–S battery configuration operating at room temperature represents a valuable option as it can provide low equivalent weight, high capacity (1672 mAh/g), low cost (about $150 per ton) and environmentally benign factors. All these characteristics cannot be accomplished with current Li-ion technology.

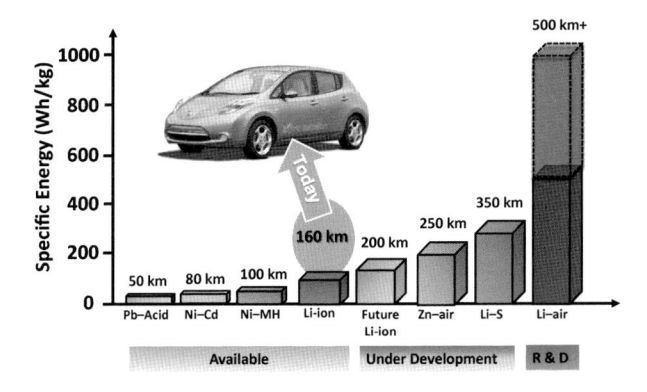

Figure 1.1 Practical, specific energies for some rechargeable batteries, along with estimated driving distances.
Source: Reproduced with permission from Ref. 1.

1.1.2. History of Li–S batteries

The history of Li–S chemistry dates back to early 1960s (even prior to the advent of rechargeable Li batteries).[4] With the patented work of Herbert and Ulam in 1962, sulfur was proposed as positive electrode and Li (or alloy of Li) as negative electrode in electric dry cells and storage batteries. Electrolyte was identified to be alkaline or alkaline-earth perchlorate, iodide, sulfocyanide, bromide, or chlorate dissolved in a primary, secondary or tertiary saturated aliphatic amine. Four years later, Herbert filed another patent,[5] which was a continuation in part of their previous patent,[4] with the electrolyte solution preferably consisting of a selected Li salt dissolved in a propyl, butyl or amyl amine. Preferably isopropyl amine was utilized as the solvent. In the same year, Rao[6] patented high-energy density metal–sulfur batteries. Electrolyte consisted of cations of light metals or ammonium ions and anions of tetrafluoroborate, tetra-chloroaluminate, perchlorate or chloride salts which were dissolved in organic solvents. The solvents were propylene carbonate, y-butyrolactone, NzN-dirnethylformamide or dimethylsulfoxide and the cells were cycled between the voltages 2.52 and 1.16 V vs. Li. Later on, in 1970, Moss and Nole,[7] represented a patent for the battery employing Li and sulfur electrodes with non-aqueous electrolyte. More information regarding the patent landscape of Li–S batteries, with analyzed and categorized total of 760 patent families, can be found in Ref. 8.

Although the concept of Li–S batteries is not new and was already intensively researched, the topic was inhibited because of missing exploitable results of the early studies.[9] Following the pioneering work published by Nazar et al.[10] in 2009, the topic was revisited and attracted drastic research interest in the field. The subsequent development of Li–S rechargeable batteries is enormous and has very high publication dynamic. The literature reports of research and review papers database from the Web of Science have shown results of over 2500 papers containing the phrase "lithium sulfur batteries" (Figure 1.2(a)), papers with more than 70,000 citations, showing the importance of this field of research (Figure 1.2(b)).

A detailed analysis of the literature studies for Li–S batteries and their topic distribution can be seen in Figure 1.3. Most of the works were devoted to the design of host matrices for sulfur impregnation and formulation of cathode composite electrode (detailed information can be found in Chapter 2). Besides all these attempts to control polysulfide dissolution with the help of different cathode architectures, recently many efforts have been performed to find suitable and effective adsorption/absorption of polysulfides within the composite cathode.

Apart from these confinement strategies, some reports have proven that those polysulfides are beneficial for the cell life. Xu et al.[11] have showed that the self-healing of Li–S batteries could be developed in the presence of polysulfide containing electrolyte by creating a dynamic equilibrium between the dissolution and precipitation of Li polysulfides at the electrode interfaces. Thus, research in the field of Li–S batteries is slightly moving from those sulfur confinements to the use of chemically synthesized dissolved polysulfides either in static condition or redox flow configuration.[12–14] Alternatively, those polysulfides are even employed as electrolyte additives for improved cycling performances.[11,15,16] More information regarding the use of polysulfide, either as electroactive component or electrolyte additives can be found in Chapter 3.

An important progress during this period was the identification of the electrolyte formulation suitable for the Li–S batteries. Many solvent/salt combinations were suggested including, sulfone based,[10,14] dimethyl sulfoxide,[17] dimethylformamide,[18] dimethoxyethane,[19] tetraethylene glycol dimethoxyethane,[20] ionic liquid,[21] mixture of those mentioned solvents at different ratios,[22] PEO polymer electrolytes,[23] or dioxolane[24] which is now

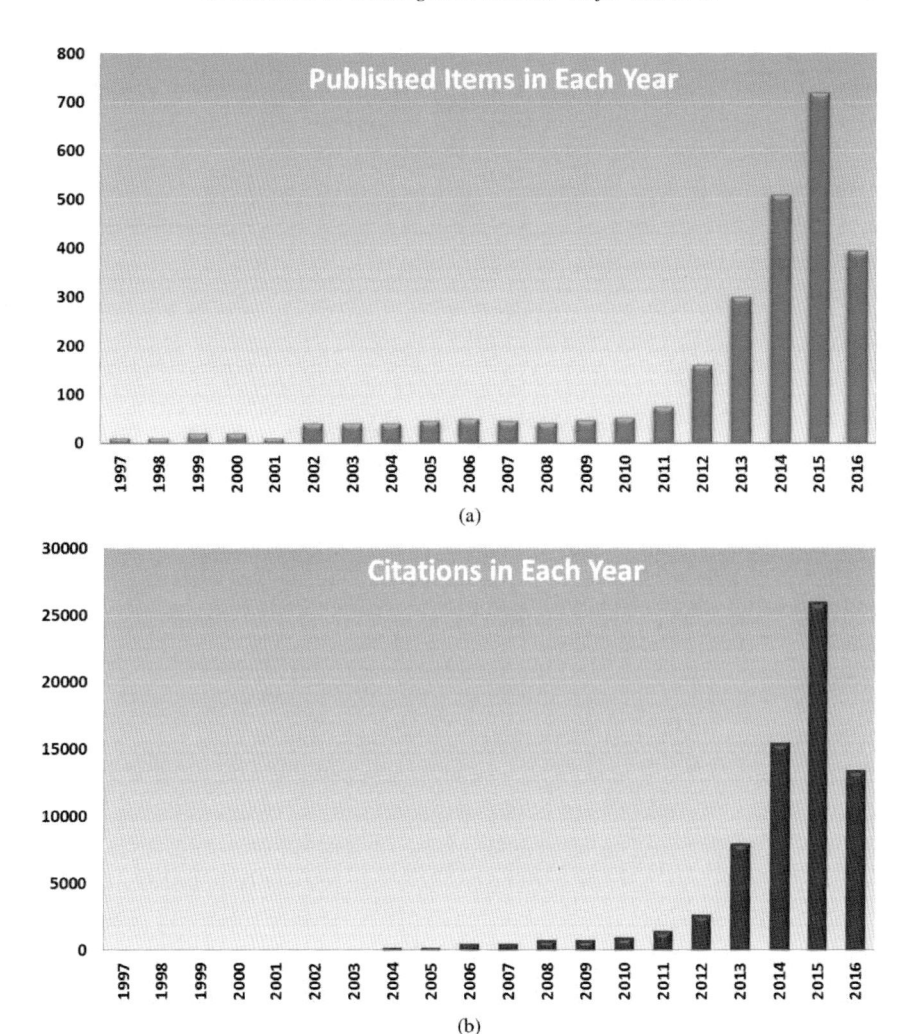

Figure 1.2 (a) Number of publications on the topic of Li–S batteries and (b) their citations over the past 20 years. Last updated in July 2016. Not subject to copyright.

one of the most used solvents in the field of Li–S batteries. For safety reasons and also to bypass the severe limitations lying on the progressive dissolution of polysulfides in liquid electrolytes, the concept of solid state electrolyte was also proposed, e.g. $Li_2S-P_2S_5$,[25] Li_2S-SiS_2,[26] thio-LISICON.[27] Solvent-in-salt, ultrahigh salt concentration, approach was also suggested for better control of the dissolution of polysulfides and

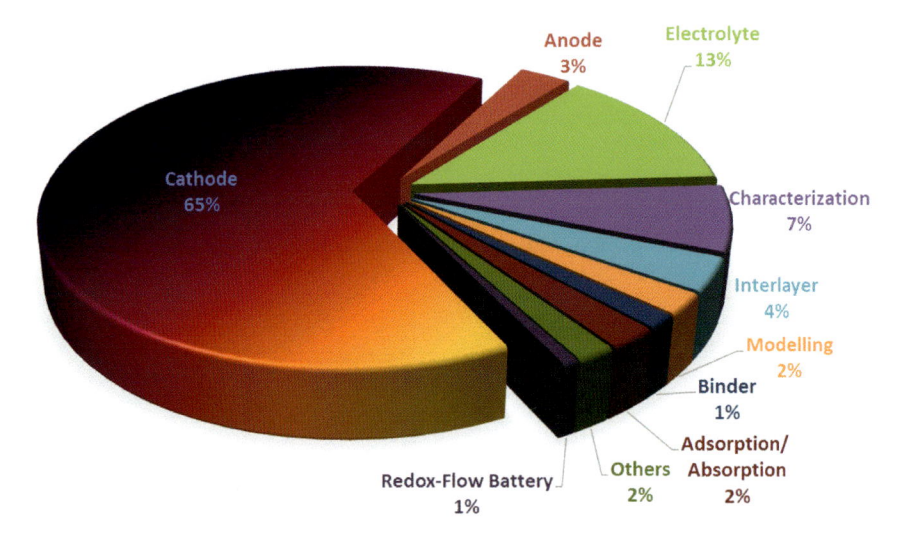

Figure 1.3 Literature distribution of Li–S batteries. Last updated in July 2016. Not subject to copyright.

safety issues of Li–S batteries via an effective suppression of Li dendrite growth.[28] Detailed information can be found in Chapter 4.

As of other metallic Li based batteries, protection of Li anode is highly important in the field of Li–S batteries. In order to protect the Li anode surface from the polysulfide species, one of the options is to use $LiNO_3$ additive which was already proven in the ether solvent to help form a uniform solid electrolyte interphase (SEI) that prevents parasitic reactions between polysulfide and lithium.[29] Certainly, apart from the strategies to protect Li surface by electrolyte additives or the use of ceramic or polymer membranes which were discussed above, another strategy would be the deep investigation of electrolyte solvents. Li anode surface chemistry in Li–S batteries still remains a mystery. Thus, at this stage a question that remains open concerns the effect of polysulfides, salts and solvents on the Li surface which is not the purpose of this chapter. A deep understanding of the surface growth mechanism and the investigation of the impact of the polysulfide species on Li anode are discussed in detail in Chapter 5.

The Li–S system has indeed a complex chemistry including many components and species in the cell, thus, its design necessitates a comprehensive approach by using different characterization tools in order to understand how changes in one of the components influence its electrochemical performances. During the past few years, several research groups embarked to gain a fundamental understanding of the operation and limitation of Li–S batteries and solving some of the problematic issues of this technology. For instance, *in situ* cyclic voltammetry (CV) was proposed by using the "4-electrode modified Swagelok cell" that can quantitatively determine the amount of polysulfides dissolved in the electrolyte by using CV.[30] The use of a 4-electrode cell by means of *in situ* CV measurements clearly demonstrated that the quantity of dissolved soluble polysulfides inside the cell allowing the choice for the best type cathode architectures or electrolyte composition at the first cycling enables faster determination of fading mechanism related to the polysulfides dissolution. For the qualitative measurements, sulfur discharge mechanism was studied by *in situ* UV–Vis spectroscopy in which the authors were able to monitor the formation of long-chain polysulfide to shorter chain during discharge (*vice versa* during charge).[31] Later on, Barchasz *et al.*[32] performed a comprehensive study to understand the Li–S battery discharge mechanism by using a combined characterization tool; HPLC, EPR, and UV–Vis spectroscopy. A sulfur discharge mechanism and the electrolyte composition at different discharge potentials were investigated in tetraethylene glycol dimethyl ether (TEGDME)-based electrolyte. *In situ* X-ray absorption spectroscopic studies were also studied by several research groups.[33–36] Gao *et al.*[33] investigated the impact of the different electrolyte composition in which they found that the electrochemical performance of the Li–S batteries are highly dependent on the type of solvents rather than the Li salts used in the electrolyte. Additionally, they elucidated the viscosity of the ethereal solvents; while soluble polysulfides remained more oxidized in viscous ethereal solvents, low viscosity resulted in a more complete reduction of soluble polysulfides. Detailed information can be found in Chapter 6.

Inspired by the Li–S and Li–oxygen chemistry, recently elemental Se which is in the same elemental group as oxygen and sulfur in the periodic table with the atomic number 34, $[Ar]4s^2 3d^{10} 4p^4$, has recently been proposed as a cathode material for rechargeable batteries. The theoretical gravimetric capacity of Se is 678 mAh/g. Se-based cathodes can be cycled to high voltages (up to 4.6 V) without failure, which is not the case for sulfur.

Li–Se battery configuration was first proposed by Amine's group.[37–39] The initial work with Se cathode resulted in a reversible capacity of ~500 mAh/g over 100 cycles. Detailed information can be seen in Chapter 7.

Recently, modeling of the Li–S cell has been considered as a new research line in the field, since it is highly important to precisely predict the cell life, performance and state-of-health. However, due to the complexity of the Li–S cell chemistry, achieving these goals is not a trivial task. Following the first model developed in 2004 by Mikhaylik and Akridge[40] to investigate the shuttle mechanism of the Li–S cell and in 2008 by Kumaresan et al.[41] for more sophisticated cell model based on the formative electrode porous theory, several studies have also been pursued which can be found in Chapter 8.

1.2. Electrochemistry and Basic Reaction Mechanism of Li–S Batteries

A conventional Li–S cell is composed of a sulfur cathode, a Li anode and a liquid electrolyte placed in between. At the open circuit voltage (OCV, Φ_{oc}), due to the difference between the electrochemical potentials of the Li anode (μ_a) and the S cathode (μ_c), the Li–S cell ensures a maximum voltage (Figure 1.4(a)). Upon reduction (discharge), molecules of elemental sulfur (S_8) are reduced by accepting electrons which leads to the formation of high-order Li polysulfides Li_2S_x ($6 < x \leq 8$) at the upper plateau (2.3–2.4 V vs. Li). As the discharge continues, further polysulfide reduction takes place progressively stepping down voltage to 2.1 V (vs. Li) and lower order Li polysulfides chains Li_2S_x ($2 < x \leq 6$) are formed. There are two discharge plateaus at 2.3 and 2.1 V with ether-based liquid electrolytes, which represent the conversions of S_8 to Li_2S_4 and Li_2S_4 to Li_2S, respectively.[42] At the end of the discharge, Li_2S is formed, which is both electronically insulating and insoluble in the electrolyte. Apart from Li_2S, the rest of sulfur reduction species are highly soluble in aprotic solvents.

The reduction process is accompanied by a decrease in cathode electrochemical potential (Figure 1.4(b)) until the battery reaches the terminal voltage (Φ, normally ≤ 1.5 V).[43] The opposite reaction (oxidation, charging) arises when an external electric field with a certain potential difference is applied, leading to the decomposition of Li_2S to Li and S. During this process, the cathode electrochemical potential gradually increases till

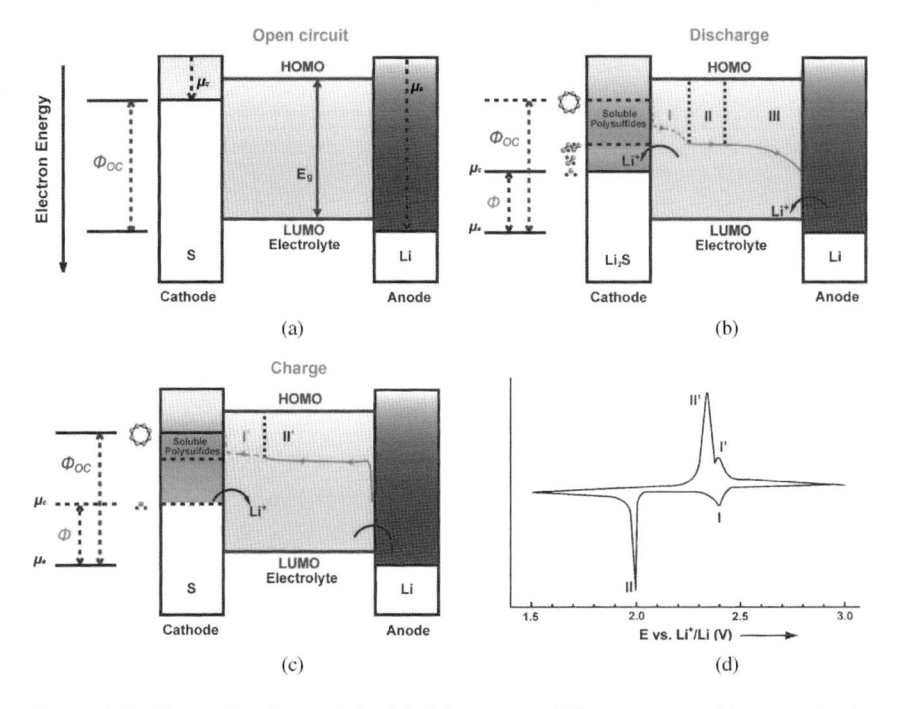

Figure 1.4 Electrochemistry of the Li–S battery at different stages: (a) open circuit. (b) discharge process. (c) charge process. (d) Typical cyclic voltammogram (CV) for an S cathode in a Li–S battery.

Source: Reproduced with permission from Ref. 43.

the battery voltage returns to Φ_{oc} (Figure 1.4(c)). The redox process displays two pairs of redox peaks, corresponds well with the CV of the Li–S battery (Figure 1.4(d)). The overall redox couple described by the reaction $S_8 + 16Li \leftrightarrow 8\,Li_2S$ lies at an average voltage of approximately 2.2 V vs. Li. This potential is around 30% less than that of conventional cathode materials in LIBs. However, the lower potential is not detrimental and compensates by its high theoretical capacity, which makes sulfur the highest energy density solid cathode material.

1.3. Energy Density of Li–S Batteries

Energy density states the energy in Watt-hours per unit mass of a battery. Calculation of an approximate energy density of a battery uses the following formula:

Energy density (Watt-hours) = Nominal Battery Voltage (V) × Capacity (Ampere-hour).

To calculate the gravimetric or volumetric energy density (Wh/kg or Wh/L), energy term needs to be divided by mass or volume, respectively.

It should be noted that energy density of all types of Li-based batteries is limited by the cathode materials.[44] For instance in the field of LIBs, even utilizing of post Li-ion cathode electrode material such as high capacity layered oxides or high voltage spinel (i.e. 4.8 V $LiMn_{1.5}Ni_{0.5}O_4$), the energy densities can only be enhanced to a few tens of percent compared to existing state of the art systems.[44] Besides that, those high voltage cathodes need to ensure the electrolyte stability as well as safety issues.

Apart from the engineering approaches to have lighter cell design leading to higher energy densities, for instance, by the use of extremely thin current collectors, light packaging foil, less electrolyte or thinner separator, a maximal cathode theoretical specific capacity up to 300 mAh/g can be achieved which cannot reach an energy density higher than 150–200 Wh/kg for an EV battery based on the best Li-ion cells.[44] Considering that standard electric cars have intrinsic limitations for the dimension and the weight of a battery, they can approximately carry several hundreds of kilograms with the maximum 200 liter volume. Thus, their driving range is limited to around 200 km between each charge (Figure 1.1).

When it comes to Li–S batteries, the Gibbs energy of the Li–S reaction is about 2600 Wh/kg. The situation of Li–S batteries in terms of specific capacity and potential (V) compared to different cathode materials for Li-ion technology is presented in the Table 1.1. Theoretically, sulfur can deliver a high specific capacity of 1675 mAh/g and a specific energy of 2567 Wh/kg, assuming complete conversion between S_8 and Li_2S. Its practical energy density is also projected to be two to three times higher than that of state-of-the-art LIBs (180 Wh/kg). As sulfur is much less expensive than the typical components of other battery systems, the Li–S technology starts with a lower material cost than Li-ion or Li-polymer batteries. Manufacturing techniques for Li–S batteries are very similar to those used in other battery chemistries.[45]

Recently, Hagen et al.[9] made a statistical Li–S literature review with 274 research papers and discussed important electrode cell design such as

Table 1.1: Properties for different cathode materials for Li-ion and Li–S batteries, not subject to copyright.

Cathode material	Theoretical gravimetric capacity (mAh/g)	Experimental capacity (mAh/g)	Theoretical volumetric capacity (mAh/cm^3)	Average voltage (V)	Level of development
LiCoO$_2$	274	130–150	1363	3.8	Commercialized
LiNi$_{0.33}$Co$_{0.33}$Mn$_{0.33}$O$_2$	280	150–170	1333	3.7	Commercialized
LiNi$_{0.8}$Co$_{0.15}$Al$_{0.05}$O$_2$	279	180–200	1284	3.7	Commercialized
LiMn$_2$O$_4$	148	100–120	596	4.1	Commercialized
Li$_2$MnO$_3$	458	180–190	1708	3.8	Research
LiFePO$_4$	170	160–165	589	3.4	Commercialized
LiCoPO$_4$	167	110–130	510	3.24	Research
Sulfur	1675	200–1400	2567	2.1	Research

sulfur fraction, sulfur load, the electrolyte/sulfur ratio and electrode thickness. They compared the results with a Panasonic NCR18650B cell which provides a capacity of ≈3.3 Ah with a nominal voltage of 3.6 V resulting in a gravimetric energy density of ≈240 Wh/kg and a volumetric energy density of ≈670 Wh/L. After disassembly of the cell inside a glove-box, they weighted the mass of electrochemically passive compartments of the cells. This information enabled the calculation of theoretical energy densities of future Li–S cells with 18650 cell form. According to their results, for superior gravimetric energy densities of a Li–S cell than the state-of-the-art, mass-produced LIBs, comparable volumetric energy densities and attractive cell costs can be obtained only if these following criteria are fulfilled: (i) the sulfur load must be ≥6 mg/cm^2, (ii) the sulfur fraction ≥70%, (iii) the sulfur utilization ≥80%, and (iv) the electrolyte/sulfur weight ratio 3:1 or less. Similar electrolyte/sulfur ratio findings were also reported from another contribution by Hagen et al.[46] Additionally, it should be noted that if optimal quantities of Li (two fold excess) and electrolyte are used for cell formulation, the typical cell life might result in dry-out of the cell.

Since manufacturing of thick electrodes with high sulfur loading (see Section 1.4.6) and lowering the electrolyte/sulfur weight ratio down to 3 still remain challenging, currently one of the biggest advantages of Li–S batteries could be their potential low cell prices.[9] According to the US Advanced Battery Consortium, a Li-ion battery pack price of $150 kW/h was targeted resulting in a probable cell price of around $100 kW/h.[1,9] Li–S cells are therefore considered particularly interesting due to the abundance of the sulfur active material and its very low initial cost which was estimated around ≈$70 kW/h for a 18650 Li–S cell.[9]

Currently, 250 Wh/kg energy densities have been achieved and reported with 80 cycles which is equivalent to volumetric energy densities of around 325 Wh/L.[9] The Li–S cell manufacturers, Sion Power and Oxis Energy expect that future Li–S cells will have a volumetric energy density comparable to that of state-of-the-art Li-ion cells (≈700 Wh/L) but more than twice the gravimetric energy density with values of 400–600 Wh/kg. Moreover, it was reported that Li–S cell had good performances at low temperature.[47] Thus, these cells have potentials for niche applications such as for unmanned aerial vehicles, space batteries or military purposes which have less requirements or long number of cycles.

1.4. Drawbacks and Key Issues of Lithium–Sulfur Batteries

Rechargeable Li–S batteries which interested the battery community for many years[4] have strong potentials to play an important role in the next generation sustainable road transport. However, the Li–S battery technology faces several drawbacks leading to a poor life cycle that hinders the practical application of the battery. Each cell compartment, namely the sulfur cathode, the metallic Li anode and the electrolyte is equally responsible for the limited performance of Li–S cells.

1.4.1. The insulating nature of sulfur

Sulfur, as well as its discharge products, is electronically insulating. Thus, the low conductivity of sulfur (5.0×10^{-30} S/cm) always requires close contact with conductive additives. In the field of Li–S batteries, research efforts were directed towards solving the challenges associated with the insulating nature of sulfur. Peled et al.[48] first described the use of conductive carbon skeletons to encapsulate the active sulfur and to create more efficient electronic contact. Then, the biggest accomplishment was from the group of Nazar who developed the concept of infiltration of sulfur into ordered porous carbon materials.[10] Two issues were simultaneously improved: (i) conductivity (due to the carbon conductive host) and (ii) soluble polysulfides trapping into the pores of conductive skeletons. Thus, different porous carbon hosts produced by hard template approaches,[49] heteroatom (sulfur, boron or nitrogen) doped carbons,[50–52] core–shell carbon/sulfur composites,[53] graphene-wrapped sulfur particles,[54,55] graphene oxides,[56] hollow carbon nanofiber,[57] carbon nanotubes[58] were already suggested (detailed information can be found in Chapter 2). Although most of these strategies proved to be feasible in slowing down the diffusion of soluble polysulfides, they do not represent viable solutions for long cycle time due to the unavoidable migration of soluble polysulfides since negatively charged polysulfides anions can easily be dragged toward Li anode side under an electrical field.

1.4.2. Formation of dissolved species and the shuttle effect

As was discussed above, one of the biggest issues in Li–S rechargeable batteries is the formation of Li polysulfides (Li$_2$S$_x$, $x \geq 2$) upon sulfur reduction which are soluble in the aprotic solvents. Since those polysulfide species generated during the battery operation are readily soluble in the electrolytes, they are prone to migrate from the cathode electrode into the separator and are deposited on the Li electrode resulting in loss of active material. In order to protect Li anode from the polysulfide poisoning effect, there are several strategies studied in the literature (detailed information can be found in Chapter 5). For instance, ceramic or polymer membranes with the all solid state configuration have been employed, which are discussed in Chapter 4. Additionally, electrolyte additives, (i.e. LiNO$_3$) are mostly used in ethereal solvent enabling a SEI type layer on the Li electrode which prevents parasitic reactions between polysulfide and Li.[29]

Another biggest problem related to the formation of soluble species is the "shuttle effect." Schematic illustration of the polysulfide shuttle mechanism is shown in Figure 3.1(b) of Chapter 3. The shuttle effect is explained by the free migration of the extremely soluble high or middle range of polysulfides Li$_2$S$_x$ ($4 < x \leq 8$) between the cathode and anode. Thus, long and middle chain polysulfides move from the sulfur cathode to the Li anode in which they are further reduced to the shorter length of polysulfides. Successively, shorter lengths of polysulfides diffuse back to the cathode electrode where they are further oxidized. As a result, a cyclic process, called shuttle mechanism, occurs that corresponds to a chemical shortcut of the cell. Unambiguous shuttle behavior is capable of resulting in an infinite recharge and poor charge efficiency. Fundamentally, the shuttle behavior is the one of the main reasons for low Coulombic efficiency of the Li–S batteries.[42] To investigate the degree of the polysulfide shuttle behavior Mikhaylik and Akridge used the simulated and experimental data.[40] In their assessment, the charge shuttle factor (f_c) can be calculated from Equation (1) in which I_c is the charge current, k_s is the shuttle constant (heterogeneous reaction constant), q_H is the specific capacity of sulfur at the upper plateau (419 mAh/g, which is a quarter of the full theoretical capacity of Li–S batteries), and [S_{total}] is the total sulfur concentration.[32,40,42]

$$f_C = \frac{k_s q_H [S_{total}]}{I_C}. \quad (1)$$

Equation (1) represents that when the $f_C < 1$ approaching zero, either the charge current should be high enough or the shuttle constant should be low, then the cell could be charged completely without having a shuttle behavior. In contrast, when the $f_C > 1$, the cell never reaches complete charge and the charge curves become horizontal. Sulfur oxidation conditions could proceed markedly which demonstrates its high overcharge protection.

Additionally, Equation (1) shows that the shuttle behavior is also proportional to the active material concentration. If the rate of reduction of high polysulfides on the Li anode surface is directly proportional to their concentration in the electrolyte, the high-order polysulfide concentration normalized to a certain cell volume or surface [S_H] in the upper plateau can be presented by Equation (2) in which t is the time, I is the charge or discharge current, q_H is the sulfur specific capacity related to the high voltage plateau, and k_s is the heterogeneous reaction constant or shuttle constant.

$$\frac{d[S_H]}{dt} = \frac{I}{q_H} - k_S [S_H]. \quad (2)$$

The general differential equation of Equation (2) can be written as

$$\frac{I - q_H k_S [S_H]}{I - q_H k_S [S_H^\circ]} = e^{-k_S t}, \quad (3)$$

where, [S_H°] is the upper plateau polysulfide concentration at $t = 0$. When oxidation reaction starts with the least reduction species of sulfur (Li$_2$S),

$[S_H^\circ]$ equals to zero and this reaction can be expressed as Equation (4) in which t_C is the charging time at the upper plateau.

$$[S_H] = \frac{I_c}{k_S q_H}(1 - e^{-k_S t_C}). \tag{4}$$

Thus, if $k_S t_C \gg 1$, the upper plateau capacity (q_H) depends on the concentration of the long-chain polysulfide. The shuttle constant k_S can be obtained when small charge currents are applied by measuring the charge capacity output from the upper plateau. As a result, the lower the shuttle constant the battery system possesses, the slighter the shuttle effect.[42]

A great amount of work is being currently devoted to the engineering of better host carbon matrices since the high solubility of polysulfides has historically been considered unfavorable in Li–S batteries since they cause a well known "shuttle" mechanism resulting in internal self-discharge and active material loss. However, those polysulfides were recently subjected for the use of cathode active materials or electrolyte additives which deviated from these approaches and explored the use of an "all liquid" system with dissolved polysulfides.[12,13] The use of dissolved polysulfides (so called catholytes) rather than sulfur impregnated composites further eliminates the formation of Li_2S within the pore structure and leads to superior performance of the conventional Li–S cell.[14] Moreover, Demir-Cakan et al.[14] discovered that the performance could be further enhanced when sulfur was directly contacted and deposited on the Li negative electrode. Even though not fully understood, such an improvement holds great promises (further information can be found in Chapter 3).

The use of electrode additives was also employed to immobilize the soluble species on the cathodic side. By introducing heteroatoms to the carbon matrices or adding metal oxides/sulfides to the sulfur cathodes, long-term cycling stability and good rate performances have been achieved.[59–64] Due to the ineffectiveness of physical interactions of nonpolar carbons with polar Li polysulfides, researchers have explored highly modified carbon surfaces in order to better adsorb the dissolved species.[65] For instance, the most studied material includes N-doped carbon[66–68] or graphene, graphene oxides were also highly used to trap dissolved species.[69–72] In addition to that MXene phases[60] or Ti_4O_7[64,73] have been reported to provide a sulfiphilic and conductive surface for enhanced redox of dissolved Li polysulfides.[65] Semiconductors metal oxide or metal organic frameworks were also suggested to adsorb sulfur reduction species, such as SiO_2,[74,75] TiO_2,[59,76] VO_x,[77] or MnO_2.[61,78]

1.4.3. Formation of insoluble species (Li_2S)

Another issue is rooted from the least sulfur discharge species, Li_2S. Contrary to the rest of the Li polysulfide species, Li_2S_x ($8 > x \geq 2$), Li_2S is insoluble in aprotic solvents and it is as well electronically insulating which increases the internal resistance of the battery. This increases results in a large polarization that reduces the energy efficiency of the battery. Moreover, Li_2S might further precipitate on the composite matrix as non-conductive slabs resulting in poor contact between electrolytes and electrodes.[14,29]

Additionally, sulfur reduction following the formation of Li_2S results in volume variations. Due to the different densities of sulfur (2.07 g/cm³) and Li_2S (1.66 g/cm³), there is around 80% calculated volume expansion. Such a Li_2S formation within the pores might pulverize and lose their electrical contacts with the conductive substrate or the current collector.[43,79] This problem is crucial for practical Li–S batteries, in which the drastic volume variation of the electrodes can lead to serious safety problems. One solution is the use of solubilizing additives to reduce or eliminate the product resistivity problem. The impact of V_2O_5[79] and P_2S_5[80] was shown to promote the dissolution of Li_2S.

1.4.4. Self-discharge phenomenon

As discussed, polysulfides are readily soluble in most of the aprotic solvents resulting in undesirable spontaneous reaction between dissolved polysulfide and the highly reactive metallic Li metal anode which is the source of the well-known "shuttle reaction." As long as a Li–S cell starts discharging, the electroactive polysulfides gradually dissolve and migrate to the anode due to the concentration gradient which continues even when the cell is at open circuit voltage so to say at resting state. This process is called self-discharge which is also known in the field like nickel–cadmium or traditional nickel–metal hydride batteries.[42] At OCV, the increase in polysulfide concentration at the sulfur electrode and reaction of sulfur via the shuttle mechanism causes a potential lessening and capacity of the cell which resembles the effect of discharging the battery by applying a current.[81]

The level of self-discharge phenomenon with the help of combination of mathematical modeling and experiments was first quantitatively determined by Mikhaylik and Akridge.[40] They made the correlation between the redox shuttle and capacity as well as charge efficiency (Equation (5)). In that work they developed the concept of the "charge-shuttle factor constant", a relationship between the upper plateau capacity and resting time to quantify the rate of self-discharge under rest conditions.

$$\frac{dlnq_H}{dt} = -k_S. \tag{5}$$

This equation demonstrates that the self-discharge behavior still closely correlates with the shuttle constant, which means that both the shuttle effect and the self-discharge effect originate from the dissolution of the active material in the Li–S battery system.[42]

Later on, Ryu et al.[82] investigated changes in the open-circuit voltage (OCV) and discharge capacity with storage time. After 30 days of storage, they observed that the OCV has fallen from 2.48 to 2.16 V (vs. Li) and the discharge capacity has decreased from 1206 to 924 mAh/g which was related to the corrosion of the stainless steel current collector.

More recently, several groups have been working on the understanding of self-discharge phenomenon and tackling it by designing different cell architectures.[81,83–90]

Al-Mahmoud et al.[81] demonstrated an experimental-based model that the rate of self-discharge was markedly affected by the distance between the sulfur and Li electrodes. The authors also showed that the rate of self-discharge was noticeably affected by the carbon content of the electrode, since the change of the electrode potential due to the increase in polysulfide concentration was buffered by the capacitive behavior of carbon. Since the self-discharge phenomenon is directly linked with the shuttle reaction, Moy et al.[90] quantified the redox shuttle at different state-of-charge by measuring the current passing through the Li–S cells. They demonstrated that at the maximum state-of-charge in which the concertation of long-chain polysulfide is higher, the shuttle current had an uppermost value. On the other hand, when the $LiNO_3$ electrolyte additive was used, an order of lower magnitude current compared with the cells without $LiNO_3$ was obtained.

Similar judgment was also presented by Lacey et al.[83] in which the redox shuttle and self-discharge behavior in electrolytes containing $LiNO_3$ was studied through semiquantitative measurements of polysulfides in the electrolyte when the cells were at resting conditions. They concluded that an improved understanding of the redox shuttle and self-discharge processes as well as the use of appropriate measurements to assess medium- and long-term charge storage was essential to the development of this system for wider practical application.

The problem of self-discharge behavior correlated with the shuttle reaction could rationally be tackled by minimizing the solubility of dissolved polysulfide species in the electrolyte as well as elimination of the polysulfide reaction with Li anode. Those strategies could be the use of polymer of ceramic membrane or use of interlayers to protect Li and promoting the formation of a favorable SEI by different electrolyte additives. A recent common approach is also to use adsorption of polysulfide by adding heteroatom doped carbonaceous materials or metal oxides which can interact with negatively charged polysulfides.

1.4.5. The use of Li anode

The use of metallic Li anode in Li–S batteries also suffers from several problems. To start with, the Fermi energy of Li is higher than the low unoccupied molecular orbital of the commonly used liquid electrolytes (Figure 1.4(a)).[91] Consequently, there is a continuous formation of a SEI layer due to the electrolytes deduction on the surface of the Li anode which results in a notable irreversible capacity loss and low deposition efficiency of Li upon charging. For practical Li–S batteries, this condition lessens the energy output of the system by requiring an excessive amount of Li to pair with S cathode.[43]

Another issue is originated from the poisoning of Li surface by dissolved polysulfide species. Chapter 5 indicated the solutions for the protection of Li metal anodes in rechargeable Li–S batteries. An imminent solution is to use Li-ion conducting glasses or membranes that segregate the active metal from detrimental side reactions.[92–97] An alternative method is to replace sulfur with its lithiated form, lithium sulfide (Li$_2$S). Thus, Li anode can be bypassed with tin, silicon or metal oxide anodes which enable Li-free anodes. The cell is called Li-ion sulfur battery[44] which was first demonstrated by the group of Scrosati[23,98] and Cui.[99] However, the use of Li$_2$S also brings some issues; low electronic conductivities, sensitivity to moisture and oxygen, limited synthesis routes and high over potential oxidation at the first cycle.[42,44]

At the Li anode side, as in any Li metal battery, the other main problem is known to be the potential dendrite formation during the charging process which can cause significant safety problems. However, some reports have shown that the Li–S system is exceptional, that dendrites do not develop during cycling.[14] It should be noted that chemical reaction of soluble polysulfides with metallic Li is directly linked with the amount of polysulfide and electrolyte ratio which is another problem of Li–S batteries. Continuous degradation of Li and consumption of electrolyte or in the forms of dissolved polysulfides require an excess of both components if we want to achieve a long cycle life from a Li–S cell. However, the excess of Li and electrolyte needs to be carefully tuned since they directly lower the energy density. Anyhow, further deep studies and knowledge are required to fully exploit the phenomenon.

1.4.6. Challenges of thick sulfur electrodes

As was described before, sulfur loads below 2 mg/cm^2 cannot compete with state of the art commercialized Li-ion technology, even at sulfur utilizations of 90%, which at present are far from being achieved at high currents. At slightly lower sulfur utilization (i.e. 70%) those calculated masses should be above 6 mg/cm^2.[29] On the other hand, in the majority of publications in the literature, the sulfur loading is well below 2 mg/cm^2 resulting in less than or equal to 1 mAh/cm^2. Thus, to achieve high-energy density Li–S cells, a cathode with high sulfur loading or so to say ensuring thick electrodes is essential.[45]

According to Hagen et al.[9] 1 mg/cm^2 sulfur loading requires an electrode thickness of 25 μm and in order to compete with the state-of-art LIBs one needs to have a sulfur loading of around 6 mg/cm^2 resulting in a thickness of 150 μm. Since the sulfur content in the final electrode and the mass loading strongly affect the overall performance of the Li–S cells, Lv et al.[45] represented the issues of the high-energy density Li–S batteries focusing on the challenges of thick sulfur electrode. The authors used integration synthesis approach by assembling nanosized amorphous carbon nanoparticles into micron-sized particles. Sulfur loading of 2–8 mg/cm^2 was tested with further controlling of the rolling pressure applied on the electrode.

However, obtaining thick electrode would not be straightforward by adopting the knowledge gained by the thin electrodes and results would not be applicable when the cathode thickness is sufficiently increased. The significant thickness difference between the thick electrodes aimed to be used by the companies and those thin-film electrodes used at the research laboratories might result in many uncertainties.[9] Xiao[100] has explained those cell performance differences between the thick and thin electrodes with three criteria. First, as the electrodes become thicker, meaning that more sulfur is packed onto the current collector, the total mass transport changes significantly. Accordingly, the concentration gradient of the dissolved polysulfide species and their spatial distribution throughout the electrode structure and electrolyte largely changes, leading to fundamentally different diffusion pathways and reaction kinetics. A second issue is associated with the Li anode. When the aerial loading of sulfur is increased to a high level, Li metal also undergoes dramatically exacerbated Li stripping and redeposition processes, which may gradually become the dominant reason for cell failure. When a thin-film electrode is used, anode side problem is relatively minor due to less contact with polysulfides which results in passivating of the Li. Thus, the anode issues become much more pronounced when coupled with a thick cathode. Possible solutions could be the optimization of the cell design promoting less contact between Li and polysulfide species, thus, polymer electrolytes or ceramic membranes could be suggested. Third issue comes from the size of the carbon/sulfur composites of the think electrodes. Since downsized nanoparticles do facilitate Li$^+$ transport by shortening the diffusion path, thick electrodes might oblige to use nanosized particles. However, large scale synthesis of nanoparticles could be the problem for the companies as we observed from the Li-ion technologies that commercially available high-energy LIBs continue to use micrometer-sized particles.

Additional issues could be the wettability problem of the electrolytes inside the thick electrodes which brings supplementary kinetic difficulties to the cell. Moreover, as was discussed before, the shuttle behavior is proportional to the active material concentration. Thus, the shuttle effect and self-discharge phenomenon would be much more pronounced in the case of thick electrodes with more aerial loading of sulfur.

1.5. Key Players in the Field

The purpose of this chapter is not to represent the entire patent analysis and product survey on Li–S batteries. A detailed patent landscape of Li–S batteries can be found in Ref. 8.

Here, in this subchapter, the idea is to give a flavor of research actions of the leading companies in the field. The information given here is only based on their patent application since most of the companies do not disclose their research actions.

USA-based company, Sion power, is one of the global leaders in the development of Li–S batteries. Their research actions date back to the early 1990s and since then they have been very active in filing patents for almost all the cell compartments as well as deep investigation of the sulfur chemistry. Some of their patents are as follows; Li anodes for electrochemical cells,[101] investigation of electrolytes[102] or separation of electrolytes with dual phase concept,[103] carbon/sulfur cathode composites[104] and separators.[105] In 2012, BASF acquired an equity ownership position in privately held Sion Power. Thus, by this joint venture they aimed to accelerate the commercialization of Sion Power's proprietary Li–S battery technology for electric and plug-in electric vehicles and other high-energy applications over the next decade.

UK-based Oxis Energy has been in this field since 2004 and holds many patents, covering mainly the positive electrode,[106] additive in the electrolyte[107] and anode made of Li metal and intercalation materials.

Both Sion Power and Oxis Energy projected that future Li–S cells will have a volumetric energy density comparable to that of state of the art Li-ion cells (~700 Wh/L) but more than twice the gravimetric energy density with values of 400–600 Wh/kg. Prototype Li–S cells obtain 350 Wh/kg, for the initial cycles, 250 Wh/kg after 80 cycles, and volumetric energy densities of around 325 Wh/L.[9]

Samsung SDI started their Li–S research in the beginning of 2000. Their patents cover electrolytes (both in organic electrolyte and aqueous electrolyte),[108] positive electrodes[109] and binder[110] for Li–S batteries.

Bosch is also very active in filing patents. They are mainly working on cathode composites such as carbon nanofiber/sulfur composites.[111]

Poly Plus is also one of the pioneers in the field back in 1996. They have started with cathode electrode[112] and later on they disclosed a patent on electrolyte solvents for ambient-temperature Li–S batteries.[113] The disclosed solvents include at least one ethoxy repeating unit compound, examples of linear solvents include the glymes ($CH_3O(CH_2CH_2)_nCH_3$). Some electrolyte solvents include donor or acceptor solvent in addition to an ethoxy compound. While donor solvents assist in solvation of Li ions, acceptor solvents include alcohols, glycols, and polyglycols which assist in solvation of the sulfide and polysulfide anions.

It should be emphasized that this subchapter part provided only brief information with a few examples of key companies and startups in the field. Indeed, it is becoming increasingly difficult to provide a full overview of this emerging field due to the intensive work that is being carried out by hundreds of research groups.

1.6. Conclusion

Here, in this chapter, a brief summary of the Li–S battery research actions aimed at expanding attractive Li–S battery systems that have a practical prospect is provided.

Research on Li–S batteries dates backed to the early 1960s, even prior to Li-ion rechargeable batteries. However, the topic regained its importance only a couple of years ago and since then a great deal of research and development contributions has flooded the literature. The field is definitely evolving and great performances were achieved at the laboratory scale in terms of cycling. In contrast, comparing energy density achievements and ranking them between the different approaches is terrifying due to the diversity in reporting data. The field will greatly gain by defining a unique way to quote numbers taking into account amount of sulfur loading, electrode surface, thickness, and so on. Whatever, the main issue presently remaining with the Li–S system is the need to operate in excess of electrolyte for achieving good cycling performances. This still inflicts a great penalty in terms of energy density to the Li–S system to the point that it cannot compete with Li-ion.

Although there is still a huge gap between the practical realization of this system and the literature reports, we do hope that the Li–S system has a great potential to fulfill the future requirements to reach "The Holy Grail" of a 500 km driving range at low cost.

Acknowledgment

The author would like to sincerely acknowledge the financial support of the Gebze Technical University and the Scientific and Technological Research Council of Turkey (TUBITAK) (Contract nos: 213M374 and 214M272). The author also acknowledges the continuous support from Dr. Mathieu Morcrette and Prof. Jean-Marie Tarascon.

Bibliography

1. P. G. Bruce, S. A. Freunberger, L. J. Hardwick and J.-M. Tarascon, *Nature Materials*, **11** (2012) 19–29.
2. M. Hook and X. Tang, *Energy Policy*, **52** (2013) 797–809.
3. S. Evers and L. F. Nazar, *Accounts of Chemical Research*, **46** (2013) 1135–1143.
4. D. Herbert and J. Ulam, US Patent (1962) No: **3043896**.
5. D. Herbert, US Patent (1966) **US3248265 A**.
6. M. L. B. Rao, US Patent (1966) **3,413,154**.
7. V. Moss and D. A. Nole, US Patent (1970) **US3532543 A**.
8. Retrieved from http://www.researchandmarkets.com/reports/2974405/lithium-sulfur-batteries-patent-landscape.
9. M. Hagen, D. Hanselmann, K. Ahlbrecht, R. Maca, D. Gerber and J. Tubke, *Advanced Energy Materials*, **5** (2015) 1401986.
10. X. Ji, K. T. Lee and L. F. Nazar, *Nature Materials*, **8** (2009) 500–506.
11. R. Xu, I. Belharouak, J. C. M. Li, X. Zhang, I. Bloom and J. Bareno, *Advanced Energy Materials*, **3** (2013) 833–838.

12. R. D. Rauh, K. M. Abraham, G. F. Pearson, J. K. Surprenant and S. B. Brummer, *Journal of the Electrochemical Society*, **126** (1979) 523–527.
13. S. S. Zhang and J. A. Read, *Journal of Power Sources*, **200** (2012) 77–82.
14. R. Demir-Cakan, M. Morcrette, Gangulibabu, A. Gueguen, R. Dedryvere and J.-M. Tarascon, *Energy & Environmental Science*, **6** (2013) 176–182.
15. S. Chen, F. Dai, M. L. Gordin and D. Wang, *RSC Advances*, **3** (2013) 3540–3543.
16. Y. Yang, G. Zheng, S. Misra, J. Nelson, M. F. Toney and Y. Gui, *Journal of the American Chemical Society*, **134** (2012) 15387–15394.
17. M. V. Merritt and D. T. Sawyer, *Inorganic Chemistry*, **9** (1970) 211–215.
18. A. Evans, M. I. Montenegro and D. Pletcher, *Electrochemistry Communications*, **3** (2001) 514–518.
19. S. Kim, Y. J. Jung and H. S. Lim, *Electrochimica Acta*, **50** (2004) 889–892.
20. H. S. Ryu, H. J. Ahn, K. W. Kim, J. H. Ahn, K. K. Cho, T. H. Nam, J. U. Kim and G. B. Cho, *Journal of Power Sources*, **163** (2006) 201–206.
21. L. X. Yuan, J. K. Feng, X. P. Ai, Y. L. Cao, S. L. Chen and H. X. Yang, *Electrochemistry Communications*, **8** (2006) 610–614.
22. V. S. Kolosnitsyn, E. V. Karaseva, D. Y. Seung and M. D. Cho, *Russian Journal of Electrochemistry*, **39** (2003) 1089–1093.
23. J. Hassoun and B. Scrosati, *Angewandte Chemie International Edition*, **49** (2010) 2371–2374.
24. E. Peled, Y. Sternberg, A. Gorenshtein and Y. Lavi, *Journal of the Electrochemical Society*, **136** (1989) 1621–1625.
25. A. Hayashi, T. Ohtomo, F. Mizuno, K. Tadanaga and M. Tatsumisago, *Electrochimica Acta*, **50** (2004) 893–897.
26. N. Machida, K. Kobayashi, Y. Nishikawa and T. Shigematsu, *Solid State Ionics*, **175** (2004) 247–250.
27. T. Kobayashi, Y. Imade, D. Shishihara, K. Homma, M. Nagao, R. Watanabe, T. Yokoi, A. Yamada, R. Kanno and T. Tatsumi, *Journal of Power Sources*, **182** (2008) 621–625.
28. L. Suo, Y.-S. Hu, H. Li, M. Armand and L. Chen, *Nature Communications*, **4** (2013) 1481–1490.
29. D. Aurbach, E. Pollak, R. Elazari, G. Salitra, C. S. Kelley and J. Affinito, *Journal of the Electrochemical Society*, **156** (2009) A694–A702.
30. R. Dominko, R. Demir-Cakan, M. Morcrette and J.-M. Tarascon, *Electrochemistry Communications*, **13** (2011) 117–120.

31. M. U. M. Patel, R. Demir-Cakan, M. Morcrette, J.-M. Tarascon, M. Gaberscek and R. Dominko, *ChemSusChem*, **6** (2013) 1177–1181.
32. C. Barchasz, F. Molton, C. Duboc, J.-C. Lepretre, S. Patoux and F. Alloin, *Analytical Chemistry*, **84** (2012) 3973–3980.
33. J. Gao, M. A. Lowe, Y. Kiya and H. D. Abruna, *The Journal of Physical Chemistry C*, **115** (2011) 25132–25137.
34. T. A. Pascal, K. H. Wujcik, J. Velasco-Velez, C. Wu, A. A. Teran, M. Kapilashrami, J. Cabana, J. Guo, M. Salmeron, N. Balsara and D. Prendergast, *Journal of Physical Chemistry Letters*, **5** (2014) 1547–1551.
35. K. H. Wujcik, J. Velasco-Velez, C. H. Wu, T. Pascal, A. A. Teran, M. A. Marcus, J. Cabana, J. Guo, D. Prendergast, M. Salmeron and N. P. Balsara, *Journal of the Electrochemical Society*, **161** (2014) A1100–A1106.
36. M. U. M. Patel, I. Arcon, G. Aquilanti, L. Stievano, G. Mali and R. Dominko, *Chemphyschem*, **15** (2014) 894–904.
37. A. Abouimrane, D. Dambournet, K. W. Chapman, P. J. Chupas, W. Weng and K. Amine, *Journal of the American Chemical Society*, **134** (2012) 4505–4508.
38. C. Luo, Y. H. Xu, Y. J. Zhu, Y. H. Liu, S. Y. Zheng, Y. Liu, A. Langrock and C. S. Wang, *ACS Nano*, **7** (2013) 8003–8010.
39. C. P. Yang, S. Xin, Y. X. Yin, H. Ye, J. Zhang and Y. G. Guo, *Angewandte Chemie International Edition*, **52** (2013) 8363–8367.
40. Y. V. Mikhaylik and J. R. Akridge, *Journal of the Electrochemical Society*, **151** (2004) A1969–A1976.
41. K. Kumaresan, Y. Mikhaylik and R. E. White, *Journal of the Electrochemical Society*, **155** (2008) A576–A582.
42. A. Manthiram, Y. Fu, S.-H. Chung, C. Zu and Y.-S. Su, *Chemical Reviews*, **114** (2014) 11751–11787.
43. Y. X. Yin, S. Xin, Y. G. Guo and L. J. Wan, *Angewandte Chemie International Edition*, **52** (2013) 13186–13200.
44. A. Rosenman, E. Markevich, G. Salitra, D. Aurbach, A. Garsuch and F. F. Chesneau, *Advanced Energy Materials*, **5** (2015) 1500212.
45. D. P. Lv, J. M. Zheng, Q. Y. Li, X. Xie, S. Ferrara, Z. M. Nie, L. B. Mehdi, N. D. Browning, J. G. Zhang, G. L. Graff, J. Liu and J. Xiao, *Advanced Energy Materials*, **5** (2015) 1402290.
46. M. Hagen, P. Fanz and J. Tubke, *Journal of Power Sources*, **264** (2014) 30–34.
47. Y. V. Mikhaylik and J. R. Akridge, *Journal of the Electrochemical Society*, **150** (2003) A306–A311.

48. E. Peled, A. Gorenshtein, M. Segal and Y. Sternberg, *Journal of Power Sources*, **26** (1989) 269–271.
49. S.-R. Chen, Y.-P. Zhai, G.-L. Xu, Y.-X. Jiang, D.-Y. Zhao, J.-T. Li, L. Huang and S.-G. Sun, *Electrochimica Acta*, **56** (2011) 9549–9555.
50. Z. Ma, S. Dou, A. Shen, L. Tao, L. Dai and S. Wang, *Angewandte Chemie International Edition*, **54** (2015) 1888–1892.
51. F. Schipper, A. Vizintin, J. Ren, R. Dominko and T.-P. Fellinger, *ChemSusChem*, **8** (2015) 3077–3083.
52. F. Chen, J. Yang, T. Bai, B. Long and X. Zhou, *Electrochimica Acta*, **192** (2016) 99–109.
53. C. Wang, J.-J. Chen, Y.-N. Shi, M.-S. Zheng and Q.-F. Dong, *Electrochimica Acta*, **55** (2010) 7010–7015.
54. H. Wang, Y. Yang, Y. Liang, J. T. Robinson, Y. Li, A. Jackson, Y. Cui and H. Dai, *Nano Letters*, **11** (2011) 2644–2647.
55. Y. Cao, X. Li, I. A. Aksay, J. Lemmon, Z. Nie, Z. Yang and J. Liu, *Physical Chemistry Chemical Physics*, **13** (2011) 7660–7665.
56. L. Ji, M. Rao, H. Zheng, L. Zhang, Y. Li, W. Duan, J. Guo, E. J. Cairns and Y. Zhang, *Journal of the American Chemical Society*, **133** (2011) 18522–18525.
57. L. Ji, M. Rao, S. Aloni, L. Wang, E. J. Cairns and Y. Zhang, *Energy & Environmental Science*, **4** (2011) 5053–5059.
58. Y.-S. Su and A. Manthiram, *Chemical Communications*, **48** (2012) 8817–8819.
59. S. Evers, T. Yim and L. F. Nazar, *The Journal of Physical Chemistry C*, **116** (2012) 19653–19658.
60. X. Liang, A. Garsuch and L. F. Nazar, *Angewandte Chemie International Edition*, **54** (2015) 3907–3911.
61. X. Liang, C. Hart, Q. Pang, A. Garsuch, T. Weiss and L. F. Nazar, *Nature Communications*, **6** (2015) 5682–5689 5682–5689.
62. X. Liang, C. Y. Kwok, F. Lodi-Marzano, Q. Pang, M. Cuisinier, H. Huang, C. J. Hart, D. Houtarde, K. Kaup, H. Sommer, T. Brezesinski, J. Janek and L. F. Nazar, *Advanced Energy Materials*, **6** (2016) 1501636.
63. X. Liang and L. F. Nazar, *ACS Nano*, **10** (2016) 4192–4198.
64. Q. Pang, D. Kundu, M. Cuisinier and L. F. Nazar, *Nature Communications*, **5** (2014) 4759–4767.
65. Q. Pang and L. F. Nazar, *ACS Nano*, **10** (2016) 4111–4118.
66. J. Song, M. L. Gordin, T. Xu, S. Chen, Z. Yu, H. Sohn, J. Lu, Y. Ren, Y. Duan and D. Wang, *Angewandte Chemie International Edition*, **54** (2015) 4325–4329.

67. Q. Pang, J. Tang, H. Huang, X. Liang, C. Hart, K. C. Tam and L. F. Nazar, *Advanced Materials*, **27** (2015) 6021–6028.
68. J. Song, T. Xu, M. L. Gordin, P. Zhu, D. Lv, Y.-B. Jiang, Y. Chen, Y. Duan and D. Wang, *Advanced Functional Materials*, **24** (2014) 1243–1250.
69. F.-F. Zhang, X.-B. Zhang, Y.-H. Dong and L.-M. Wang, *Journal of Materials Chemistry*, **22** (2012) 11452–11454.
70. G. Zhou, L.-C. Yin, D.-W. Wang, L. Li, S. Pei, I. R. Gentle, F. Li and H.-M. Cheng, *ACS Nano*, **7** (2013) 5367–5375.
71. C. Wang, K. Su, W. Wan, H. Guo, H. Zhou, J. Chen, X. Zhang and Y. Huang, *Journal of Materials Chemistry A*, **2** (2014) 5018–5023.
72. G. Zhou, L. Li, C. Ma, S. Wang, Y. Shi, N. Koratkar, W. Ren, F. Li and H.-M. Cheng, *Nano Energy*, **11** (2015) 356–365.
73. X. Tao, J. Wang, Z. Ying, Q. Cai, G. Zheng, Y. Gan, H. Huang, Y. Xia, C. Liang, W. Zhang and Y. Cui, *Nano Letters*, **14** (2014) 5288–5294.
74. X. Ji, S. Evers, R. Black and L. F. Nazar, *Nature Communications*, **2** (2011) 325–331.
75. R. Demir-Cakan, M. Morcrette, F. Nouar, C. Davoisne, T. Devic, D. Gonbeau, R. Dominko, C. Serre, G. Ferey and J.-M. Tarascon, *Journal of the American Chemical Society*, **133** (2011) 16154–16160.
76. Z. W. Seh, W. Li, J. J. Cha, G. Zheng, Y. Yang, M. T. McDowell, P.-C. Hsu and Y. Cui, *Nature Communications*, **4** (2013) 1331–1336.
77. K. T. Lee, R. Black, T. Yim, X. Ji and L. F. Nazar, *Advanced Energy Materials*, **2** (2012) 1490–1496.
78. Z. Li, J. Zhang and X. W. Lou, *Angewandte Chemie International Edition*, **54** (2015) 12886–12890.
79. R. Demir-Cakan, *Journal of Power Sources*, **282** (2015) 437–443.
80. Z. Lin, Z. C. Liu, W. J. Fu, N. J. Dudney and C. D. Liang, *Advanced Functional Materials*, **23** (2013) 1064–1069.
81. S. M. Al-Mahmoud, J. W. Dibden, J. R. Owen, G. Denuault and N. Garcia-Araez, *Journal of Power Sources*, **306** (2016) 323–328.
82. H. S. Ryu, H. J. Ahn, K. W. Kim, J. H. Ahn, J. Y. Lee and E. J. Cairns, *Journal of Power Sources*, **140** (2005) 365–369.
83. M. J. Lacey, A. Yalamanchili, J. Maibach, C. Tengstedt, K. Edstrom and D. Brandell, *RSC Advances*, **6** (2016) 3632–3641.
84. M. L. Gordin, F. Dai, S. R. Chen, T. Xu, J. X. Song, D. H. Tang, N. Azimi, Z. C. Zhang and D. H. Wang, *ACS Applied Materials & Interfaces*, **6** (2014) 8006–8010.

85. N. Azimi, Z. Xue, N. D. Rago, C. Takoudis, M. L. Gordin, J. X. Song, D. H. Wang and Z. C. Zhang, *Journal of the Electrochemical Society*, **162** (2015) A64–A68.
86. J. Q. Huang, T. Z. Zhuang, Q. Zhang, H. J. Peng, C. M. Chen and F. Wei, *ACS Nano*, **9** (2015) 3002–3011.
87. M. Kazazi, M. R. Vaezi and A. Kazemzadeh, *Ionics*, **20** (2014) 1291–1300.
88. W. T. Xu, H. J. Peng, J. Q. Huang, C. Z. Zhao, X. B. Cheng and Q. Zhang, *ChemSusChem*, **8** (2015) 2892–2901.
89. L. N. Wang, J. Y. Liu, S. Y. Yuan, Y. G. Wang and Y. Y. Xia, *Energy & Environmental Science*, **9** (2016) 224–231.
90. D. Moy, A. Manivannan and S. R. Narayanan, *Journal of the Electrochemical Society*, **162** (2015) A1–A7.
91. J. B. Goodenough and K. S. Park, *Journal of the American Chemical Society*, **135** (2013) 1167–1176.
92. M. Buonaiuto, S. Neuhold, D. J. Schroeder, C. M. Lopez and J. T. Vaughey, *ChemPlusChem*, **80** (2015) 363–367.
93. S.-H. Chung and A. Manthiram, *Electrochemistry Communications*, **38** (2014) 91–95.
94. Y. Jung and S. Kim, *Electrochemistry Communications*, **9** (2007) 249–254.
95. D. J. Lee, H. Lee, J. Song, M.-H. Ryou, Y. M. Lee, H.-T. Kim and J.-K. Park, *Electrochemistry Communications*, **40** (2014) 45–48.
96. S. H. Lee, J. R. Harding, D. S. Liu, J. M. D'Arcy, Y. Shao-Horn and P. T. Hammond, *Chemistry of Materials*, **26** (2014) 2579–2585.
97. R. S. Thompson, D. J. Schroeder, C. M. Lopez, S. Neuhold and J. T. Vaughey, *Electrochemistry Communications*, **13** (2011) 1369–1372.
98. J. Hassoun, Y.-K. Sun and B. Scrosati, *Journal of Power Sources*, **196** (2011) 343–348.
99. Y. Yang, M. T. McDowell, A. Jackson, J. J. Cha, S. S. Hong and Y. Cui, *Nano Letters*, **10** (2010) 1486–1491.
100. J. Xiao, *Advanced Energy Materials*, **5** (2015) 1501102.
101. T. A. Skotheim, C. J. Sheehan, Y. V. Mikhaylik and J. Affinito, US Patent, (2007) **US7247408 B2**.
102. Y. V. Mikhaylik, US Patent, (2008) **US7354680 B2**.
103. Y. V. Mikhaylik, C. Scordilis-Kelley, I. Kovalev and C. Burgess, US Patent, (2013) **US8617748 B2**.
104. S. P. Mukherjee and T. A. Skotheim, US Patent, (2001) **US6238821 B1**.

105. S. A. Carlson, Z. Deng, Q. Ying and T. A. Skotheim, World Patent, (1999) **WO1999033125 A1**.
106. V. Kolosnitsyn and E. Karaseva, World Patent, (2007) **WO2007034243 A1**.
107. V. Kolosnitsyn and E. Karaseva, US Patent, (2015) **US9196929 B2**.
108. S. S. Choi, Y. S. Choi, Y. J. Jung, J. D. Kim and S. Kim, European Patent, (2003) **EP1302997 A2**.
109. S. Kim, Y. Jung, J. S. Han and J. D. Kim, US Patent, (2007) **US7291424 B2**.
110. S. Kim, Y. Jung, J. S. Han and J. D. Kim, US Patent, (2007) **US7179563 B2**.
111. G. Deromelaere, R. Aumayer, U. Eisele, B. Schumann and M. H. Koenigsmann, European Patent, (2011) **EP2339674 A1**.
112. M. Y. Chu, US Patent, (1996) **US5523179 A**.
113. M. Y. Chu, J. L. C. De, S. J. Visco and B. D. Katz, World Patent, (1999) **WO1999019931 A1**.

Chapter 2

Sulfur Cathode

Elton J. Cairns[*,†,‡] and Yoon Hwa[*,†,§]

*Energy Storage and Distributed Resources Division,
Lawrence Berkeley National Laboratory, 1 Cyclotron Road MS
Berkeley, CA 94720, USA
†Chemical and Biomolecular Engineering Department,
University of California, Berkeley, CA 94720, USA
‡ejcairns@lbl.gov
§yhwa@lbl.gov

2.1. Introduction

Since the 1990s, human life has greatly changed because commercial lithium (Li)-ion batteries (LIBs) were first released by Sony[1] and became widely used for small electronic devices, which allow people to use their electronics without power cables, i.e. popularization of portable electronics. More recently, the application of LIBs has been expanding to large systems such as electric vehicles (EVs), advanced portable electronics and large scale stationary energy storage systems, because environmental pollution and limited deposits of fossil fuel issues came to the fore, which makes the development of the next generation of energy sources essential. Unfortunately, it is a great challenge for conventional Li-ion cells because the performance of Li-ion cells has not improved as fast as the increase of demand for high-performance portable electronics and is not satisfactory for emerging market demands such as electric automobiles and aircraft. For example, an EV that is fully operated by battery power requires a high specific energy of ~350 Wh/kg at the three-hour discharge rate with reasonably low cost.[2] The specific energy (Wh/kg) of a cell is technically determined by the operating voltage (V) and the specific capacity (Ah/kg) of the cell, however, conventional Li-ion cells composed of a carbon anode and a transition metal oxide cathode can offer only about 100–200 Wh/kg of practical specific energy, due mainly to low specific capacity of the transition metal oxide cathode (theoretical specific capacity of $LiCoO_2$ cathode[3]: 274 Ah/kg). Even the theoretical specific energy of conventional Li-ion cells (500–600 Wh/kg) is not far from the practical requirement for EV applications, so it seems very unlikely that any Li-ion cell can meet the 350 Wh/kg goal. For these reasons, seeking the next generation of rechargeable cells with higher specific energy has become essential.

Even though elemental sulfur was first investigated in the 1960s as a cathode material for light metal (Li, Na) based electrochemical cells,[4,5] sulfur has recently attracted a great deal of interest and been intensively researched due to its advantages as cathode material for Li rechargeable cells such as a large theoretical specific capacity, low cost, abundance, and low environmental impact. The hundreds of research articles and reviews published since 2010 reflect how much the Li–S cell has attracted researchers' interest.[6–20] In the Li–S cell system, the sulfur cathode can accommodate 2 Li-ions per S atom by the electrochemical reaction shown below:

$$S_8 + 16Li^+ + 16e^- \leftrightarrow 8Li_2S \qquad (E^0 = 2.15 \text{ V vs. Li/Li}^+). \qquad (1)$$

The large theoretical specific capacity (1675 mAh/gS) of the sulfur cathode obtained by Reaction (1) can compensate for its relatively low redox potential compared to that of transition metal oxide cathodes, so the Li–S cell can offer a theoretical specific energy of ~2600 Wh/kg and a theoretical energy density of 2800 Wh/L under the assumption of complete Li_2S formation during the discharge process, which is much larger than that of the Li-ion cell ($LiMO_2$/graphite system: ~500 Wh/kg).[21] Besides, sulfur is one of the most plentiful and widespread elements in the world, so it is much less expensive (~$150/ton) than the cobalt and nickel that are employed as cathode materials for Li-ion cells. In addition, the common sulfur cathode has technical similarity to the fabrication of Li-ion cell cathodes, therefore, the technological entry barrier for being adapted to practical rechargeable cell manufacture is relatively low. These advantages of the sulfur cathode allow the Li–S cell to be a strong candidate for the next generation rechargeable cell.

Despite the promising advantages, multiple scientific and technical challenges for the sulfur cathode need to be overcome in order to develop a practical Li–S cell; a large volumetric change of about 80% occurs upon the formation and decomposition of Li_2S during the discharge and charge processes, resulting in mechanical failure of the sulfur cathode[22,23]; the electronically insulating nature of S and Li_2S limits complete conversion of sulfur to Li_2S. Especially, formation and growth of Li_2S on the surface of the sulfur cathode during discharge impedes further lithiation of sulfur owing to its poor electronic and Li ion diffusivity[24]; and the dissolution of Li polysulfides (Li_2S_n = 2–8) into most liquid electrolytes leads to both active material loss from the cathode and the shuttle effect, where dissolved Li polysulfides diffuse to the surface of the Li metal anode and form an insoluble Li_2S (or Li_2S_2) layer on its surface resulting in low Coulombic efficiency of the cell.[25] By understanding reaction and failure mechanisms of the sulfur cathode, important keys for the design of the sulfur cathode can be deduced: (1) mechanical stability of the sulfur cathode during cycling; (2) improvement of electronic and ionic conductivities of the sulfur cathode; (3) protection in order to suppress the dissolution of Li polysulfides into the liquid electrolyte; (4) physical or chemical trapping of Li polysulfides in order to suppress the shuttle effect. These must be addressed for rational design of a sulfur cathode to achieve a high specific energy Li–S cell, and the design of the cathode should not only be aimed at developing sulfur-based active materials, but should also consider the other components of the electrode such as binder, additives and current collector.

A goal of this chapter is to provide an overview of recent research on the sulfur cathode that represents the notable achievements and to discuss the scientific and technical challenges for developing practical Li–S cells. Various effective approaches and concepts to improve the electrochemical performance of the sulfur cathode with regard to all cathode components including sulfur-based active materials, binder, additives as well as current collectors will be discussed. In particular, remarkable strategies used to improve the cell performance significantly will be highlighted. The additional interlayer commonly placed between the cathode and the separator in the cell will be included in this chapter because of its importance although it is not a part of the cathode. Finally, perspectives for further advances toward practical, high specific energy Li–S cells will be provided.

2.2. Nanostructured Sulfur as Active Material

2.2.1. Introduction

Since nanotechnology has been used in battery engineering, electrode materials can be manipulated and engineered at the nanoscale, which may greatly change the physical and chemical properties. Many kinds of materials have resurfaced and been studied as electrode materials because significantly different electrochemical behaviors of the materials are exhibited when they are nanosized. In the electrochemical cell, because the Li-ion diffusion rate in solid particles is much more sluggish than the rate of Li-ion movement in a liquid electrolyte, the impedance due to Li insertion/extraction can be significantly lowered by reducing the length of the Li-ion diffusion pathway through the particles.[26,27] In particular, this approach can be very effective if the lithiated phase of the host has poor electronic and ionic conductivity. During the electrochemical lithiation process, a new phase first forms at the surface of the particle and grows into the interior part, but poor conductivity of the new phase can impede Li-ion insertion, which results in limited utilization of the active material. So the active material can be more fully utilized as the particle size of the active material decreases. In addition, nanosized particles can have better mechanical stability against fracture during electrochemical cell cycling. Volume change commonly occurs when the host material forms an alloy phase with Li during lithiation and the degree of volume change depends on the crystal structure of the host material and the amount of Li accommodated. But some materials such as silicon, tin and sulfur expand too much compared to the original volume and that causes pulverization of the particles (or electrode). If pulverization of the particles occurs, the fragments (or some part of the electrode laminate) will lose electrical contact with the current collector rendering it no longer available for electrochemical reaction, resulting in capacity degradation. The pulverization problem is a very critical issue for some anode materials for Li-ion cells such as silicon and tin. Recent advances show significantly improved electrochemical performance of the cell by using nanosized particles of active material. It is revealed that nanosized particles are mechanically more stable than micron-size particles when the volume changes by reaction with Li due to facile strain relaxation.[28–30] As is well known, sulfur and the lithiated solid phase (Li_2S) have very poor conductivity and sulfur particles expand by up to 80% compared to the original volume of the S particle. So, based on the discussion above, the slow lithiation of sulfur could be addressed in the same manner, so the particle size of sulfur should necessarily be reduced in order to have high utilization of sulfur and improve the mechanical stability of the electrode.

The electrochemical performance of nanoengineered sulfur can be further improved when it forms unique composite structures with complementary materials such as additional media that can improve physical and chemical properties or provide new advantages to the sulfur cathode. Sulfur/porous material nanocomposites is one of the most popular designs because nanopores not only provide the space to accommodate the volume expansion of sulfur during lithiation, but also physically trap polysulfides in its space, which is called the 'polysulfide reservoir' concept. Thus, long cycle life with good Coulombic efficiency of the cell can be achieved with this concept. Porous carbons or metal oxides are commonly employed for porous media. The core–shell nanostructure is another representative design, consisting of a core material (main component) surrounded by a shell material. In general, the shell material acts as a protection layer to prevent the main material from being physically or chemically degraded, or provide a new property. The shell can also prevent direct contact between the core material and the electrolyte to avoid unnecessary side reactions and sulfur loss.[31] This is very effective for the sulfur cathode because polysulfide dissolution into liquid electrolyte is prohibited if the sulfur particle and the liquid electrolyte are physically separated by the shell. So, the core–shell structured nanocomposite with sulfur as the core and a second material that has good electrical conductivity, Li permeability, and mechanical stability as the shell can be considered as a promising approach for enhancing the electrochemical performance of the sulfur cathode.

Due to the structural benefits of nanocomposites, most sulfur cathodes that have shown promising cell performance mainly employ sulfur-based nanocomposites instead of pure sulfur powder as the active material. Carbonaceous materials such as porous carbon, graphene (oxide), and carbon nanotubes (CNTs) have been extensively researched as second media for the sulfur cathode due to its reasonably good conductivity, Li permeability, and additional benefits depending on its physical and chemical structure. Polymers that may be flexible or conductive are also strongly considered as second media to improve electrochemical performance of the sulfur cathode. Besides carbonaceous materials and polymers, some inorganic compounds such as oxides and sulfides and metal-organic frameworks (MOFs) have been studied as well. These physical mixture concepts could be further improved when combined with chemical surface functionalization that can chemically trap polysulfide species to suppress the polysulfide shuttle effect.

The concepts briefly discussed above have significantly improved the electrochemical performance of the sulfur cathode. In the following sections, important previous accomplishments will be introduced and discussed in detail, to help trace recent progress on developing sulfur nanocomposites as active materials and understand how they work. Organosulfur compounds as cathode material will also be briefly discussed.

2.2.2. Sulfur–carbon nanocomposites

Sulfur–carbon nanocomposites have been intensively researched for developing advanced sulfur cathodes because most carbonaceous materials have good electronic conductivity so it can compensate for the inadequate electronic conductivity of sulfur (5×10^{-30} S cm^{-1} at 25°C) resulting in improvement of sulfur utilization, especially at high rates. Various carbonaceous materials have been investigated in order to improve the electrochemical performance of sulfur cathodes and they can be classified according to their physical structure; (1) nanoporous carbon, (2) one dimensional (1D) carbons such as CNTs and carbon nanofibers (CNFs) and (3) two-dimensional (2D) carbons such as graphene and GO. Basically, carbonaceous materials have reasonably good electrical conductivity and Li-ion transport, so they can provide a stable conductive pathway, which could influence the electrochemical performance of the sulfur cathode positively. Furthermore, those carbonaceous materials offer their own advantages depending on their physical and chemical structure, so various sulfur–carbon nanocomposites can be designed.

There are several methods to fabricate sulfur–carbon nanocomposites such as the infiltration method, high-energy mechanical ball-milling (HEMM), precipitation of sulfur onto carbon surfaces, the hydrothermal method, and wrapping sulfur with 2D carbonaceous materials. Unfortunately, the chemical vapor deposition (CVD) method which is one of the most popular and effective methods for synthesizing core–shell nanostructures cannot be used with sulfur due to the low melting temperature of sulfur (~115°C) that is much lower than the common operating temperature of the CVD process (above 450°C depending on the carbon precursor gas). Instead, the infiltration method by heat treatment (commonly conducted at 155°C, where the viscosity of liquid sulfur is the lowest) is used to embed sulfur into the inner spaces of carbon particles. A vacuum atmosphere can help to insert sulfur into carbon particles during heat treatment.

2.2.2.1. *Sulfur/porous carbon nanocomposites*

Porous materials are solid media containing pores and voids in their structure and are commonly categorized into microporous (pore size: <2 nm), mesoporous (pore size: 2–20 nm) and macroporous materials (pore size: >50 nm) depending on pore size, which was defined by Sing *et al.* in 1985.[32] In addition to these classifications of porous structure, the term 'nanoporous material' has been popularly used, which refers to materials with a pore size of <100 nm in diameter. Porous materials have been intensively studied as a component of sulfur-based composite materials for Li–S cells because the pores in the structure of the materials can play an important role in improving the Li–S cell lifetime by accommodating volume expansion of up to 80%, so mechanical fracture of the sulfur cathode can be avoided as a result. Furthermore, pores can physically trap Li polysulfides that are formed by the electrochemical reaction of sulfur with Li (and can dissolve into liquid organic electrolytes), resulting in better cycle life of the Li–S cells. Among many porous materials, porous carbonaceous materials necessarily attract researchers' attention, due to the diverse methods to prepare a porous structure and a good electrical conductivity in order to improve the electrochemical performance of the Li–S cells. Traditional methods to synthesize porous carbon are categorized by Hyeon et al. as follows[33]; (1) chemical, physical activation or their combination[34,35]; (2) catalytic activation of carbon precursors using metal salts or organometallic compounds[36,37]; (3) carbonization of polymer blends composed of a carbonizable polymer and a pyrolyzable polymer that does not leave carbon residue[38]; (4) carbonization of a polymer aerogel synthesized under supercritical drying conditions.[39] In addition to the above mentioned methods, uniform and ordered porous structures can be obtained by the template synthesis method with an inorganic template that is rigid and designable. Porous carbon with desired pore sizes and shapes can be obtained with the help of template materials. Template inorganic materials are necessarily stable under the process conditions of carbon formation and removable under a condition which is safe for carbon. Zeolites[40] and silica[41] are representative template materials for the fabrication of porous carbons.

In the early 2000s, a few papers reported on sulfur/porous carbon composites fabricated by heat treatment of a sulfur/porous activated carbon (pore size of 2.5 nm) mixture for the sulfur cathode materials.[42,43] In the early works, the sulfur/porous carbon cathodes exhibited an initial specific capacity of about 800 mAh/g that is about a half of its theoretical value, and then the specific capacity rapidly decreased to about 400 mAh/g during the following cycles at 0.1 C discharge–charge.[42,43] Significant progress on sulfur/porous carbon nanocomposite cathodes with high sulfur utilization were reported by Liang et al.[44] and Ji et al.[45] in 2009. In Liang's work, a hierarchically structured sulfur–carbon nanocomposite was prepared with micro and mesoporous carbon (MPC) obtained by the postactivation process of MPC using KOH.

MPC has a uniform mesopore size of about 7.3 nm, and a Brunauer–Emmett–Teller (BET) surface area of 368.5 m^2/g. After the KOH activation, the BET surface area increased to 1566.1 m^2/g and the volume contributed by small pores from the size of 2 to 4 nm increased without morphological change, which indicates that micro and small mesopores were generated on the walls of MPC. The sulfur/micro- and MPC nanocomposite cathodes exhibited 1584 and 818 mAh/g of initial discharge capacity and exhibited 700 and 220 mAh/g after 50 cycles at the test specific current of 2.5 A/g when the sulfur contents in the sulfur/micro and MPC composite were 11.7 and 51.5 wt.%, respectively. In this chapter, all specific capacity and applied currents for testing are based on the mass of sulfur unless it is specified otherwise. For comparison, sulfur/microporous carbon and sulfur/MPC nanocomposite cathodes were also prepared, however, they exhibited a very low initial specific capacity and very poor capacity stability during 50 cycles. It is suggested that the synergetic effect of different sizes of pores occurs, wherein the micropores provide the volume for retaining sulfur species in the cathode and the mesopores provide an efficient Li-ion diffusion pathway and thus confer a high ionic conductivity to the cathode.

Nazar and coworkers reported a sulfur/highly ordered MPC (CMK-3) nanocomposite that is comprised of sulfur embedded in the conductive MPC framework as shown in Figure 2.1(a).[45] CMK-3 was fabricated using a mesoporous silica (SBA-15) template and sucrose as carbon precursor and the sulfur/CMK-3 nanocomposite was prepared by following a melt-diffusion method at 155°C. The nanoarchitecture of the sulfur/CMK-3 nanocomposite was intended to act as an electronic and ionic conduit to the nanosized sulfur encapsulated within and provides pores that not only act as the Li polysulfide reservoir but also enhance mechanical stability by accommodating the volume expansion of sulfur during lithiation. For further improvement, polyethylene glycol (PEG) was linked to the external surface of the composite, which acts as an additional barrier to suppress lithium polysulfide diffusion out of the cathode structure, resulting in improvement of cycle life. As shown in Figure 2.1(b), the sulfur/CMK-3 and sulfur/CMK-3/PEG nanocomposite cathodes exhibited a high initial discharge capacity of about 1000 and 1320 mAh/g respectively at a discharge specific current of 168

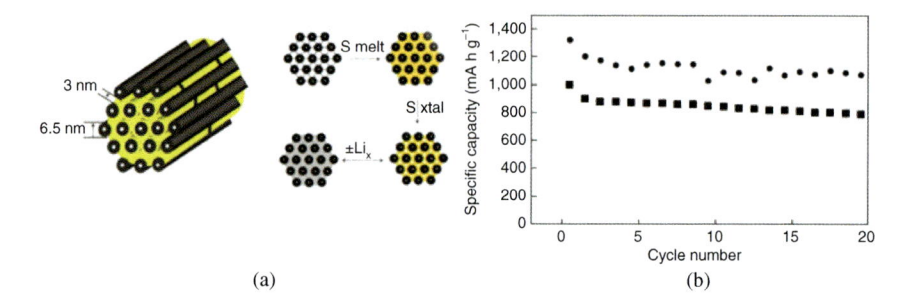

Figure 2.1 (a) A schematic diagram of sulfur (yellow) confined within the interconnected pore structure of MPC (CMK-3) formed from carbon tubes that are held together by CNFs. (b) Cycling stability comparison of CMK-3/S-PEG (upper points, in black) vs. CMK-3/S (lower points, in black square) at 168 mA/g at room temperature.[45]

mA/g. To verify the effect of CMK-3 as a Li polysulfide reservoir during cycling, the sulfur contents (%) in the sulfur-free electrolyte of the cells employing sulfur/acetylene black composite, sulfur/CMK-3 nanocomposite and sulfur/CMK-3/PEG nanocomposite as cathode materials and the morphology changes of the cathodes were investigated after 30 cycles. It was demonstrated that the mesoporous structured sulfur/CMK-3 cathode lost only about 25% of the total active mass of sulfur, while the sulfur/acetylene black cathode lost 96% of the total active mass of sulfur into the electrolyte after 30 cycles. The sulfur/CMK-3/PEG cathode showed a minimal amount of dissolved sulfur in the electrolyte, which indicates that the PEG-functionalized surface can serve to trap Li polysulfides by providing a highly hydrophilic surface chemical gradient that preferentially solubilizes them in relation to the electrolyte. The morphology changes of the electrodes after 30 cycles were shown in scanning electron microscopy (SEM) images, in which the sulfur/acetylene black composite cathode showed significant surface change caused by deposition of insoluble species while the other two cathodes showed very little change. This work is worth highlighting because it presents a high capacity of about 1320 mAh/g with a relatively higher sulfur loading of about 1.1 mg/cm^2 than that of earlier works and a twofold protection concept was proposed in order to enhance the electrochemical performance by combining two different strategies; i.e. porous structure acting as lithium polysulfide reservoir with functional polymer outer layer.

Although Liang et al.[44] and Ji et al.[45] reported great improvement in terms of specific capacity by employing unique nanostructures, long-term cycle life was not demonstrated, which is one of the essential properties for practical Li–S cells. Therefore, much more effort has been devoted to improvement of the cycle life of Li–S cells, since reasonably high sulfur utilization was achieved.[46–66] In 2010, Zhang et al. reported on sulfur/microporous carbon nanocomposites with a narrow pore size distribution (~0.7 nm) as cathode material for Li–S cells and it exhibited good rate capability and cycle life.[46] The specific capacity of ~890 and 730 mAh/g were obtained at the test specific currents of 200 mA/g and 1200 mA/g, respectively. The specific capacity of the sulfur/microporous carbon nanocomposite cathode was almost recovered when the test specific current was decreased back to 200 mA/g, indicating a good rate capability. The cathode also exhibited an excellent capacity retention of about 80% for 500 cycles with a high Coulombic efficiency at the test specific current of 400 mA/g, as shown in Figure 2.2(a). The excellent electrochemical performance of the sulfur/microporous carbon nanocomposite cathode was due to good conductivity of the carbon matrix and the unique structure that is comprised of fine sulfur embedded in the micropores of the carbon matrix. As described in Figure 2.2(b), fine sulfur particles were highly dispersed and well encapsulated in the micropores (~0.7 nm) of the carbon matrix, therefore, the embedded sulfur and formed Li polysulfides were captured by micropores during cycling. As a result, active sulfur loss caused by dissolution of Li polysulfides and its shuttle reaction and formation of a thick Li_2S layer on the cathode surface can be prevented, resulting in good cycle life and excellent Coulombic efficiency. Interestingly, the sulfur/microporous carbon nanocomposite cathode showed only one obvious plateau between 1.5 and 2.8 V, whereas the common sulfur cathodes typically show two-plateau behavior during the discharge process. It is suggested that the sulfur that stayed in the micropores is no longer in the S_8 cyclical structure, instead sulfur and carbon are mixed at the atomic level. Consequently, the corresponding Gibbs free energy of the reaction could change, which is reflected in the voltage profile of the sulfur/microporous carbon nanocomposite cathode.

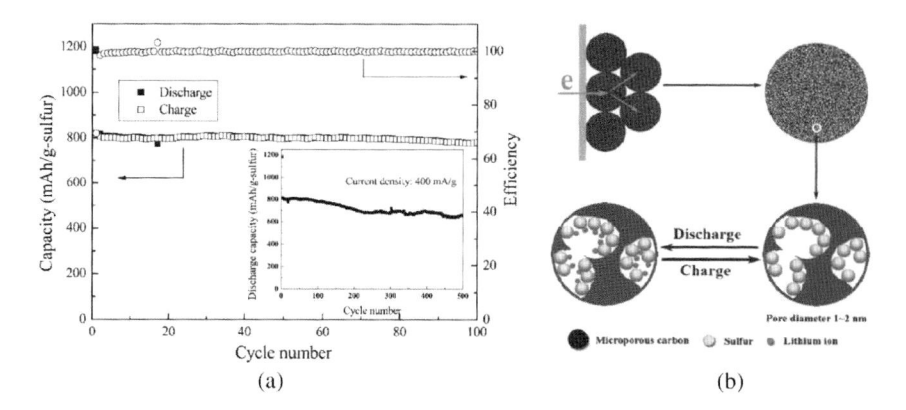

Figure 2.2 (a) The cycling performance of a sulfur–carbon composite with 42 wt.% sulfur the constant specific current of 400 mA/g (inset: The long cycle life of the composite at the specific current of 400 mA/g). (b) A scheme of the constrained electrochemical reaction process inside the micropores of the sulfur–carbon sphere composite cathode.[46]

According to the above discussion, the properties of porous carbon such as pore size distribution and defect concentration, which mainly depend on the sources of carbon and the synthesis conditions play an important role in determining the amount of sulfur infiltrated into the pores and to improve the electrochemical performance of the Li–S cell. Lee *et al.* reported a sulfur/micro and MPC nanocomposite obtained by controlling the chlorination temperature during the synthesis process of silicon carbide derived porous carbon (CDC) in order to obtain a different pore size distribution.[56] According to the SEM and BET test results, the chlorination temperature has a very minor impact on the morphology of the particles, the shape, and the size of the isotherm plots, suggesting that all samples have similar porosity. Although the porosity of three samples are similar, small increases in the volume of the sub-nm micropores, small decreases in the volume of 1.3–3.0 nm pores and small increases in the size of the largest (3–4 nm) mesopores were calculated using a quenched solid DFT model. Interestingly, Raman spectra of three samples demonstrated that defect concentration decreases as chlorination temperature increases; for instance, CDC-900°C has the lowest defect concentration and the highest volume of the smallest pores, and has the highest S content among the CDC-700, 800, and 900°C samples. The comparisons of electrochemical cycling performance of the sulfur/CDC-700, 800, and 900°C cathodes at the 0.2 C rate showed that the sulfur/CDC-900°C cathode exhibited the highest reversible specific capacity of about 600 mAh/g, while the other cathodes exhibited that of about 400 mAh/g for 100 cycles. Although the reasons for the improvement were not completely revealed, some possible reasons were suggested such as faster Li-ion diffusion due to the smaller concentration of defects and side reactions caused by chemical residues in the CDC.

As discussed in the introduction to this section, the core–shell nanostructure is a promising strategy to enhance Li–S cell performance. Commonly, core material/carbon shell nanocomposites can easily be synthesized using CVD carbon coatings or carbonization of a carbon precursor shell (commonly a polymer) at a heating temperature of between 450 and 1000°C. However, the low melting point of sulfur does not allow the use of a high temperature method. Instead, core–shell-like nanostructures can be synthesized by inserting sulfur into macropores or the hollow inner space of micro- or MPC shells using the sulfur melt-diffusion method.[49-55] He *et al.*[49] and Jayaprakash *et al.*[52] reported the sulfur/hollow carbon core–shell nanostructured composite for the Li–S cell cathode material. Hollow carbon shells were produced using a silica template, and then sulfur/porous carbon core–shell nanostructured composites were prepared using the melt diffusion method. Sulfur can be diffused into the inner space of carbon through micro- or mesopores of the shell during the melt-diffusion process. With the sulfur/hollow carbon core–shell nanostructured composite cathode, high reversible capacity of about 1000 mAh/g at the 0.5 C discharge rate for 100 cycles was achieved.[49,52] In addition, He *et al.*[49] reported comparisons that show the trade-off relationship between shell porosity and capacity retention. The highest initial capacity of 1300 mAh/g was obtained with the sulfur/carbon core–shell nanocomposite that has the highest porosity of the shell, presumably owing to the open structure of the shell, but the cycle life of the cell is very poor. In contrast, the lowest initial capacity was obtained with the least porous carbon shell, but it showed the best cycle stability for 100 cycles. The results might indicate that the open structure of the shell helps to supply enough lithium ions to the surface of the sulfur particles, but it allows lithium polysulfides to escape from the core–shell structure.

44 Li–S Batteries: The Challenges, Chemistry, Materials and Future Perspectives

Figure 2.3 shows the advanced design concept of sulfur/porous carbon core–shell structures that employs a nanoporous carbon matrix with micropores in the outer shell, surrounding the inner meso and macropores.[53,54,57] The micropores existing in the outer shell and in the walls between the meso- and macropores prevent migration of the highly soluble Li polysulfides into the liquid organic electrolyte due to the strong adsorption of the lithium polysulfides. Furthermore, the desolvation of the electrolyte ions in the micropores might be able to suppress the lithium polysulfide dissolution as the solvent concentration is very low or likely to be close to zero in these micropores (Figure 2.3(a)).[61] By contrast, highly soluble Li polysulfides could form near the surface of sulfur/porous carbon nanocomposite when porous carbon has irregular pore structure as shown in Figure 2.3(b). Thus, the polysulfide dissolution into the liquid electrolyte and the shuttling effect occur continuously during cycling. Jung et al.[54] reported a specific capacity of about 540 mAh/g with excellent capacity retention of 77% with respect to its specific capacity of the

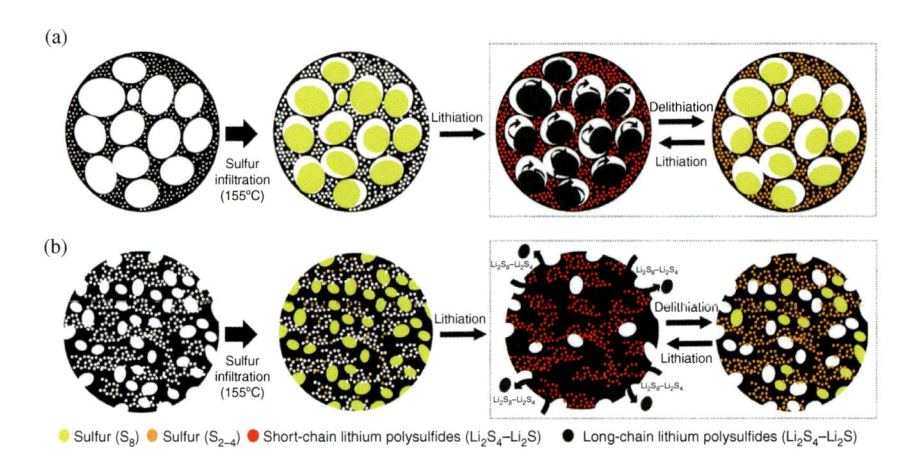

Figure 2.3 Schematic illustration of the electrode structures and their electrochemical processes. (a) Hierarchical porous carbon particles have micropores in the outer shell, surrounding the inner meso and macropores. The dissolution of the soluble long-chain Li polysulfides in the inner macro and mesopores is suppressed by the outer micropores serving as a barricade. (b) Conventional activated carbon (AC1600) containing micropores and mesopores in a random geometry. The long-chain Li polysulfides are liable to dissolution through the open pore ends.[54]

fifth cycle after 500 cycles at the 2.4 C test rate, which indicates the sulfur/hierarchical porous carbon nanocomposite is structurally stable during cycling. A sulfur/ordered meso@microporous carbon nanocomposite reported by Li *et al.* also showed good capacity retention of up to 81% with the specific capacity of about 837 mAh/g at 0.5 C after 200 cycles.[57]

In summary, some design strategies of sulfur/porous carbon nanocomposites in order to enhance the electrochemical performance of the sulfur cathode can be successful according to the discussion above; (1) Optimization of pore size distribution and uniform distribution of sulfur in the pore space, (2) Large pore volume in order to increase the amount accommodated sulfur, which is important to achieve high specific energy of a practical Li–S cell. (3) Highly porous carbon that has a closed structure near the outer surface can trap lithium polysulfides more effectively.

2.2.2.2. Sulfur–graphene (oxide) nanocomposite

Graphene is one allotrope of carbonaceous material that is two-dimensionally structured with a continuous array of the hexagonal carbon lattice. Graphene is fundamentally a single layer of graphite, however, it offers extraordinary properties, i.e. very high electronic and thermal conductivity, transparency and excellent mechanical stability, which are good features for enhancing the electrochemical performance of the Li–S cell. Graphene can be normally synthesized using the CVD method and epitaxial growth. Unfortunately, mass production of high quality graphene at low cost, and in a reproducible manner with those methods is still a challenge, which limits the practical rechargeable battery application of pure graphene. As an alternative, reduced graphene oxide (RGO), the reduced product of graphene oxide (GO) by thermal, chemical or electrochemical methods has become interesting due to its relatively high capability of mass production. GO is a compound of carbon, oxygen and hydrogen and is normally produced by exfoliation of graphite that is oxidized using a strong oxidizing agent. GO commonly does not have a perfect graphene structure due to the presence of oxygen-containing functional groups such as epoxy and hydroxyl functional groups on the surface. Because of the structural characteristics of GO, RGO also commonly has many pores that are generated at the site occupied by functional groups after the reduction process. Because of the imperfect structure of GO and RGO, they usually have relatively poor properties compared to pure graphene, however, some properties can almost be identical to those of pure graphene depending on the reduction process.

For the design of sulfur–graphene (oxide) nanocomposites, graphene (oxide) is employed as deposition site for sulfur to form a thin sulfur layer or small sulfur particles on the surface or wrapping media that forms a nanostructure similar to the core–shell concept.[26,67–90] Wang et al. reported a sulfur–graphene nanosheet (GNS) composite produced by the melt-diffusion method to form a sulfur layer on the surface of GNS.[73] The initial discharge capacity of pure sulfur and the sulfur/GNS composite cathodes of about 1100 and 1611 mAh/g, respectively, at a test specific current of 50 mA/g were demonstrated and the sulfur/GNS composite cathode showed a discharge capacity of about 580 mAh/g after 40 cycles while the pure sulfur cathode showed only about 200 mAh/g. The electrochemical impedance spectroscopy results indicated that the charge-transfer resistance of the cell employing sulfur/GNS composite cathode was lower than that of the cell with pure sulfur cathode, which presumably is able to explain the improvement of the reversible capacity of the cell. Despite the improvement in terms of specific capacity, the capacity retention of the sulfur/CNS composite cathode was only about 36% after 40 cycles, even with very low sulfur content of about 22 wt.% in the sulfur–CNS composite, which is not comparable with conventional Li-ion cells. The poor cycle life of the sulfur–CNS composite cathode might be due to direct contact of sulfur with liquid organic electrolyte and a lack of a physical or chemical barrier against polysulfide dissolution. This result emphasizes that a simple mixture of sulfur and graphene cannot offer long cycle life, therefore, the sulfur–graphene nanocomposite needs to employ an additional strategy to maintain its enhanced specific capacity.

A graphene (oxide)-wrapped sulfur nanocomposite is one of the unique nanostructures that is intended to physically trap the sulfur and Li polysulfides by surrounding sulfur particles with graphene (oxide) sheets.[68–71,81] Wang et al. reported GO-wrapped submicron sulfur particles including carbon black (CB) in the structure to improve the conductivity of the cathode.[68] Triton X-100 which is a surfactant containing a PEG chain was used as a capping agent for sulfur particles to limit the size of sulfur particles in the nanoscale during the synthesis. The strategies of the GO wrapped sulfur nanocomposite in order to improve the cell performance are: (1) the outer GO layer on the sulfur nanoparticles not only can help to trap the Li polysulfides as they form, but also provide a flexible cushion to accommodate the volumetric strain caused by volume expansion of sulfur particles during lithiation process, (2) PEG containing surfactant coating provides additional chemical barrier against lithium polysulfides dissolution into liquid electrolyte, (3) the GO and the CB improve the electronic conductivity of the sulfur based cathode. The sulfur–PEG, the sulfur–GO without PEG and the sulfur–GO with PEG cathodes were prepared for comparison and they exhibited similar initial specific capacity of 700–800 mAh/g at the 0.2 C discharge rate. However, the sulfur–GO cathode with PEG only showed a stable cycle life for 100 cycles, while the other two cathodes showed very poor cycle life. The sulfur–GO cathode with PEG still delivered a specific capacity of about 500 mAh/g after 100 cycles, but the specific capacity of the other two cathodes dramatically decreased to about 300 mAh/g after only 50 cycles, which indicates that sulfur was more effectively retained in the cathode when two types of protection were combined. On the other hand, Evers and Nazar used RGO as both an electrical conduit for insulating sulfur and as a barrier to retard polysulfide dissolution.[69] This work is highlighted by the high sulfur content of about 78% in the sulfur cathode achieved by excluding conductive carbon additives (RGO used as wrapping media was the only carbonaceous material in the cathode), which is suitable for practical level of sulfur content. The RGO wrapped sulfur cathode achieved a specific capacity of about 700 and 500 mAh/g at the first and 50th cycle, respectively. The results suggest that rationally engineered sulfur–graphene (oxide) nanocomposites can provide an opportunity for increasing the sulfur content in the cathode by decreasing the content of conductive carbon additives in order to achieve a high specific energy and energy density of a practical Li–S cell.

A unique core–shell nanostructure that is comprised of hollow graphene nanoshells (HGNs) and inserted sulfur was reported by

Peng et al.,[71] which delivers an extended lifetime of 1000 cycles at 1.0 C with a slow specific capacity decay rate of 0.06% per cycle (Figure 2.4(a)). The initial specific capacity of the sulfur–HGNs cathode was up to 1100 mAh/g and it still showed 420 mAh/g after 1000 cycles. The rate capability test results of the sulfur/HGNs cathode showed high discharge capacities of 1520, 1058, and 737 mAh/g at 0.1, 2.0, and 5.0 C, respectively, and then the specific capacity recovered quickly to 1200 mAh/g when the C-rate was decreased to 0.1 C, which indicates the excellent rate capability of the sulfur/HGNs cathode. The excellent electrochemical performance of the sulfur–HGNs cathode was due to the three-dimensional (3D) graphene framework that provides: (1) free space provided by the hollow structure of the graphene nanoshells for accommodating volume expansion of sulfur particles during discharge,

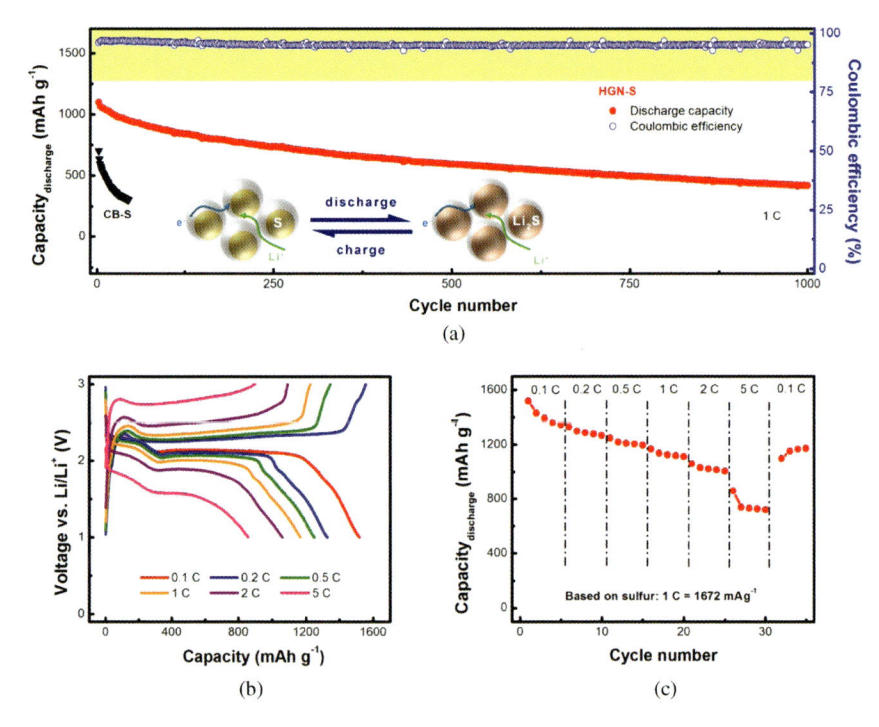

Figure 2.4 Electrochemical performance of HGN–S. (a) Cycling performance at a rate of 1 C and schematic illustration of HGN–S during discharge and charge (inset), (b) galvanostatic charge–discharge profiles at different rates, and (c) rate performance.[71]

(2) a physical protection layer that suppresses Li polysulfide dissolution into the liquid electrolyte during cycling and (3) stable electron pathways that improve the electronic conductivity of the cathode.

In 2011, a notable discovery was reported by Ji *et al.* which demonstrated that the functional groups on the surface of GO have strong adsorbing ability to anchor S atoms and to effectively prevent the subsequently formed Li polysulfides from dissolving into the electrolyte during cycling.[67] Figure 2.5(a) shows the carbon K-edge absorption spectra for both GO and sulfur–GO nanocomposites, which revealed that the oxygen containing functional groups on GO can enhance the binding of S to the C atoms. The sulfur–GO nanocomposite was prepared by the chemical deposition of sulfur onto GO in a micro emulsion system followed by heat treatment in order to remove some bulk sulfur. This unique nanostructured sulfur–GO nanocomposite cathode showed a very high initial specific capacity of about 1320 mAh/g at 0.02 C and a reversible specific capacity of about 950 mAh/g at 0.1 C for 50 cycles as shown in Figure 2.5(b). The excellent electrochemical performance of the sulfur–GO nanocomposite cathode is due to: (1) improvement of conductivity by the GO in the heat-treated composites, (2) The oxygen containing functional groups on the GO surface immobilize sulfur and Li polysulfides, (3) very thin and conformal sulfur coating layer shortens the

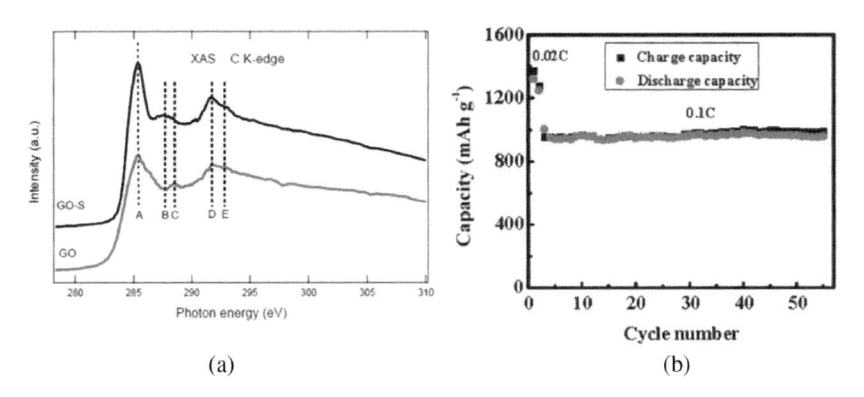

Figure 2.5 (a) C K-edge XAS spectra of GO and GO–S nanocomposites after heat treatment in Ar at 155°C for 12 hours. (b) Cycling performance of GO–sulfur nanocomposite cathode at a constant rate of 0.1 C after initial activation processes at 0.02 C for two cycles.[67]

conduction pathway through sulfur and (4) the GO network accommodates the volume expansion caused by the lithiation process of sulfur. A further study on the electronic structure and chemical bonding between sulfur and GO was conducted by Zhang *et al.* using X-ray spectroscopies such as X-ray photoelectron spectroscopy (XPS), near-edge X-ray absorption fine structure (NEXAFS) and X-ray emission spectroscopy (XES).[76] The results revealed that; (1) the incorporation of sulfur can partially reduce the GO and thus improve the conductivity of the GO; (2) the mild interaction between GO and sulfur can not only preserve the fundamental electronic properties of GO but also stabilize the sulfur by direct bonding with the GO sheet, which may prevent the diffusion of Li polysulfides formed during the discharge–charge cycling into the electrolyte. Wang *et al.* conducted a theoretical study of the effect of oxygen-containing functional groups to stabilize the polysulfides through molecular dynamic simulations and density functional theory calculations.[91] The results suggested that oxygen-containing functional groups, especially oxygen bonded to vacancies and edges, may be used to stabilize the polysulfides, which can explain the experimentally observed improved cycling stability when graphene-based materials are introduced in the cathode material.

In order to reinforce the chemical barrier against Li polysulfide dissolution or to provide a new opportunity for trapping sulfur and Li polysulfides, additional functionalities were provided to the surface of graphene (oxide) using various methods such as nitrogen doping,[81,82] hydroxylation,[80] employing cationic surfactants[88] and so on.[74,85,89,90] Nitrogen doping on the surface of carbonaceous materials is a promising method to enhance electronic conductivity because nitrogen atoms substituting for carbon atoms in the graphite matrix are electron donors and promote *n*-type conductivity.[92] It is also beneficial for the Li–S cells because nitrogen-containing functional groups on the surface of the carbonaceous materials can trap lithium polysulfides, resulting in improvement of the cycle life of the Li–S cells. For these reasons, various nitrogen-doped carbonaceous materials such as porous carbon,[60,63] graphene (oxide)[81,82] and CNTs (or fibers)[93,94] have been employed in the sulfur cathode in order to enhance the electrochemical performance of Li–S cells. Qui *et al.* demonstrated an ultralong cycle life exceeding

2000 cycles at the 2.0 C rate and an extremely low capacity-decay rate (0.028% per cycle) using nitrogen-doped RGO (NG) produced by a thermal nitridation process of GO in a NH_3 atmosphere at 750°C as wrapping media on the surface of sulfur nanoparticles.[81] Three types of nitrogen functional group such as pyridinic N, pyrrolic N, and graphitic N on NG were demonstrated by N 1s XPS and especially, the first two types are dominant in the product, which are believed to be more effective in alleviating dissolution of Li polysulfides into the liquid electrolyte and improving their re-deposition process during cycling.[95,96] The sulfur–NG cathode that is comprised of sulfur (60%) without any additional carbon additive delivered reversible specific capacities of about 730 and 492 mAh/g at the 100th and 1000th cycle, respectively, and showed a capacity retention of about 44% after 2000 cycles. The excellent performance of the sulfur–NG cathode was attributed to the NG which not only provides an efficient electrical pathway, but also acts as a sulfur immobilizer using the nitrogen-containing functional groups on the surface of NG. The cetyltrimethyl ammonium bromide (CTAB)-modified sulfur–GO nanocomposite cathode reported by Cairns *et al.* shows an ultralong service life exceeding 1500 cycles at the 1.0 C discharge rate (extremely low decay rate of 0.039% per cycle) with excellent specific capacity of about 846 mAh/g at 0.05 C after 1000 cycles at 1.0 C and about 740 mAh/g at 0.02 C after 1500 cycles at 1.0 C as shown in Figure 2.6.[88] The excellent electrochemical performance of

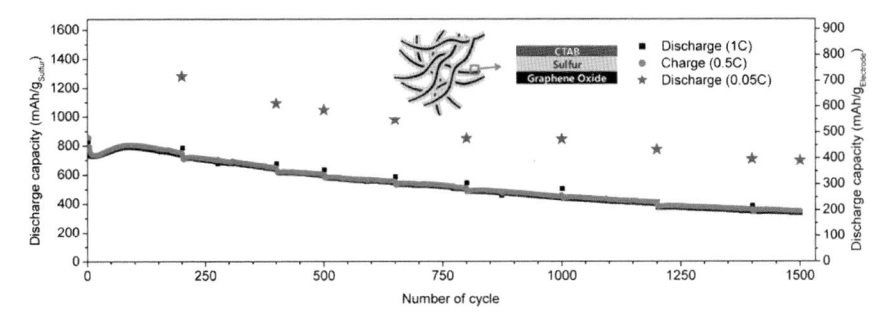

Figure 2.6 Longterm cycling test results of a Li–S cell with a CTAB-modified sulfur–GO nanocomposite cathode. (inset: schematic illustration of CTAB-modified sulfur–GO nanocomposite.[88]

CTAB-modified sulfur–GO nanocomposite cathode was achieved by a multifaceted approach: (1) GO acts as a sulfur immobilizer, (2) CTAB-modified sulfur anchored on the functional groups of GO, (3) elastomeric SBR/CMC binder, and (4) an ionic liquid-based novel electrolyte containing $LiNO_3$ additive.

Most synthesis methods of sulfur–graphene (oxide) nanocomposites discussed above use graphene (oxide) as a sulfur deposition site or wrapping medium on sulfur particles, therefore, the properties of graphene (oxide) such as number of layers and surface area strongly influence the nanostructure of the sulfur–graphene (oxide) composite, which means that the preparation process of graphene (oxide) needs to be conducted with care. Facile synthesis methods for sulfur–graphene (oxide) were developed by Xu et al.[83] and Lin et al.[78] using HEMM. The HEMM method is very attractive for practical applications because it allows a scalable one-step synthesis and commonly doesn't require harmful chemicals, so it is feasible for mass production without severe impact on the environment. The "sandwich-like" layered meso/macroporous sulfur–graphene nanoplatelets (GnPs) were prepared by HEMM with the sulfur-GnP ratio of 7:3 (0.7S–0.3GnP, the final sulfur content in 0.7S–0.3GnP was about 65.5%), indicating that the presence of sulfur in the ball-milled graphite facilitated the formation of 3D nanostructured carbon foams, presumably due to the strong S–S interaction between the edges of the functionalized S–GnP.[83] The 0.7S–0.3GnP cathode showed excellent high C-rate performance with average discharge capacities of 1043.1, 885.6, 756.8, 610.4, 404.8, and 186.5 mAh/g at 0.2, 0.5, 1.0, 2.0, 5.0, and 10.0 C, respectively. Long cycle life of the 0.7S–0.3GnP cathode was demonstrated with a discharge specific capacity retention of about 50% (966 mAh/g at the first cycle and 485 mAh/g at 500th cycle) after 500 cycles at 2.0 C, which is due to the "sandwich-like" layered meso–macroporous nanostructure of S–GnP. The electrochemical performance of the S–GnP cathode was further improved by employing additional carbon paper (CP) between the separator and the cathode in order to suppress the Li polysulfide shuttle to the anode. This is called the 'Interlayer concept' and will be discussed in separate section. The 0.7S–0.3GnP–CP cell delivered an initial discharge capacity of 970.9 and 679.7 mAh/g at the 500th cycle at the 2.0 C rate.

In summation, graphene (oxide) has significant advantages as a component of the sulfur cathode, such as a high electrical conductivity, the immobilizing effect on sulfur and lithium polysulfides depending on its functional groups on the surface, mechanical flexibility as well as a large surface area. It is commonly used as sulfur deposition surface or wrapping media for sulfur nanoparticles and it compensates for the low electronic conductivity of sulfur, and physically or chemically suppresses Li polysulfide dissolution into liquid electrolytes. However, the effectiveness of graphene (oxide) for enhancing the electrochemical performance of the Li–S cell strongly depends on its physical and chemical structure such as defect concentration and degree of oxidation, so it should be engineered very carefully in order to achieve the most promising and reliable performance of the Li–S cell.

2.2.2.3. Sulfur/1D structured carbon nanocomposites

1D structured carbonaceous materials such as CNFs and CNTs have been employed in many technologies due to their extraordinary thermal and electrical conductivity and mechanical properties. CNT is regarded as a CNF with stacking of rolled graphene layers at specific rolling angles, thus, it forms into perfect cylinders, consisting of a cylindrical wall with a hollow inner space. The properties of CNTs strongly depend on the radius, the rolling angle, and especially, the number of rolled graphene layers which categorize the CNTs into single-walled carbon nanotubes (SWCNTs) and multi-walled carbon nanotubes (MWCNTs). For the Li–S cell application, one-dimensional (1D) structured carbonaceous materials can be employed in the cathode as a conductive additive or a component of a composite active material, because it can provide a continuous and efficient electrical pathway, thus the reaction kinetics of the sulfur cathode can be enhanced, resulting in an improvement of Li–S cell performance. In the early work conducted by Han *et al.* MWCNTs were employed as a conductive additive in the sulfur cathode to improve the electrical conductivity.[97] Two different sulfur cathodes were prepared, with and without MWCNT addition, respectively and were electrochemically tested for comparison. It was demonstrated by cyclic voltammetry (CV) and electrochemical impedance spectroscopic (EIS) techniques that the charge and discharge overpotentials were reduced when the MWCNTs were employed. In addition, the sulfur cathode with the MWCNT additive showed an initial capacity of about 500 mAh/g while that of the sulfur cathode without the MWCNT additive yielded about 400 mAh/g, which verified the effect of the MWCNT as a conductive additive. However, the specific capacity and cycle life of the Li–S cell employing the sulfur cathode with the MWCNT additive were still poor due to the large particle size of the sulfur powder (~80 μm) and no protection against lithium polysulfide dissolution into the electrolyte.

Yuan *et al.* fabricated a sulfur-coated MWCNT composite cathode using the melt-diffusion method.[98] The MWCNTs were used as a component of the composite that provides a large surface area to encourage the formation of a very thin sulfur layer, so the sulfur-coated MWCNT composite can have benefits in terms of electrical conductivity by providing a continuous electrical pathway through the 1D structured MWCNT network, reducing the length of the Li-ion diffusion pathway through insulating sulfur. As a result of the composite material preparation, the 1D structured sulfur-coated MWCNT composite had an average thickness range of sulfur from several to tens of nanometers, depending on the sulfur content in the composite. To verify the benefits of the nanostructure, a sulfur-coated CB cathode and a MWCNT mixed sulfur cathode were prepared for comparison. The results of the electrochemical cycling test at a specific current of 100 mA/g showed that the sulfur-coated MWCNT composite cathode maintained a reversible capacity of about 670 mAh/g after 60 cycles, whereas the other two cathodes exhibited reversible capacities of only about 270 mAh/g after 40 cycles, which confirmed the effect of the nanostructure that improved the electrochemical performance. SEM examination of the cycled cathodes showed a significant difference between the sulfur-coated MWCNT composite cathode and the other two cathodes. In the SEM images, the sulfur-coated MWCNT composite maintained a uniform distribution of sulfur on the cathode whereas the other two cathodes showed obvious aggregation of sulfur (or Li sulfide) after cycling. Based on the results, it was suggested that the nanostructure of the sulfur-coated MWCNT composite not only helped to enhance the reaction kinetics, but also to stabilize the structure of the sulfur cathode during cycling, resulting in enhancement of the Li–S cell's performance.

Although the sulfur-coated CNT nanocomposite cathodes improved the electrochemical properties of the Li–S cells, no effective protection barriers against Li polysulfide dissolution into the liquid organic electrolyte and its shuttle effect were available, thus the specific capacity continuously decreased during cycling. To overcome the lack of protection, a porous material concept discussed above was adopted in order to physically trap Li polysulfides during cycling.[63,99–104] Ji et al. prepared a CNF-sulfur composite, comprising embedded sulfur in the pores of CNF generated by elimination of poly (methyl methacrylate) during the carbonization of poly acrylonitrile (PAN) wires.[99] The large surface area of the pores in the CNFs was filled with deposited sulfur *via* a solution-based chemical reaction, and that was confirmed by electron microscopy and BET analysis. Because sulfur was dispersed well in the pores on the highly electronically conductive CNFs, a high initial specific capacity of nearly 1400 mAh/g at 0.05 C could be delivered with ~85% capacity retention after 30 cycles. In addition, the specific capacity of 987 mAh/g was obtained at the 0.2 C after 30 cycles at various C rates, indicating good rate capability of the Li/S cells. Xin et al. prepared CNT@MPC by coating a porous layer on the surface of MWCNT using D-glucose and sodium dodecyl sulfate (SDS) as carbon precursors followed by a carbonization process.[102] Sulfur was inserted into the pores of the outer porous carbon layer using the melt diffusion method to form a coaxial nanostructure consisting of a CNT core and MPC/S sheath. It was demonstrated that sulfur embedded in the pores of MPC is metastable small sulfur molecules of S_{2-4} (S_2, S_3, and S_4) that have a chain-like structure instead of S_8 rings. Consequently, the S/(CNT@MPC) cathode showed a unique electrochemical behavior that exhibited a single discharge plateau at ~1.85 V (Figure 2.7(a)), instead of two distinguishable discharge plateaus, which is a similar phenomenon to that shown in the previous work discussed above.[46] As shown in Figure 2.7(b), the S/(CNT@MPC) cathode showed a high initial specific capacity of 1670 mAh/g, an impressive cycling stability of 1149 mAh/g after 200 cycles, and a favorable high-rate capability of 800 mAh/g at 5 C. It was suggested that the excellent

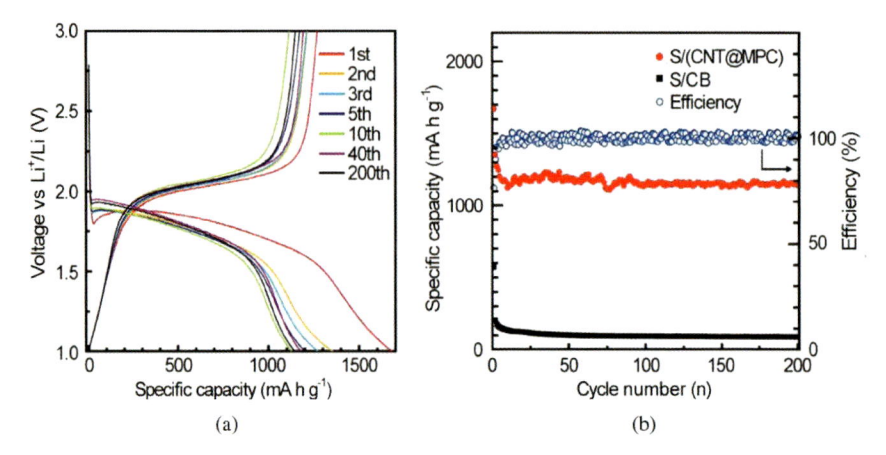

Figure 2.7 Electrochemical properties in carbonate-based electrolyte. (a) Voltage profiles of S– (CNT@MPC) at 0.1 C. (b) Cycling performance of S– (CNT@MPC) and S–CB at 0.1 C (open circles show the Coulombic efficiency of S– (CNT@MPC)).[102]

electrochemical properties of the S/(CNT@MPC) cathode could be attributed not only to an efficient 3D network of electrical pathways for enhancing the reaction kinetics, but also to the space confinement of the carbon micropores on sulfur that prevents the formation of highly soluble Li polysulfides (Li_2S_n, $n = 4–8$) by avoiding the transition between S_8 and S_4^{2-}.

As was discussed above, a core–shell-like structure that provides a closed environment for sulfur is one of the most promising concepts to improve the cycle life of Li–S cells by isolating the sulfur from the liquid organic electrolyte. Hollow 1D structured carbonaceous materials such as hollow CNFs and CNTs can be used as a host material that accommodates sulfur into their inner empty space by solution-based sulfur precipitation methods or a melt diffusion method with the help of capillary forces.[105–112] Moon et al. prepared hollow carbon nanowire (NW) arrays using anodic aluminum oxide (AAO) as a template material and then sulfur was inserted into the hollow space of the NWs, as shown in Figures 2.8(a)–(e).[107] As a result, the uncommon monoclinic phase of sulfur encapsulated in hollow carbon wires (S@C NW) with a very thin carbon shell of about 3 nm was obtained with a high sulfur content of up to 81 wt.%. To fabricate the cathode, the prepared 1D structured S@C NW composite array was attached to a substrate of 316 stainless steel (SS) by using conductive

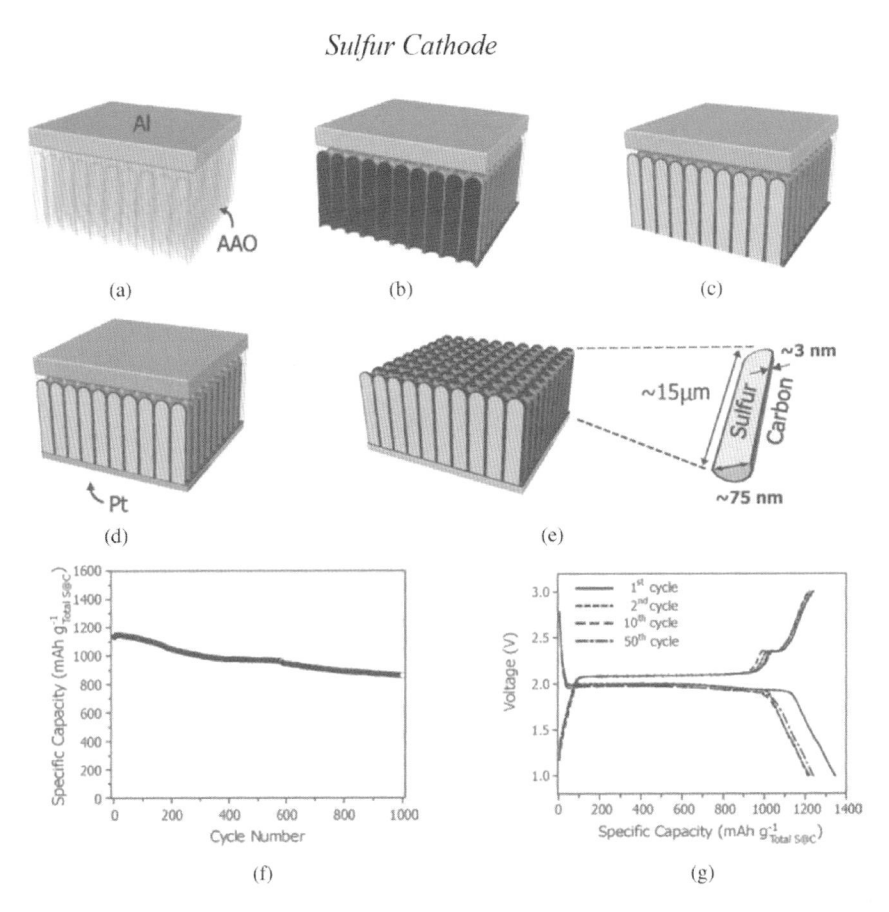

Figure 2.8 Schematic illustration of the fabrication procedure of the S@C NW electrode. (a) Preparation of porous alumina template by a two-step anodization process. (b) Carbon layer deposition by a CVD process to form an array of CNTs. (c) Sulfur infiltration utilizing capillary force through a heat treatment at 155°C. (d) Pt deposition on the opposite side and a heat treatment at 400°C. (e) The final electrode structure after removal of the membrane by a wet etching step, alongside a schematic of each S@C NW with the dimensions of the key components denoted. Electrochemical characterization of the S@C NW electrode. (f) Capacity retention at the discharge rate of 5.0 and the charge rate of 2.0 C for 1000 cycles. (g) Galvanostatic profiles of the S@C electrode at different cycle numbers when measured at 0.5 C for both charge and discharge (1 C = 1675 mA/g).[107]

epoxy and an additional conductive platinum (Pt) layer between the S@C NW composite and the 316 SS substrate. A high discharge specific capacity of about 1200 mAh/g at the very high rate of 20 C was measured and a capacity retention of more than 75% (with specific capacity of 1078 mAh/g at 1000th cycle) was shown even after 1000 cycles at 2.0 and 5.0 C (Figure 2.8(f)). It was suggested that the excellent electrochemical performance of the S@C NW composite cathode is due to: (1) the unique structure of the electrode that prevents the embedded sulfur from having direct contact with the electrolyte. The absence of the first discharge plateau related to the formation of soluble lithium polysulfide (shown in Figure 2.8(g)) may support this idea because soluble Li polysulfide cannot form unless there is direct contact between sulfur and electrolyte; (2) the short electrical diffusion pathway through sulfur overcomes the insulating nature of sulfur. The Pt layer may help to enhance the electronic conductivity of the cathode as well; (3) a robust carbon coating layer which endures the volume expansion of sulfur during discharge. Despite the excellent electrochemical performance of the 1D structured S@C NW composite cathode, some remaining issues due to the uncommon electrode structure can potentially limit the practical usage of the 1D structured S@C NW composite cathode. Especially, it was emphasized that the adhesion between the novel Pt current collector and the 1D structured S@C NW nanocomposite array can significantly influence the electrochemical performance of the cathode. Of course, the high price of a Pt current collector can significantly increase the cost of the Li–S cell.

In the case of the 1D structured sulfur/C core–shell nanocomposites with common electrode microstructure, the reversible capacity of the cells trended downward for the first 100–200 cycles [105,106,108] unless additional strategies to stabilize the cycling of these Li–S cells were employed. Zheng *et al.* revealed one possible capacity fading mechanism of the 1D structured sulfur/C core–shell nanocomposite cathode using *ex situ* TEM observations, and overcame it by introducing an amphiphilic polymer that modifies the carbon surface, rendering strong interactions between the non-polar carbon and the polar Li_xS clusters.[108] Figure 2.9 shows the TEM images of the pristine and the discharged (at 0.2 C to 1.7 V) sulfur/hollow CNF cathode without and with functionalization of the carbon surface using polyvinylpyrrolidone (PVP). As shown in Figures 2.9(a) and 2.9(b), the inner core (identified as Li_2S) of the sulfur/hollow CNF cathode without modification shrank away from the carbon wall along the length of the hollow CNF compared to that of the pristine

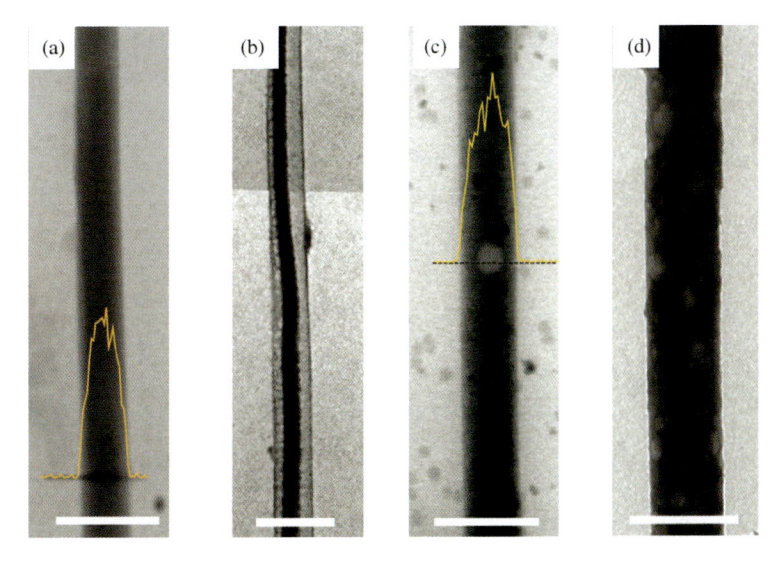

Figure 2.9 *Ex situ* study of hollow CNF encapsulated sulfur cathode. (a) TEM image of the sulfur cathode before discharge. The line represents the EDS counts of the sulfur signal along the dark line. (b) TEM image of the sulfur cathode after full discharge to 1.7 V. (c) TEM image of the sulfur cathode after functionalization with polymer and infusion of sulfur. The line represents the EDS count of the sulfur signal along the dashed line. (d) TEM image of the sulfur cathode after full discharge. The scale bars are 500 nm.[108]

hollow carbon–sulfur cathode after the first discharge. This result was unexpected because the volume of sulfur should be increased instead of shrinking, when sulfur is lithiated (~80% when Li_2S forms). To explain this phenomenon, it was suggested that the separation of Li_2S from the carbon wall was caused due to the leakage of intermediate polysulfides from the hollow CNFs into the electrolyte through the openings, resulting in the loss of electrical contact and capacity decay. On the other hand, the sulfur–PVP modified hollow CNF cathode did not show the severe detachment of Li_2S from the carbon wall; instead, some small spots were observed that might be formed by localized detachment of Li_2S (Figure 2.9(d)). Consequently, the sulfur–PVP modified hollow CNF cathode showed significantly improved electrochemical performance. The initial specific capacity of the unmodified and the modified cathodes were similar (about 800 mAh/g) at 0.5 C, but the unmodified cathode showed a dramatic specific capacity decay with a discharge capacity of a little more than 500 mAh/g after 100 cycles, while the PVP-modified cathode showed a more stable capacity retention of over 80% for more than 300 cycles. DFT simulation results suggested that the hydrophobic groups of PVP allow anchoring of the polysulfide species within the carbon matrix, which supports the *ex situ* TEM observation and the electrochemical test results. According to the improved electrochemical performance of the sulfur–PVP modified hollow CNF cathode and a few more reports,[113–115] the polymer assisted sulfur/1D structured carbon nanocomposite concepts are fairly effective for improving the cycle life of the Li–S cell. The polymer assisted sulfur cathode will be discussed in a separate section.

A tube-in-tube carbon nanostructure (TTCN) with MWCNTs confined within hollow porous CNTs was prepared by Zhao *et al.*[112] using a SiO_2 template coated on the surface of MWCNTs followed by carbon coating on the surface of SiO_2 and etching of SiO_2, in sequence. The prepared TTCN was used as the host material accommodating sulfur in the space between the outer hollow porous CNTs, which was simply prepared by a melt diffusion method. This unique structure was intended not only to provide an efficient electron pathway through both inner MWNTs and the outer porous carbon wall, but also to help suppressing the dissolution of lithium polysulfides into the electrolyte using the micro- and mesopores of the outer CNTs. The S–TTCN nanocomposite cathode delivered a specific capacity of about 700 mAh/g at the third cycle (the first and second cycles were cycled at 0.5 A/g for activation of sulfur) and 647 mAh/g after 200 cycles with a low decay rate of 0.089% per cycle at the discharge specific current of 2 A/g. Rate capability test results showed discharge specific capacities of about 800, 750, 650, and 550 mAh/g when the cathode was cycled at 1.0, 2.0, 4.0, and 6.0 A/g, respectively, and then the discharge capacity recovered to 850 mAh/g when the test specific current was reduced back to 0.5 A/g, indicating good rate capability of the S–TTCN cathode.

As was discussed above, sulfur/1D structured carbon nanocomposites have a structural advantage in terms of conductivity, because the continuous 1D structure of conductive carbon is efficient for delivering electrons to the active sulfur particles in the whole cathode area. From the structural design point of view, sulfur can not only be placed on the surface of 1D structured carbon, but can also be inserted into the inner space of 1D structured carbon, in case of CNTs or hollow CNFs that have a hollow inner space. The nanocomposite with embedded sulfur inside the 1D structured carbon might be more effective for protecting sulfur against lithium polysulfide dissolution during cycling rather than the nanocomposite with coated sulfur onto the 1D structured carbon because of their less open structure. However, even if sulfur is embedded in the inner space of the 1D structured carbon, lithium polysulfides formed during cycling can diffuse into the electrolyte through openings in the carbon outer layer that are used for the sulfur infiltration process. Thus, it seems that both designs need an additional strategy to improve the cycle life of the Li–S cell; i.e. employing micro or mesoporous material or a polymer as protection layer, or doping 1D structured carbon with nitrogen.

2.2.3. Sulfur–polymer nanocomposites

As discussed above, sulfur-carbonaceous material composites are very promising due to the attractive physical and chemical properties of carbonaceous materials that can help to overcome the drawbacks of the sulfur cathode. However, the synthesis processes of carbon are normally conducted at a relatively high temperature (>450°C) which is much higher than the melting temperature of sulfur (~115°C), thus the opportunity for synthesis of carbon structures without damage to the sulfur particles, especially closed carbon structures that encapsulate sulfur particles, is limited. Alternatively, a melt diffusion method is commonly used for sulfur insertion into the empty space of carbon structures through pores or an open structure of carbon, however the pores or open structure of carbon can potentially serve as a Li polysulfide diffusion pathway into the liquid electrolyte during cycling, resulting in capacity decay of the Li–S cell. Because of the technical difficulty in synthesizing a sulfur encapsulating carbon nanocomposite, alternate composite systems that can encapsulate sulfur have been intensively studied. Polymers have been considered as promising materials for sulfur-composite cathodes because various polymer processing methods, especially solvent based processes are feasible at processing temperatures below 100°C allowing sulfur to maintain its structure as long as the solvent does not dissolve sulfur during the process. The advantage of an ambient-temperature polymerization process is that it allows the polymer to be useful for fabrication of a sulfur-encapsulated nanostructured composite without structural damage or loss of sulfur.

Generally, polymers are mechanically soft, so they are suitable for use as a matrix material for sulfur-based composites that will accommodate the volume expansion caused by the lithiation of sulfur particles. The mechanical stability of a polymer matrix helps to physically confine sulfur in the structure during cycling without mechanical failure of the composite, so physical contact between encapsulated sulfur particles and liquid electrolyte can be avoided, which means that sulfur can be confined in the cathode effectively. Moreover, some polymers have functional groups or atoms such as nitrogen and oxygen, so they can act as sulfur immobilizers that chemically trap sulfur and Li polysulfides.[90,108,116–118] For these reasons, rational design of polymer-coated or –assisted sulfur based nanocomposites with proper choice of polymers should be helpful in overcoming the drawbacks of the sulfur cathode. Among many polymers, conducting polymers such as polyaniline (PANi),[113,116,119–121] polythiophene (PTH),[122,123] polypyrrole (PPy),[115,121,124–131] poly(3,4-ethylenedioxythiophene) (PEDOT),[121,132,133] and so on[74,134,135] have been explored for use in the sulfur cathode. Compared to traditional polymers that are used for electrically passive applications due to their electrically insulating nature, conducting polymers are beneficial for the sulfur cathode, because the insulating nature of sulfur (and Li_2S) is one of the major drawbacks of the sulfur cathode that limits utilization of sulfur as an electrode material. Conjugated polymers are the most popular class of polymeric conductors that is represented by poly(p-phenylene),[136] PPy,[137] PTs[138] and its 3-methoxy-derivative[139] and PANi.[140] The common feature of the structure of conjugated polymers is polyconjugation in the π-system of their backbone (for PANi, this holds only in the case of a doped polymer) and they possess the electronic properties of metals while retaining the mechanical properties and processability of conventional polymers.[141] The scientific details of conducting polymers are available in review articles.[141–147]

A sulfur–PANi nanotube polymer backbone/sulfur nanocomposite (SPANi–NT/S) was reported by Xiao et al.[116] The self-assembled PANi nanotubes were treated at 280°C with elemental sulfur, so elemental sulfur reacted with the unsaturated bonds in the polymer chains to form cross-linked, stereo-network structures. This is known as the "vulcanization reaction".[148–150] The usual views on the mechanism of the vulcanization process are that the activity of sulfur liberated in *statu nascendi* is high enough to enable it to react with rubber and to create the spatial structure of the vulcanizate.[149,150] As a result, a SPANi–NT–S nanocomposite that is composed of a SPANi network with both inter and/or intrachain disulfide bond interconnectivity (chemically confined sulfur) and infused sulfur into the hollow voids (physically confined sulfur) was prepared. A possible structure for the SPANi–NT–S nanocomposite is shown in Figure 2.10(a). The prepared SPANi–NT–S nanocomposite cathode was electrochemically tested at various C rates and it exhibited specific capacities of 837, 614 and 568 mAh/g after 100 cycles at the 0.1, 0.5, and 1.0 C rates, respectively, indicating a good rate capability of the SPANi–NT–S nanocomposite cathode (Figure 2.10(b)). A long cycle life of a Li–S cell with good capacity retention of 76% with respect to its specific capacity of the 100th cycle (432 mAh/g) after 500 cycles at the 1.0 C rate was also demonstrated. The author suggested several factors that might contribute to the good electrochemical performance of SPANi–NT–S nanocomposite cathode such as: (1) the vulcanization process produces a 3D, cross-linked SPANi network that provides molecular-level encapsulation of the sulfur compounds; (2) the polymer matrix SPANi functions as a self-breathing, flexible framework during charge/discharge to reduce

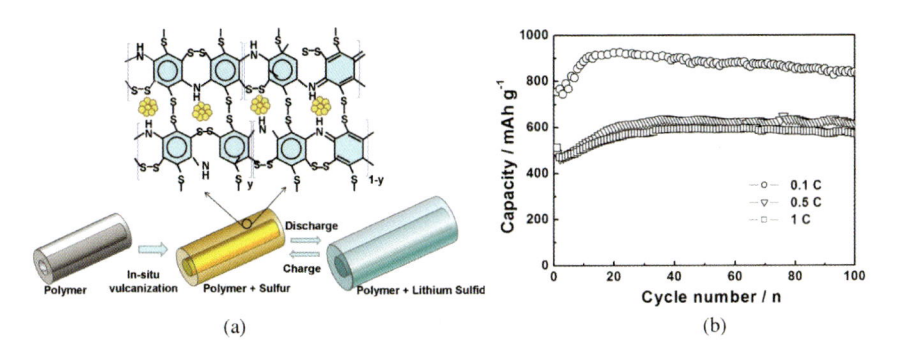

Figure 2.10 (a) Schematic illustration of the construction and discharge–charge process of the SPANi–NT/S composite. (b) Discharge capacity vs. cycle number of the SPANi–NT/S composite electrode at different rates as labeled.[116]

stress and structural degradation; (3) the electropositive amine and imine groups on the SPANi chains further attract polysulfides through electrostatic forces, thereby reducing the displacement of sulfur during repeated cycling.[116]

A yolk–shell nanoarchitecture consisting of PANi as the shell and sulfur as the core was synthesized through the coating of PANi on sulfur nanoparticles followed by a heating process to create void space in the PANi shell by reducing the size of sulfur particles using the vulcanization reaction that consumes some of the elemental sulfur from inside, and partial evaporation of the elemental sulfur.[120] The empty space in the PANi shell can help to accommodate the volumetric expansion of the sulfur during the lithiation process, thus, a yolk–shell nanoarchitecture can exhibit better mechanical stability during cycling as compared to a common core–shell nanostructure. The electrochemical performances of the pure sulfur, the common core–shell S–PANi and the yolk–shell S–PANi nanocomposite cathodes were compared using a galvanostatic discharge–charge test at 0.2 C. The cycling stability of the common core–shell S–PANi cathode was slightly improved, compared to that of the pure sulfur cathode, however, the specific capacities of both pure sulfur and S–PANi core–shell nanocomposite cathodes decayed very quickly and only delivered specific capacities of 124 and 280 mAh/g, respectively after 125 cycles. These results indicate that the PANi coating on the sulfur particles can improve the electrochemical performance of the sulfur cathode by enhancing the electrical conductivity of the cathode and by reducing Li polysulfide dissolution into the liquid electrolyte, however, the protective coating integrity of the S–PANi core–shell nanocomposite was not preserved during the volumetric expansion caused by lithiation of the sulfur core, thus polysulfides can eventually escape during cycling. In contrast, the yolk–shell S/PANi nanocomposite cathode retained a stable specific capacity of 765 mAh/g after 200 cycles, corresponding to a capacity retention of 69.5%, which was much higher than that of both pure sulfur and S–PANi core–shell nanocomposite cathodes. The improved cycling stability verified that the mechanical stability of the PANi shell provided by the empty space to accommodate volume expansion of sulfur particles can help to stabilize the specific capacity more effectively during cycling. The structural stability of the PANi shell of the S–PANi yolk–shell nanocomposite was confirmed by *ex situ* SEM observation of the cycled S–PANi core–shell and S–PANi yolk–shell nanocomposite cathodes. In the SEM images, the PANi shells of the S–PANi yolk–shell nanocomposite were still well preserved, whereas most PANi shells of the S–PANi core–shell nanocomposite were cracked after only 5 cycles.

Another conducting polymer, PTH was also used as a conductive composite matrix to form a sulfur–polythiophene (S–PTH) core–shell nanostructured composite.[122] After the introduction of PTH, a flake-like PTH shell with a thickness of about 20–30 nm formed on the surface of the sulfur particles, which was observed by SEM and TEM. To verify the effect of the PTH coating on the electrochemical performance of the sulfur cathode, galvanostatic charge-discharge cycle tests were conducted and the results revealed that the cycle life and rate capability of the Li–S cell were significantly improved after PTH coating. The initial capacities of S–PTH core–shell nanostructured cathode and bare sulfur cathode were 1119 and 1019 mAh/g, respectively, however, the reversible capacity of the bare sulfur cathode decreased to only 282 mAh/g after 80 cycles, corresponding to a capacity retention of about 28%. In contrast, the S–PTH core–shell nanostructured cathode showed a much higher specific capacity of 830.2 mAh/g after 80 cycles, corresponding to an improved capacity retention of 74.2%. Because PTH itself can be used as an active material for a Li rechargeable cell,[151] a galvanostatic test was conducted for a PTH cathode under the same test conditions to demonstrate the contribution of PTH to the specific capacity of the S/PTH core-shell nanostructured cathode. The measured specific capacity of the Li–PTH cell was only about 10 mAh/g, indicating the negligible capacity contribution of PTH to the specific capacity of the S–PTH core–shell nanostructured cathode. The specific capacity of the Li–PTH cell was much smaller than that reported previously because the charge reaction between PTH and Li–ions mainly occurs between about 3.0 and 4.0 V,[151] which is higher than the upper limit of the voltage window for the galvanostatic cycling test for the S–PTH core–shell nanostructured cathode. To evaluate the rate capability of the bare sulfur and the S–PTH core–shell nanostructured cathodes, a rate capability test was conducted at various test specific currents from 100 to 1600 mA/g and then back to 100 mA/g. At a low specific current of 100 mA/g, the specific capacities of the bare sulfur and the

S–PTH core–shell nanostructured composite cathodes were comparable, however, as the test specific current and the cycle number increased, the specific capacity of the bare sulfur cathode became much smaller than that of the S–PTH core–shell nanostructured composite cathode. After a rate test from 100 to 1600 mA/g, the specific capacity of the S/PTH core–shell nanostructured composite cathode remained at 811 mAh/g when the specific current returned to 100 mA/g, indicating improved cycling stability of the Li–S cell after PTH coating. The improved sulfur utilization, cycle life and rate capability of the S/PTH core–shell nanostructured composite cathode was mainly due to the conductive nature of the PTH shell that reduced the particle-to-particle contact resistance.

Ultrafine sulfur nanoparticles with a size of about 10–20 nm were prepared using a membrane to form microdroplets of S–CS_2 solution and PVP in ethanol (precipitant) as a wrapping ligand to prevent the sulfur aggregation.[132] As the particle size of sulfur is reduced, the length of both electron and Li–ion conducting pathways through insulating sulfur are also reduced, thus reaction kinetics can be improved. Next, PEDOT was coated on the sulfur nanoparticles to prevent Li polysulfide dissolution into the liquid electrolyte by stably-limiting direct contact between sulfur and liquid electrolyte and improving the electrical conductivity of the electrode.[132] The TEM images (Figures 2.11(a) and 2.11(b)) showed that the obtained sulfur nanoparticles have a spherical shape with diameters ranging from 10 to 20 nm and an amorphous shell with a thickness of about 5 nm is formed on the sulfur nanoparticles after PEDOT coating. Figure 2.11(c) shows the voltage profiles of the commercial sulfur (cp–S), the synthesized sulfur nanoparticles (nano-S) and the sulfur encapsulated in PEDOT matrix (nano-S@PEDOT) cathodes. As shown in the voltage profiles, the nano-S@PEDOT cathode exhibited a much more flat and well-defined lower plateau at 2.0 V, while the other two cathodes showed slopes between 2.1 and 1.5 V. The author suggested that the shape change of the voltage profile after PEDOT coating was because the PEDOT matrix can help to trap freshly formed long-chain polysulfides in their structure, thus the formation of insoluble Li_2S_2 and Li_2S layers that impede electrical conduction at the sites in contact with the conducting network can be suppressed before the complete transition from long-chain Li polysulfide to short-chain Li polysulfide. On the contrary, when the sulfur

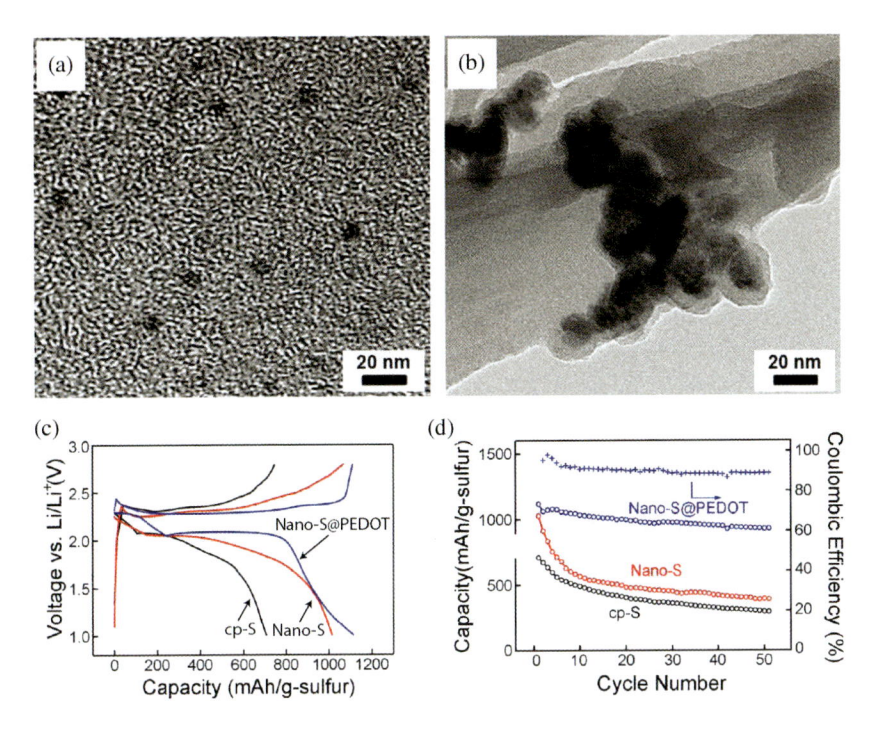

Figure 2.11 TEM images of (a) pure S nanoparticles and (b) S/PEDOT core/shell nanoparticles. (c) Initial discharge/charge curves of nano-S@PEDOT, nano-S, and CP-S cathodes at a specific current of 400 mA/g (0.25 C) (d) Discharge capacity and Coulombic efficiency cycling stability for nano-S@PEDOT, nano-S, and CP-S cathodes at a specific current of 400 mA/g.[132]

particles directly contacted the liquid electrolyte due to the lack of a protection layer, long-chain polysulfides may be reduced to short-chain polysulfides and even to insoluble and insulating Li_2S_2 and Li_2S near the contact sites, which may hinder the reduction of polysulfides far away from the contact sites, and thus lead to a rapid increase of polarization as the discharge proceeds.[132] The comparison of cycling results shown in Figure 2.11(d) indicates that the nano-S cell exhibited an initial discharge specific capacity of up to 1000 mAh/g, more than that of the CP–S cell (about 700 mAh/g), however both the CP–S and the nano-S cells showed drastic capacity fading at a test specific current of 400 mA/g. After 10 cycles, the specific capacities of the CP–S and the nano-S cathodes decreased to 69 and 55% of their initial specific capacities, respectively.

On the other hand, the nano-S@PEDOT cathode showed the highest initial specific capacity of about 1117 mAh/g and the best capacity retention of about 83% after 50 cycles. The test results verified that the strategies of the nano-S@PEDOT were fairly effective for improving the electrochemical performance of the Li/S cell.

Li *et al.* studied the influence of different conducting polymers on the electrochemical performance of the sulfur cathode.[121] PEDOT, PPy, and PANi, three of the most well-known conducting polymers, were coated, respectively, onto monodisperse hollow sulfur nanospheres through a simple polymerization process. The structures of the PANi–S, the PPy–S, and the PEDOT–S electrodes were the same except for the polymer coatings on the hollow sulfur nanospheres (the coating thicknesses were also maintained the same, the electrochemical performances of conducting polymer coating with 20 nm thickness are discussed here) to compare the effect of each conducting polymer on the electrochemical performance of Li–S cells, and the results of electrochemical tests are shown in Figure 2.12. The PANi–S, the PPy–S and the PEDOT–S electrodes exhibited high initial specific discharge capacities of 1140, 1201, and 1165 mAh/g, respectively, at the 0.5 C rate, however, the PANi–S cell showed a relatively poor capacity retention of about 45%, whereas the PPy-S and PEDOT-S cathodes exhibited capacity retentions of 60 and 67%, respectively (Figure 2.12(a)). Because the functional groups in the polymers can affect the cycling stability of the sulfur cathode positively, *ab initio* simulations in the framework of DFT were performed to investigate the interaction between Li polysulfide (Li$_x$S, $0 < x \leq 2$) species and the conducting polymers. The simulation results indicate that PEDOT has a much stronger binding energy (1.08 eV) with the Li atom in Li–S than PANi (0.59 eV) and PPy (0.50 eV). Although the simulation results might not give us an absolute quantification of the binding strength, the experimentally demonstrated cycle life of the PANi–S, the PPy–S and the PEDOT–S electrodes can be supported by a qualitative understanding of the influence of chemical bonding on the cycling stability of the sulfur cathode. The rate capability test results and voltage profiles of the three cathodes (Figures 2.12(b) and 2.12(c)) clearly reveal that the rate capability of the cells were ordered as follows: PEDOT–S > PPy–S > PANi–S. At a high

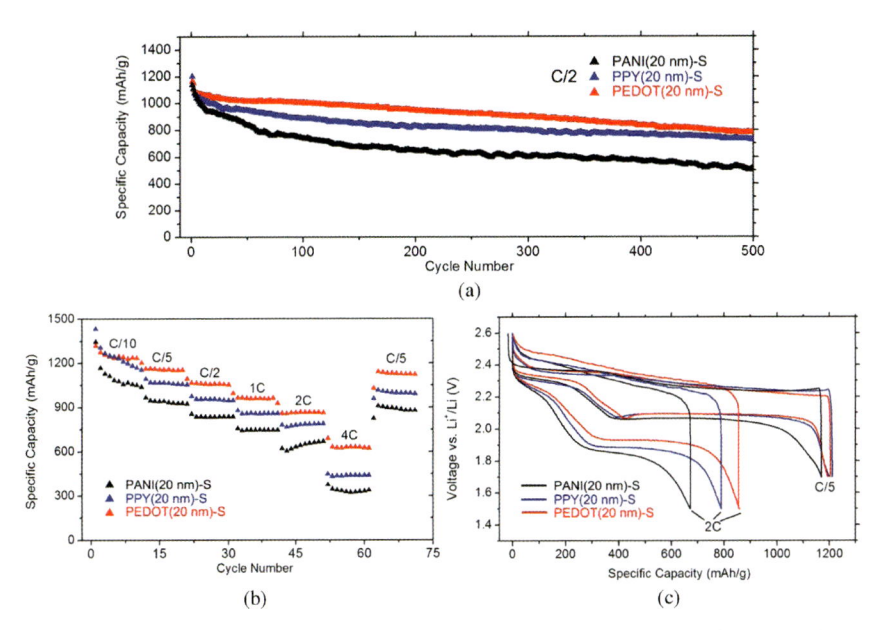

Figure 2.12 (a) Cycling performance of the cells made from hollow sulfur nanospheres with PANi, PPy and PEDOT coatings of ~20 nm at C/2 rate for 500 cycles. (b) Rate capability of the cells discharged at various rates from C/10 to 4 C. (c) Typical discharge–charge voltage profiles of the cells at 0.2 and 2.0 C (1 C = 1673 mA/g).[121]

C–rate of 2.0 C, the PEDOT–S and PPy–S delivered reversible capacities of 858 and 789 mAh/g, which was higher than that of the PANi–S (666 mAh/g). As shown in Figure 2.12(c), the voltage hysteresis between the discharge and the charge curves decreased in the order of PANi–S > PPy–S > PEDOT–S at both 0.2 and 2.0 C. Because the structures of the three materials were the same, it was suggested that the rate capability of the cathodes was mainly determined by the conductivity of the polymer shell. The EIS test results supported the electrochemical test results, where the charge transfer resistance of the cathodes in order of PEDOT–S < PPy–S < PANi–S at different states of charge during the first discharge. Consequently, in this work, PEDOT was found to be the best choice to achieve a long cycle life and high-rate capability among the three conductive polymers, due to its strong binding energy with the lithium atom in Li–S, and better conductivity than those of the other conductive polymers.

It was emphasized previously that electrically conductive matrix materials for sulfur based composites are desired due to the insulating nature of sulfur and Li$_2$S. This is the reason that conducting polymers have been mainly considered for the sulfur cathode, among many polymers. However, as was discussed above, non-conductive PVP can also improve the cycle life of the Li–S cell by chemically trapping Li polysulfides in the cathode during cycling.[108] A few works have been recently reported on developing polymer–sulfur cathodes with the non-conducting polymers such as PVP,[108] amylopectin[90] and polydopamine (PD)[117,118] that are capable of chemically trapping Li polysulfides. Because these polymers are not beneficial in terms of electronic conductivity, they were used as supporting materials to further improve the electrochemical performance of sulfur–carbon composite cathodes. Zhou et al. reported a PD coated, nitrogen-doped hollow carbon–sulfur double-layered core–shell structure that has multiple barriers to confine sulfur in the cathode.[117] As discussed in the sulfur–carbon nanocomposites section, a nitrogen doped porous carbon shell can physically and chemically trap Li polysulfides during cycling. However, pores in the carbon shell used for sulfur insertion into the hollow space can serve as a Li polysulfide diffusion pathway to the electrolyte, thus a potential risk to lose sulfur from the electrode exists during cycling. To overcome this drawback, a nitrogen doped hollow carbon–sulfur cathode (NHC–S) was surrounded by an additional PD shell layer (PDA-NHC-S), so the immobilization of Li polysulfides was facilitated. Based on the results of galvanostatic charge–discharge testing at the 0.2 C rate, both the NHC–S and the PDA–NHC–S nanocomposite cathodes delivered similar initial discharge capacities of 1141 and 1070 mAh/g, respectively. However, the specific capacity of the NHC–S cathode gradually decreased and showed about 600 mAh/g after 150 cycles. In contrast, the PDA–NHC–S composite cathode exhibited the best capacity retention of about 84% after 150 cycles with a capacity of 900 mAh/g. To verify a long cycle life of the PDA–NHC–S composite cathode, it was also galvanostatically cycled at 0.6 C. The excellent cycle life of the PDA–NHC–S composite cathode was demonstrated with the specific capacity of 630 mAh/g after 600 cycles, corresponding to a capacity retention of 85.1%.

In summary, polymers are useful as matrix materials for sulfur based nanocomposites because many polymerization or polymer coating processes can be conducted at a relatively low temperature of less than 100°C, which allows sulfur to maintain its structure as long as the solvent for the polymer processing does not dissolve sulfur. Therefore, sulfur encapsulated nanostructures with a polymer matrix can easily be prepared without structural damage or loss of sulfur. Among many polymers, electrically conducting polymers have been extensively studied for sulfur cathodes to facilitate ion and charge transport. Moreover, various polymers can chemically trap sulfur and Li polysulfides using their functional groups, thus, the cycle life of Li–S cells can be improved. In addition, polymers are mechanically soft, thus the volume expansion of sulfur particles caused by the lithiation process can be effectively accommodated, and stable protection against mechanical failure of the electrode, and Li polysulfide dissolution can be provided. However, chemical and thermal stability of these polymers in organic electrolytes needs to be addressed for any commercial Li–S cell.

2.2.4. Other sulfur based nanocomposites

As discussed above, uniquely structured sulfur/carbonaceous material and sulfur–polymer nanocomposites with promising electrochemical performance have been proposed as cathode materials for the Li–S cell in numerous papers. It is unquestioned that nanostructured sulfur/carbonaceous material and sulfur–polymer composites are in the mainstream of research on developing advanced Li–S cell cathodes, however, there are some other notable achievements for developing sulfur-based composite cathode systems such as sulfur metal oxides[152–165] and sulfur MOF[166] nanocomposites and organosulfur compounds[167–177] for Li–S cells. These classes of sulfur-based nanocomposite cathodes can offer new opportunities to improve the electrochemical properties of the Li–S cell.

Among them, sulfur/metal oxide nanocomposites have attracted attention from researchers with a few different concepts to improve the cycle life of Li–S cells, e.g. absorbing materials to trap Li polysulfides[152–157];

porous structured metal oxide as Li polysulfide reservoirs[158–162] and insulating protection layers to suppress Li polysulfide dissolution into the liquid electrolyte.[163–165] In 2013, binary and ternary metal oxides such as La$_2$O$_3$,[152] Al$_2$O$_3$,[153] and Mg$_{0.6}$Ni$_{0.4}$O[154] were proposed as absorbers of Li polysulfides which were homogeneously distributed in the sulfur based composite. The metal oxides were intended to support the sulfur–carbon or sulfur/polymer nanocomposites for further improvement in the electrochemical performance of Li–S cells, so they can also be regarded as additives. Although the working mechanisms of a Li polysulfide absorber are not clearly revealed in the papers, sulfur based nanocomposite cathodes containing metal oxide additives showed higher reversible discharge capacity and better capacity retention than those of the sulfur based nanocomposite cathodes without the metal oxides. The beneficial effect of Mg$_{0.6}$Ni$_{0.4}$O as polysulfide absorber was demonstrated by SEM observation of both the cycled sulfur–PAN and the sulfur–PAN–Mg$_{0.6}$Ni$_{0.4}$O cathodes. In the SEM image of the sulfur–PAN cathode, agglomeration of the composite particles in the cathode was clearly observed, whereas agglomeration was suppressed when Mg$_{0.6}$Ni$_{0.4}$O was employed in the sulfur–PAN composite.[154]

Recently, the metallic oxide, Ti$_4$O$_7$ was reported as a component of a sulfur-based nanocomposite by Pang et al.[156] and Tao et al.[157] Magnéli Ti$_4$O$_7$ is a highly conductive oxide that exhibits a bulk metallic conductivity of 2×10^3 S/cm at 298 K[178] and it contains polar O–Ti–O units that have a high affinity for Li polysulfides, thus it can bind Li polysulfides.[156] Pang et al. prepared a highly porous Magnéli Ti$_4$O$_7$, exhibiting a very high BET surface area of 290 m^2/g for efficient utilization of a polar surface for a strong Li polysulfide binding effect as well as a physical confinement of sulfur and Li polysulfide using its pores. Consequently, the sulfur/porous Ti$_4$O$_7$ nanocomposite has multiple features for improving the electrochemical performance of the Li–S cell. To verify the effect of the porous Magnéli Ti$_4$O$_7$ host, a sulfur–Ti$_4$O$_7$ nanocomposite cathode was prepared and galvanostatically tested between 1.8 and 3.0 V. The voltage profiles of the sulfur–Ti$_4$O$_7$ nanocomposite cathode at different C-rates from 0.05 to 1 C only showed a little increase in polarization with increasing rate, which indicates that the high electronic conductivity of the Ti$_4$O$_7$ host improved the rate capability without contribution to the capacity of the cell (only 6 mAh/g at the same test condition). The long cycle life of the sulfur–Ti$_4$O$_7$ (60% S, Ti$_4$O$_7$–S/60) nanocomposite cathode was demonstrated at the 2.0 C rate, with an initial specific capacity of 850 mAh/g and a capacity fade rate of only 0.06% per cycle over 500 cycles. The excellent cycle life of the Ti$_4$O$_7$–S/60 cathode was attributed to not only physical confinement of sulfur and lithium polysulfides in the pores of the mesoporous Ti$_4$O$_7$, but also a polar surface for strong lithium polysulfide binding. To verify a polysulfide binding effect of Ti$_4$O$_7$, a sulfur/porous carbon nanocomposite cathode was prepared, which has a similar BET surface area of about 260 m^2/g and sulfur content (60%) in the composite. Compared to the fast capacity decay rate of the sulfur–carbon cathode (0.16% per cycle), the Ti$_4$O$_7$–S-60 cell showed good capacity retention with a fade rate as low as 0.08% per cycle over 250 cycles at 0.5 C. This result was supported by SEM observations of the electrodes after the first discharge that showed conformal Li$_2$S deposition on the particle surface of the Ti$_4$O$_7$/S-60 cathode, whereas the Li$_2$S on the carbon composite was broadly and non-specifically distributed. DFT calculations and experimental characterizations such as XPS, and TEM with EDS conducted by X. Tao et al. also proved that the surface of Ti$_4$O$_7$ with abundant low coordinated Ti sites is favorable for the adsorption and selective deposition of the sulfur species, resulting in improved cycle life of the Li–S cell.[157]

A highly efficient polysulfide mediator, manganese dioxide (MnO$_2$) was reported by Nazar et al. in 2015.[155] They found that ultrathin MnO$_2$ nanosheets can entrap Li polysulfide by the unique surface reaction chemistry of MnO$_2$, thus sulfur can be confined in the cathode during electrochemical cycling, resulting in good cycle life for the Li–S cell. For visual confirmation of polysulfide entrapment at specific discharge depths, electrochemical cells with sulfur–ketjen black (S–KB) and S–MnO$_2$ nanosheet cathodes were assembled in clear vials (Figure 2.13(a)) and the cells were galvanostatically discharged to 1.8 V at 0.05 C. As a result, the clear electrolyte in the S–KB cell changed to bright yellow-green on partial discharge of the cell over 4 hours and the color of the electrolyte remained yellow until the end of discharge (12 hours), indicating the existence of the solubilized Li polysulfide in the electrolyte. In contrast, in the S/MnO$_2$ cell, the electrolyte exhibited only a faint yellow color at 4 hours, demonstrating a relatively better trapping of Li polysulfide compared to the

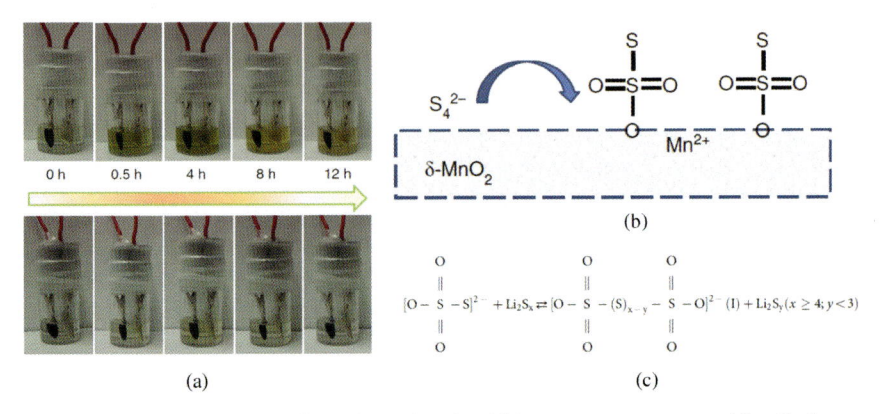

Figure 2.13 (a) Visual confirmation of polysulfide entrapment at specific discharge depths (upper: S/KB, lower: S/MnO$_2$ cells). (b) Schematic showing the oxidation of initially formed polysulfide by d-MnO$_2$ to form thiosulfate on the surface, concomitant with the reduction of Mn^{4+} to Mn^{2+}. (c) Reaction chemistry of thiosulfate on the surface of MnO$_2$ to anchor soluble long-chain polysulfides and its conversion to insoluble lower polysulfide that is Li$_2$S$_2$ or Li$_2$S.[155]

S–KB cells. Interestingly, the electrolyte was rendered completely colorless at the fully discharged state, which means that the soluble Li polysulfide was effectively converted to insoluble reduced species such as Li$_2$S$_2$ and Li$_2$S. A unique two step reaction chemistry of MnO$_2$ to mediate Li polysulfide was proposed as follows: (1) thiosulfate groups are first created *in situ* by oxidation of initially formed soluble Li polysulfide species on the surface of ultrathin MnO$_2$ nanosheets (Figure 2.13(b)); (2) As reduction proceeds, the surface thiosulfate groups are proposed to anchor newly formed soluble 'higher' polysulfides by catenating them to form polythionates and converting them to insoluble 'lower' polysulfides (Figure 2.13(c)). The S–MnO$_2$ cell exhibited a long cycle life with an initial discharge capacity of up to 1300 and 380 mAh/g after 1200 cycles at 0.2 C, which was attributed to active polythionate complexes on the surface of MnO$_2$ serving as an anchor and transfer mediator to confine sulfur in the cathode.

Porous nanostructured metal oxides such as silica[162] and titania,[158–161] which have been widely studied for many applications (e.g. microelectronics, chemical sensors and catalyst substrates) are also notable as a component of sulfur based nanocomposites for Li–S cell cathodes. The strategy of using porous metal oxides to improve the cycle life of the Li–S cell is technically the same as that of porous carbonaceous materials, which can be briefly described as physical confinement of sulfur and Li polysulfides in micro, meso and macropores during electrochemical cycling, the so-called polysulfide reservoir concept. One of the representative works was conducted by Ji *et al.* wherein mesoporous silica (SBA–15) was used as a Li polysulfide reservoir. SBA–15 is well-ordered hexagonal silica with tunable large uniform pore sizes that exhibit high surface area, a large pore volume and highly hydrophilic surface properties.[179] SBA–15 was mixed with a MPC–sulfur (SCM–S) nanocomposite to form a SBA–15–SCM–S nanocomposite and the composite was used as the active material for a sulfur cathode. The SCM–S and the SBA–15–SCM/S cathodes were galvanostatically cycled at 0.2 C and the results showed that similar initial capacities of the SMC–S and the SBA–15–SCM–S cells were obtained (920 and 960 mAh/g, respectively), however, the capacity of the SMC–S cathode decayed to less than 500 mAh/g after 40 cycles, whereas that of the SBA–15–SCM–S cathode was above 650 mAh/g. The Coulombic efficiency of the Li–S cell was also improved to about 98% when SBA-15 was added, indicating that the existence of the SBA-15 can further improve cycle life of the SMC/S cell, as a result of sulfur polysulfide confinement.

Recently, a sulfur MOF nanocomposite cathode was reported by Zheng *et al.*[166] MOFs are an emerging class of porous materials constructed from metal-containing nodes (also known as secondary building units, or SBUs) and organic linkers.[180] MOFs are structurally and functionally tunable by organic ligand design and postsynthetic modification of the linker as well as the metal clusters and their coordination bonds. These inorganic–organic hybrid materials commonly offer very high porosity and large surface area, thus it is reasonable that MOFs are considered as a porous structured host material for use in sulfur based nanocomposites. In Zheng's work, a hierarchical porous structured Ni–MOF was synthesized with an extremely high BET surface area of 5243 m^2/g and a sulfur–Ni–MOF nanocomposite that consisted of about 60 wt%. sulfur was obtained using a melt-diffusion method at 155°C. After sulfur insertion, the BET surface area decreased to 514 m^2/g, which may indicate that the pores of Ni–MOF were filled with elemental sulfur. The prepared sulfur–Ni–MOF nanocomposite was electrochemically evaluated as a Li–S cell cathode material and the results showed an initial specific capacity of about 689 mAh/g and a specific capacity of 611 mAh/g after 100 cycles at 0.1 C. Although the specific capacity of the sulfur–Ni–MOF nanocomposite cathode was lower than that of sulfur–carbonaceous material nanocomposite cathodes reported in the previous literature, which is probably due to the poor conductivity of Ni–MOF, it showed a good capacity retention of about 89% after 100 cycles. The improved sulfur–Ni–MOF cell cycle life could be attributed to not only the porous structure of Ni–MOF, but also the strong interactions between the Lewis acidic Ni(II) center and the polysulfide soft Lewis base, which was confirmed by a DFT study.

Use of metal oxide materials as a protective coating layer to inhibit polysulfide dissolution into liquid electrolytes has been reported using alumina (Al_2O_3)[163,164] or silica (SiO_x).[165] H. Kim et al. coated an Al_2O_3 layer onto the surface of a sulfur–carbon composite using rapid plasma enhanced atomic layer deposition (PEALD). Sulfur–activated carbon fibers (S–ACFs) were first prepared using a melt-diffusion method and they were cast onto a Ni current collector with a polyacrylic acid (PAA) binder. After casting and vacuum drying of the electrode, PEALD was conducted to form an Al_2O_3 layer on the electrode surface. Because of the low operating temperature of PEALD, evaporation of sulfur could be avoided during the Al_2O_3 coating. To verify the effect of the Al_2O_3 protective layer on the electrochemical performance of the Li–S cell, both the uncoated and coated S–ACFs nanocomposite cathodes were galvanostatically cycled at 0.5 C at 70°C to accelerate the polysulfide dissolution into the liquid electrolyte, intending faster cell degradation. As a result, the uncoated S–ACFs nanocomposite cathode showed an initial specific capacity of about 900 mAh/g, whereas that of the Al_2O_3 coated S–ACFs nanocomposite cathode (50-Al_2O_3 with thickness of Al_2O_3 in the range from 3 to 5 nm.) was around 400 mAh/g at 0.2 C, which might be caused by the Al_2O_3 layer increasing the electrode resistance. Although the initial capacity of the Al_2O_3 coated S–ACFs nanocomposite cathode was smaller than that of the uncoated S–ACFs nanocomposite cathode, it maintained a reversible capacity of above 300 mAh/g for 370–470 high temperature cycles, whereas that of the uncoated S–ACFs nanocomposite cathode decreased to below 50 mAh/g after 200 cycles. The stable cycling performance of the

Al$_2$O$_3$ coated S–ACFs nanocomposite cathode was supported by SEM observation of both the anode and the cathode in the Al$_2$O$_3$ coated S–ACFs cell after 200 cycles, which showed a uniform morphology of cathode and a smooth surface of the Li foil anode. In contrast, the morphology of both anode and cathode in the uncoated S–ACFs cell were covered by large chunks of Li sulfide. These results indicated that the Al$_2$O$_3$ coating can reduce Li sulfide dissolution from the cathode, thus the shuttling effect of Li polysulfides to the Li anode can be suppressed.

Organosulfur compounds which are one class of organic materials that contain sulfur in their chemical structure offer a novel class of sulfur cathode for the Li–S cell. Organosulfur compounds were reported by S. J. Visco *et al.* as alternative electrode materials for the high temperature sodium–sulfur cell to lower the operating temperature of the cell, in order to overcome the corrosion of the positive electrode container caused by the molten polysulfides at the operation temperature of about 320–350°C.[181] Since the Li rechargeable cell became popular in the 1990s, various organosulfur compound cathodes have been investigated as active materials for the Li rechargeable cell cathode.[167–177] The basic overall cell reaction of Li/organosulfur compound cathodes can be expressed as[167]:

$$\text{Positive electrode: RSSR} + 2e^- \rightarrow 2\text{RS}^-$$

$$\text{Negative electrode: Li} \rightarrow \text{Li}^+ + e^-$$

$$\text{Overall: 2Li} + \text{RSSR} \rightarrow 2\text{LiSR}$$

where R is an organic moiety.

In the context of searching for a new sulfur cathode system, organosulfur compound cathodes are attractive due to the advantages of polymer engineering such as a potentially low production cost and flexibility of material design for functionalization, because the physical and chemical properties or electrochemical behavior of organosulfur compound cathodes can vary depending on their molecular structure. However, some issues regarding the use of organosulfur compounds as cathode materials need to be addressed: (1) poor electrochemical kinetics that increase the polarization overpotential, resulting in loss of cell voltage. Thus, employing electronically conductive materials such as carbonaceous materials, conducting polymers and highly conductive metal powders might be needed to improve the electrochemical rate capability of organosulfur cathodes; (2) a low percentage of sulfur in the organosulfur compound causes a low specific energy of the Li–S cell, even if the organosulfur compound cathode exhibits a high specific capacity based on the mass of active sulfur. The weight corresponding to the electrochemically inactive portion of the molecules needs to be minimized to achieve a high specific energy of the Li–S cell; (3) physical and chemical stability of the organosulfur compounds in electrochemical cell in the desired voltage window, which is necessary to obtain long cycle life. Some examples of organosulfur cathodes that have been studied for Li rechargeable cells are given in Table 2.1.

Recently, a notable organosulfur compound cathode was reported by Chung *et al.* which was prepared by a facile 'inverse vulcanization' process to prepare chemically stable and processable polymeric materials

Table 2.1: Some organosulfur compounds investigated as positive electrodes.

Compound	Gram per equivalent (g/eq)	Chemical structure	Theoretical specific capacity (mAh/g)
2,5-Dimercapto-1,3,4-thiadiazole (DMcT)	74		362
Dimethyl ethylenediamine	75		357
2-Mercapto-5-methyl-1,3,4-thiadiazole (McMT)	131		205
Anthra [1′,9′,8′-b,c,d,e] [4′,10′,5′-b′ c′ d′ e′] bis-[1,6,6a (6a-S^{IV}) trithia]pentalene (ABTH)	60.7		442
Poly(5,8-dihydro-1 H,4H-2,3,6,7-tetrathiaanthracene) (PDTTA)	64		419
Polydithiodianiline (poly (DTDA))	121		222

through the direct copolymerization of elemental sulfur with vinylic monomers.[175] The inverse vulcanization process was described as the stabilization of polymeric sulfur against depolymerization by copolymerizing a large excess of sulfur with a modest amount of small-molecule dienes, whereas polydienes are cross-linked with a small fraction of sulfur to form synthetic rubber in the conventional vulcanization process. The copolymerization process of S_8 with 1,3-diisopropenylbenzene (DIB) via inverse vulcanization is described in Figure 2.14(a). Briefly, sulfur powder was heated at 185°C for the ring-opening polymerization (ROP) of S_8 and DIB was added directly to the molten sulfur medium while the temperature was maintained at 185°C. The prepared poly(sulfur-*random*-1,3-diisopropenylbenzene) (poly(S–r–DIB)) material was employed as a cathode material for the Li–S cell. Among the

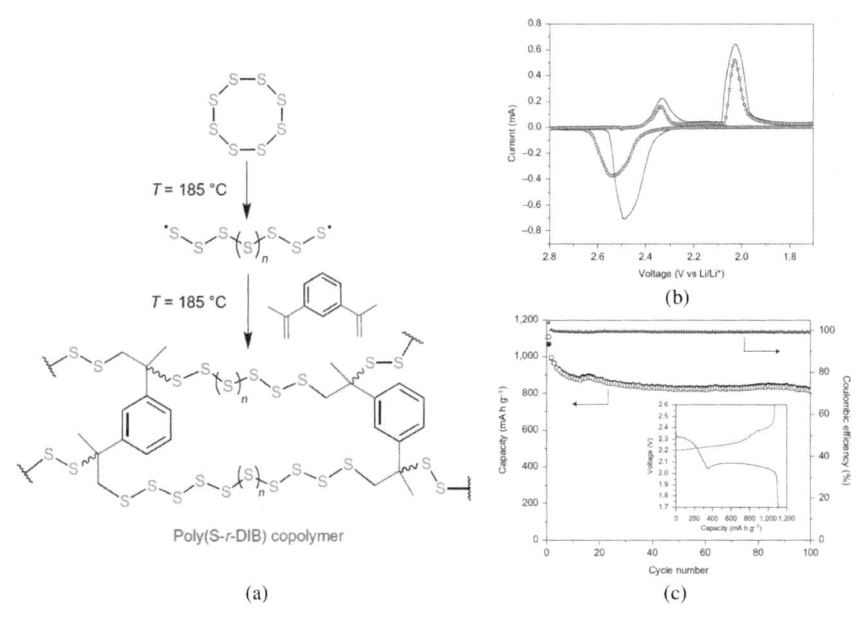

Figure 2.14 (a) Synthetic scheme for the copolymerization of S_8 with DIB to form chemically stable sulfur copolymers. Electrochemical performance of the poly(S–r–DIB) cathode. (b) CV of the S_8 (solid line) and the poly(S–r–DIB) cathodes at a scan rate of 20 μV/s. (c) Cell cycling data for the poly(S–r–DIB) cathode showing the discharge capacity (open circles), charge capacity (filled circles) and Coulombic efficiency (triangles) with the inset showing a typical charge/discharge profile.[175]

poly(S–r–DIB) materials which had sulfur contents from 50 to 90%, poly(S–r–DIB) with a very high sulfur content of 90% was chosen as the Li–S cell cathode material. As shown in Figure 2.14(b), CV curves of both the S_8 and the poly(S–r–DIB) cathodes were very similar, which showed two distinguishable discharge peaks, indicating similar electrochemical behavior between the poly(S–r–DIB) and the S_8 cathodes. The poly(S–r–DIB) cathode (Figure 2.14(c)) exhibited a good cycle life with specific discharge capacities of about 1100 and 823 mAh/g at the first and the 100th cycle, respectively, at 0.1 C. Rate capability test results showed reasonable capacity retention between 900–700 mAh/g from the range of 0.2 to 1.0 C rates and a specific capacity of 400 mAh/g was obtained at the 2.0 C rate. The poly(S–r–DIB) cathode should be highlighted among many organosulfur compound cathodes due to its good electrochemical performance with an exceptionally high sulfur content of 90% in the compound as well as simple, inexpensive and scalable synthesis procedure.

In summary, various kinds of metal oxides have been explored as a component of sulfur based composites in order to enhance the cycle life of the Li–S cell. The role of metal oxides in sulfur-based composite cathodes depends on the physical and chemical structure of the metal oxides, for example, Ti_4O_7 or MnO_2 can entrap sulfur and Li polysulfides on their surface which is attributed to their unique surface chemistry; Al_2O_3 or SiO_x can be used as an insulating protection layer that suppress polysulfide dissolution into the liquid electrolyte; porous structured metal oxides can act as polysulfide reservoirs using the pores to physically trap Li polysulfides. However, it should be noted that the nanostructured metal oxides which can be regarded as 'additives' not only can potentially increase the production cost of the cell, but also decrease the specific energy of the Li–S cell owing to the weight of metal oxide that does not contribute to specific capacity. In the case of organosulfur compound cathodes, although they can offer more alternative cathode systems for the Li–S cell, they need to be further investigated for practical use due to their poor reaction kinetics, uncertainty of thermal or chemical stability of organosulfur compounds in the temperature range for many battery applications and undesired 'dead weight' due to all of the remaining portions of the molecules except for the active sulfur.

2.3. Binders, Conducting Additives and Current Collectors

Most investigations for developing advanced sulfur cathodes have mainly been focused on the design of active sulfur-based nanocomposites. A significant amount of previous work successfully demonstrated that the design of the active sulfur-based nanocomposites plays an important role in improving the electrochemical performance of the Li–S cell. In the sulfur cathode, however, 1–20 wt.% of the active material layer on the current collector is generally devoted to binder and conductive carbon, which means that they also greatly influence the cell performance. Therefore, it is necessary to extensively study the binder and conductive carbon additive as well as the current collector to optimize the architecture of the sulfur cathode. In the following section, various kinds of binder, conductive carbon additives as well as current collectors and their recent progress will be discussed.

2.3.1. Binders

The binder plays an important role in creating a durable electronically conductive network by maintaining connection among all electrode components during electrochemical cycling while allowing Li-ion transport into the internal electrode space. Mechanical failure of the sulfur cathode can occur during electrochemical cycling due to the volume expansion of sulfur particles up to 80% (when Li_2S forms), which can cause a loss of electrical contact between active sulfur and the current collector, resulting in a decrease of specific capacity. Thus, the binder needs to act as a buffer material to accommodate the volume change of sulfur particles during cycling to maintain the electronic contact of all electrode components in order to achieve good utilization and cycle life. The binders can be categorized depending on the solvent that is used for dissolving the binder: (1) binders used in aqueous systems; (2) binders used in organic solvent systems. The aqueous systems have advantages for mass production compared to organic solvent systems, because they use water as solvent which is inexpensive, safe and environmental friendly, whereas organic solvents are commonly toxic, flammable and expensive. Polyvinylidene difluoride

(PVDF) is the most common binder for both anode and cathode in conventional Li-ion cells due to its good adhesion properties with other electrode components,[182] and it has also been popularly used for sulfur cathodes. Nevertheless, some critical drawbacks have been noted that cause degradation of cell performance, e.g. PVDF cannot successfully accommodate the stress–strain caused by the volume change of sulfur particles due to its stiff nature; use of N-methyl-2-pyrrolidone (NMP) which is toxic and hard to evaporate due to its low vapor pressure (40 pa while that of water is 2300 pa at 20°C) and high boiling point (~202°C); the electrically insulating nature of PVDF increases the resistance of the cathode. For these reasons, many attempts have been made to develop alternative binder systems for the sulfur cathode.

Polyethylene oxide (PEO) is one of the representative binders, which is often used for Li–S cells[183,184] instead of PVDF, because it can be dissolved in acetonitrile (ACN) or propyl alcohol which are less toxic and easier to evaporate than NMP. However, electrode swelling was found to be extreme and the cause of capacity decrease due to a loss of electrical contact between sulfur and the current collector.[185] Recently, Lacey *et al.* reported that the sulfur cathode with a PEO–PVP composite binder system can improve the electrochemical performance over that of cathodes with the PVP binder or the PEO single binder systems.[186] It was previously demonstrated by Seh *et al.* that the PVP binder has a strong affinity with both Li_2S and Li polysulfide, thus it can help to improve cycle life of the Li–S cell.[187] In Lacey's work, the sulfur cathode with the PEO–PVP (8:2 by weight) composite binder system showed significant improvement in electrochemical performance with a specific capacity of about 1000 mAh/g after 50 cycles at 0.2 C, whereas the cathodes with the PEO and PVP binders showed about 900 mAh/g. Besides the chemical entrapment of Li polysulfides, addition of PVP into the PEO binder allows PEO to be used in aqueous systems. Commonly, water-based slurries containing only PEO as the binder are extremely viscous,[186] so it is difficult to obtain a good laminate. However, the viscosity of the slurry can be greatly improved by adding PVP because PVP is a well-known polymer dispersant.

In general, aqueous system binders have great advantages such as their environmentally benign nature, low cost and easy engineering of aqueous processes, which does not interfere with the advantages of the sulfur cathode. So, various aqueous binder systems such as gelatine[188,189] nafion[190] PAA[118,191] and styrene–butadiene rubber (SBR)[88,192,193] have been explored for Li–S cell. In 2013, Wang et al. designed a long-life sulfur cathode using a cross-linking reaction between PD buffer on the surface of S–RGO nanocomposite and PAA binder.[118] The PD coating layer on S–RGO nanocomposite not only chemically traps Li polysulfides during cycling, but also forms a strong covalent bond through the dehydration of the carboxyl group of PAA and the amino group of PD to produce a tertiary amide, resulting in improvement of the electrode stability. As a result, the PD coated S–RGO cathode with PAA binder exhibited a long cycle life with a specific capacity of 728 mAh/g after 500 cycles at a specific current of 0.5 A/g.

SBR is commonly used in combination with a thickening agent, carboxymethylcellulose (CMC), and the laminate with SBR–CMC composite binder shows strong adhesion to the current collector. The elastomeric nature of SBR is beneficial for accommodating stress (or strain) caused by the volume change of sulfur particles during cycling, thus the structural integrity of the sulfur cathode can be enhanced, indicating that the electronic network of the sulfur cathode can be effectively maintained.[192] Consequently, the cycle life of the Li–S cell can be improved by replacing PVDF binder with the SBR–CMC composite binder, which was demonstrated in the previous research on the CTAB modified S–GO cathode with SBR–CMC binder that exhibited a long cycle life of a Li–S cell over 1500 cycles with an extremely low capacity decay rate (0.039% per cycle) at 1.0 C discharge.[88]

Another water soluble binder, a carbonyl–β–cyclodextrin (C–β–CD) which exhibited a high solubility in water and strong bonding capability was reported by Wang et al.[194] By modifying the β–CD through a partial oxidation reaction in H_2O_2 solution, the solubility of the C–β–CD in water at room temperature increased to approximately 100 times higher than that of the β–CD. The C–β–CD binder also showed strong adhesion strength and formed a gel film tightly wrapping the surface of the sulfur-based composite (based on the SEM images), which suppressed the aggregation of the sulfur composite during cycling. The sulfur cathode with C–β–CD binder showed a high initial capacity of about 1543 mAh/g and remained at 1456 mAh/g after 50 cycles at 0.2 C, whereas that of the sulfur cathodes with PVDF and PTFE binders decreased to about 900 mAh/g.

According to many researches on binder systems, binders certainly play an important part in achieving good electrochemical performance. Binders necessarily enable the maintenance of electronic contact between sulfur, conductive carbon and current collector against the volume change of sulfur particles during cycling. Among many binders, aqueous binder systems are beneficial for manufacturing of sulfur cathodes due to their environmentally benign nature, low cost and easy engineering of the process. Despite the advantages of new binder systems, none of them have completely replaced the PVDF binder for the sulfur cathode yet, which might be due to a lack of understanding of new binders. Thus, additional effort for developing an advanced binder system is required for a practical Li–S cell.

2.3.2. Carbon additives

Carbonaceous materials such as graphite, CB, CNT, and graphene are widely used as conductive additives for sulfur cathodes due to its high electrical conductivity, physical and chemical stability and environmentally benign nature. In general, carbon additives are not involved in the electrochemical reaction, but act as an electronic pathway to compensate for the poor electronic conductivity of sulfur, resulting in improvement of sulfur utilization and rate capability. However, because the weight of the carbon additive is 'dead weight' for the specific energy of the cell, the carbon content in the sulfur cathode needs to be minimized (typically less than 10%). Therefore, the carbonaceous materials that have a high surface area or unique structure (e.g. wired structure) are favorable for providing an effective electronic pathway with a limited carbon amount.

Four different CBs (Ketjen black EC-600JD, Super C65, Printex-XE2, and Printex-A) were employed as conductive carbon in sulfur cathode and electrochemically evaluated by A. Jozwiuk et al.[195] Each CB has a different BET surface area (Printex-A: 30 m^2/g, Super C65: 65 m^2/g, Printex XE-2: 1000 m^2/g, and Ketjen black EC-600JD: 1400 m^2/g), thus the slurry concentration was varied to optimize the consistency of the blend. As a result, it was found that the carbon additive with a higher specific surface area can provide a higher specific capacity, but a brittle electrode was produced as the surface area of carbon increased. It was suggested that it is beneficial to mix the CBs with different surface areas for better performance and at the same time high stability.

Unique structured carbonaceous materials such as CNT,[55,97] CNF,[184,196] and graphene[55] were also explored as conductive carbon additives. Because of their 1D or 2D continuous structure, they can provide very stable and effective electron conduction pathways. For example, a sulfur cathode with addition of CNF exhibited a higher specific capacity and better capacity retention than that of a sulfur cathode with carbon particles, because the CNFs provide a good conductive network with structural stability.[196] In the same manner, MWCNT is also helpful for improving the cycling performance and rate capability of sulfur electrodes.[97] When MWCNT was employed in the sulfur cathode as a conductive additive, lower charge transfer resistance was measured by EIS than for the sulfur cathode without MWCNT addition. The sulfur cathode with a graphene and CNT mixture exhibited about two times higher specific capacity at 0.2 C than the sulfur cathode with Super P.[55]

There is no doubt that the use of carbon additives in the sulfur cathode can significantly improve the electrochemical performance of Li–S cells. Therefore, it is important to choose the proper conductive carbon additives that can work well with the other cathode components in order to form a structurally homogeneous microstructure. Even though the use of unique structured carbonaceous materials such as CNT and graphene can improve the electrochemical performance of the sulfur cathode, these materials should be considered carefully because they can potentially increase the production cost of the sulfur cathode due to their high price.

2.3.3. Current collector

The current collector is used for electronic connection between the active sulfur particles and the external terminals of the cell. The sulfur-based active material is generally pasted onto the current collector together with a polymer binder and conductive carbon additive, thus the geometry of the sulfur cathode is strongly influenced by that of the current collector. For various cell packaging types such as cylindrical, rectangular, pouch and coin type cells, the current collector needs to offer not only good surface conditions for obtaining strong adhesion strength to the laminate, but also the proper geometry with flexibility. Typically, aluminum (Al) thin foil is employed as the current collector for the sulfur cathode due to its chemical stability at the potential of the sulfur cathode, good electrical conductivity[197] and the relative ease of manufacturing thin foil. To improve the electrochemical performance of the sulfur cathode, various kinds of modified Al foils have been studied such as carbon coated Al foil, expanded metal mesh and surface etched Al foil.

Recently, 3D structured current collectors which can accommodate active sulfur particles and conductive carbon additives in their interior space have been studied for use in the sulfur cathode. In the case of the sulfur cathode with a conventional Al foil current collector, the laminate composed of active sulfur powder, conductive carbon additive and polymer binder just sits on the Al foil, thus the only efficient channel for conducting electrons is associated with the conductive carbon. On the other hand, 3D structured current collectors are beneficial for delivering electrons to (or from) the entire volume of the sulfur cathode because active sulfur particles are generally well dispersed in the interior space of the current collector. Moreover, the porous structure of the sulfur cathode with a 3D structured current collector can be used as a direct channel for the electrolyte to reach the inner space of the sulfur cathode, so Li-ions in the electrolyte can reach the sulfur particles easily, resulting in the improvement of charge transfer. Furthermore, the void space of the sulfur cathode with a 3D structured current collector can accommodate the volume change of the sulfur particles, so better structural stability of the sulfur cathode can be obtained during electrochemical cycling. In 2013, A nanocellular carbon (NC) current collector which has micro-meso-macroporous architecture was reported by Manthiram and co-worker.[198] In that work, it was demonstrated that the sulfur cathode with a NC current collector provided a high initial specific capacity of 1249 mAh/g with a good capacity retention of 71% at 0.2 C after 50 cycles, whereas the sulfur cathode with conventional Al foil showed an initial capacity of 871 mAh/g with a capacity retention of less than 35% after 50 cycles. It was suggested that the 3D and porous structure of the NC current collector improves the electrical conductivity of the sulfur cathode, and further inhibits active material loss and the polysulfide shuttle effect via its porous structure that acts as the Li polysulfide reservoir.

A 3D structured Al foil current collector was employed for a sulfur cathode that accommodates a very high sulfur loading of about 7.0 mg/cm^2.[199] An initial capacity of this sulfur cathode with 3D structured Al foil current collector was about 860 mAh/g at 0.1 C, whereas that of the conventional sulfur cathode was only about 534 mAh/g, even with a lower sulfur loading of about 4.6 mg/cm^2. A long cycle life of more than 1000 cycles was demonstrated with a graphene foam (GF) based 3D structured flexible sulfur electrode.[200] GF was supported by a poly(dimethyl siloxane) (PDMS) coating on the surface of the GF to form a network while providing flexibility to the cathode. This flexible and bendable sulfur cathode exhibited much better cycleability and rate capability than those of a typical sulfur cathode.

Recent progress on 3D current collectors is notable for improving the electrochemical performance of the sulfur cathode. They successfully accommodate a large amount of sulfur in their structure and the cathode showed improvement in sulfur utilization. However, suitability of the current collector design for various cell casings, cost issues, and thickness or weight of the 3D structured current collectors should be considered carefully for their use in practical high specific energy Li–S cells.

2.4. The Interlayer Concept

In 2012, A. Manthiram and coworker reported a notable new concept of a cell component to improve the cycle life of the Li–S cell, which is called the 'interlayer'.[201] The interlayer is not a conventional component of the Li–S cell, however, this electrolyte-permeable interlayer is generally placed between the sulfur cathode and the separator, so it is in direct contact with the sulfur cathode. In Manthiram's work, a bifunctional microporous CP was added between the sulfur cathode and the separator, which can act not only as an "upper current collector", but also a "polysulfide stockroom". Because the interlayer is electronically connected to the sulfur cathode, electrons can be supplied to the Li polysulfides trapped in the interlayer, thus Li polysulfide can be reutilized at the surface of the interlayer. As a result, both the sulfur utilization and the capacity retention can be improved. The Li–S cell with a porous carbon interlayer showed a specific capacity over 1000 mAh/g after 100 cycles at the 1.0 C rate with an average Coulombic efficiency of 97.6%, corresponding to specific capacity retention of 85%.

Since Manthiram's work was reported, various interlayer materials such as polymer interlayers,[114,202] carbonaceous material interlayers,[63,83,203–208] metallic interlayers[208,209] and composite systems[208, 210] have been studied. In addition to the porous structure of the interlayer, a hydroxyl-functionalized porous CP was proposed to further improve the functionality of the interlayer to trap Li polysulfides during cycling.[204] It was suggested that the hydrophilic groups and the microcrack enhanced surface adsorption properties of the carbon promote bonding of sulfur species to the surface of the interlayer. A conductive polymer, PPy interlayer which was fabricated onto the surface of the sulfur cathode was also reported by Ma et al. The functional interlayer, composed of PPy nanoparticles, can suppress Li polysulfide dissolution into the liquid electrolyte and its shuttle owing to the adsorption effect of PPy for Li polysulfides, thus the cycleability of the Li–S cell was greatly improved. The discharge specific capacity of the Li–S cell with the PPy interlayer on the sulfur cathode was 846 mAh/g after 200 cycles with an average Columbic efficiency of 94.2% at the 0.2 C rate, while the specific capacity of a Li–S cell without the PPy interlayer was only 587 mAh/g after 100 cycles with an average Coulombic efficiency of 83.4%.

Conductive carbon interlayers coated on the surface of the separator, facing the cathode side were also reported.[208] A doctor blade method with a slurry containing PVDF binder was employed to coat the separator with a conductive layer. Super P, Ketjen black and MWCNT were employed as interlayer materials and all the cells that employed the modified separators with those carbonaceous materials showed significant improvement as compared to the cell with an untreated separator. Among the Li–S cells with different types of CBS the cells with separators coated by Ketjen black or MWCNT showed better electrochemical performance than that of the cell with a separator coated by Super P. These carbons contributed to a higher electronic conductivity, resulting in enhancement of the polysulfide reutilization at the surface of the interlayer on the separator. The cells with separators coated by Ketjen black or MWCNT exhibited about 760 mAh/g after 150 cycles at 0.2 C, whereas the cell with the separator coated by Super P and the cell with the untreated separator showed specific capacities of only little more than 500 and 300 mAh/g, respectively.

According to the previous work on the interlayer concept, employing the interlayer can significantly improve the electrochemical performance of Li–S cells. Their porous structure or chemical functionality can suppress the polysulfide shuttle effect by reutilizing the polysulfides at the surface of the interlayer, resulting in the improvement of the electrochemical performance of the cell. However, the increase of cell weight caused by adding the interlayer and the additional electrolyte filling the pores should be taken into account in designing a high specific energy Li–S cell.

2.5. Summary and Outlook

As discussed in this chapter, the Li–S cell has been widely regarded as a front runner in the search for the next generation rechargeable battery because the Li–S cell offers a theoretical specific energy of ~2600 Wh/kg under the assumption of complete Li_2S formation during the discharge process, which is much larger than that of the Li-ion cell (500–600 Wh/kg). Moreover, the advantages of the Li–S cell such as the low cost of sulfur, a low environmental impact and a technical similarity to the fabrication of the Li-ion cell are very attractive features for rechargeable cells beyond Li-ion cells. However, the drawbacks associated with the sulfur cathode (e.g. a poor electrical conductivity of sulfur or Li_2S; dissolution of polysulfides into liquid electrolytes and the shuttle effect; the volume expansion of sulfur particles up to 80%) limit the sulfur utilization and cycle life of the Li–S cell. Thus extensive studies on the continuing development of more advanced sulfur electrodes must be done in order to produce a practical Li–S cell.

Recently, significant progress in the sulfur cathode has been achieved by developing a rational design of the sulfur cathode composed of sulfur-based nanocomposites as active materials, conductive additives and binders. Many strategies for the design of sulfur based nanocomposites have been proposed to improve the electrochemical performance of sulfur cathodes, which can be summarized as follows: (1) fabrication of nanosized sulfur not only to reduce the length of both electron and Li-ion conducting pathways through insulating sulfur, but also to improve the structural stability against the volume change during electrochemical cycling; (2) sulfur-based nanocomposites with conductive materials to improve the conductivity of the sulfur cathode; (3) sulfur based nanocomposites with physical or chemical immobilizers to trap sulfur and polysulfides within the cathode during cycling; (4) suppression of stress (or strain) caused by the volume change of sulfur particles during cycling using voids in the porous structures or mechanically stable matrix materials. The sulfur based nanocomposites designed with these strategies and their multiple combinations show great improvement in the electrochemical performance of the sulfur cathode and sulfur–carbonaceous materials and sulfur conducting polymer nanocomposite are represented among them. In addition to the sulfur-based active materials, there is no doubt that the conductive carbon additive, binder and current collector play important roles in improving the electrochemical performance of the sulfur cathode by constructing a good microstructure of the sulfur cathode that can offer a good electrical conductivity and structural stability, although most of the attention of researchers has been focused on developing nanostructured sulfur active materials. The addition of an interlayer in the Li–S cell is also worth considering further for improving the cell performance.

Although many promising sulfur cathode systems have been reported recently, some important issues still remain to be addressed before a practical Li–S cell is possible. First of all, both the sulfur loading and the sulfur content need to be significantly increased while maintaining good sulfur utilization and cycle life of the Li–S cell. For example, the specific capacity of the sulfur cathode estimated using the weight of the laminate is only half of the specific capacity normalized by the sulfur mass on the cathode, if the sulfur content in the laminate is 50%, because the other 50% is 'dead weight'. Moreover, the weights of the other cell components such as electrolyte, separator and Li foil are important, so the weight ratio between active sulfur and the sum of the other component weights has to be high enough to achieve a high specific energy Li–S cell, which means that a high weight of sulfur per unit area is essential. Unfortunately, most of the results reported in the previous works were obtained with low sulfur loading (commonly around 1 mg/cm^2 or less) or low sulfur contents (below 70%), or the cell performance was evaluated without consideration of the weight of the other cell components, even in cases where the sulfur loading the sulfur cathodes was high. Figure 2.15 shows the performance of the Li–S cells reported in selected references, and indicates that only a few works have achieved high sulfur loading cathodes while maintaining reasonably high sulfur utilization, although the cycle life of the cells still needs to be improved for practical use. The green zone of the figure indicates the performance necessary for achieving a specific energy greater than that of a Li-ion cell.

In addition to the sulfur loading issue, the production cost of the sulfur cathode is important, especially if it requires expensive or toxic materials. Scalability, reliability and a simple production environment for the sulfur-based nanocomposite also need to be addressed for the design of the sulfur cathode. In addition, the cell performance at various temperatures and the safety properties need to be evaluated and improved for a commercial

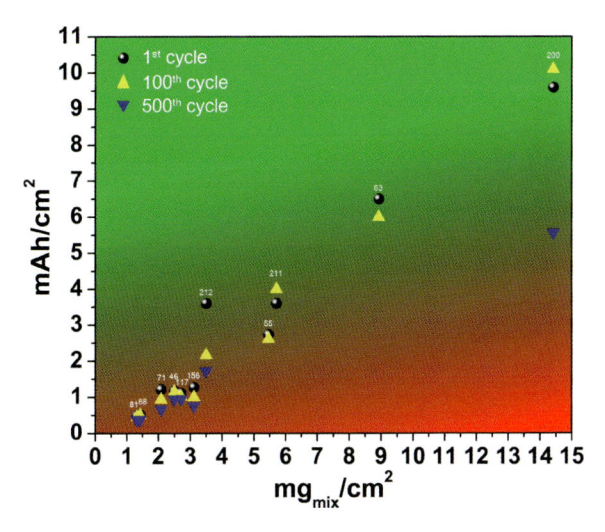

Figure 2.15 Estimated cell capacity/cm^2 plot as a function of the mixture weight/cm^2 of the cathode. The capacity/cm^2 of the cathodes for the first cycle (circle), 100th cycle (triangle) and 500th cycle (inverted triangle) are marked.

Li–S cell. Therefore, continuing research for the further development of the sulfur cathode system is necessary to address all of these issues. However, we believe that the intensive effort of the battery community will overcome the existing challenges and eventually produce practical high specific energy Li–S cells.

Bibliography

1. T. Nagaura and K. Tozawa, *Progress in Batteries and Solar Cells*, **9** (1990) 209.
2. *Electrochemical Energy Storage Technical Team Roadmap*, U.S. DRIVE Partnership/U.S. Department of Energy: Washington, DC, 2013.
3. K. Mizushima, P. C. Jones, P. J. Wiseman and J. B. Goodenough, *Solid State Ionics*, **3–4** (1981) 171–174.
4. H. Danuta and U. Juliusz, **US3043896 A**, 1962.
5. E. J. Cairns and H. Shimotake, *Science*, **164** (1969) 1347–1355.
6. X. Ji and L. F. Nazar, *Journal of Materials Chemistry*, **20** (2010) 9821–9826.
7. G. Jeong, Y.-U. Kim, H. Kim, Y.-J. Kim and H.-J. Sohn, *Energy & Environmental Science*, **4** (2011) 1986–2002.
8. Z. Liu, W. Fu and C. Liang, In *Handbook of Battery Materials*, Wiley-VCH Verlag GmbH & Co. KGaA, DOI: 10.1002/9783527637188, pp. 811–840 (2011).
9. P. G. Bruce, S. A. Freunberger, L. J. Hardwick and J.-M. Tarascon, *Nature Materials*, **11** (2012) 19–29.
10. S. Evers and L. F. Nazar, *Accounts of Chemical Research*, **46** (2013) 1135–1143.
11. Y. Yang, G. Zheng and Y. Cui, *Chemical Society Reviews*, **42** (2013) 3018–3032.
12. L. F. Nazar, M. Cuisinier and Q. Pang, *MRS Bulletin*, **39** (2014) 436–442.
13. G. Xu, B. Ding, J. Pan, P. Nie, L. Shen and X. Zhang, *Journal of Materials Chemistry A*, **2** (2014) 12662–12676.
14. Z. Lin and C. Liang, *Journal of Materials Chemistry A*, **3** (2015) 936–958.
15. L. Ma, K. E. Hendrickson, S. Wei and L. A. Archer, *Nano Today*, **10** (2015) 315–338.
16. S. Wu, R. Ge, M. Lu, R. Xu and Z. Zhang, *Nano Energy*, **15** (2015) 379–405.

17. M.-K. Song, E. J. Cairns and Y. Zhang, *Nanoscale*, **5** (2013) 2186–2204.
18. D.-W. Wang, Q. Zeng, G. Zhou, L. Yin, F. Li, H.-M. Cheng, I. R. Gentle and G. Q. M. Lu, *Journal of Materials Chemistry A*, **1** (2013) 9382–9394.
19. L. Chen and L. L. Shaw, *Journal of Power Sources*, **267** (2014) 770–783.
20. D. Bresser, S. Passerini and B. Scrosati, *Chemical Communications*, **49** (2013) 10545–10562.
21. J. R. Akridge, Y. V. Mikhaylik and N. White, *Solid State Ionics*, **175** (2004) 243–245.
22. X. He, J. Ren, L. Wang, W. Pu, C. Jiang and C. Wan, *Journal of Power Sources*, **190** (2009) 154–156.
23. R. Elazari, G. Salitra, Y. Talyosef, J. Grinblat, C. Scordilis-Kelley, A. Xiao, J. Affinito and D. Aurbach, *Journal of the Electrochemical Society*, **157** (2010) A1131–A1138.
24. Y. Yang, G. Zheng, S. Misra, J. Nelson, M. F. Toney and Y. Cui, *Journal of the American Chemical Society*, **134** (2012) 15387–15394.
25. Y. V. Mikhaylik and J. R. Akridge, *Journal of the Electrochemical Society*, **151** (2004) A1969–A1976.
26. H. Chen, C. Wang, W. Dong, W. Lu, Z. Du and L. Chen, *Nano Letters*, **15** (2015) 798–802.
27. C. Nan, Z. Lin, H. Liao, M.-K. Song, Y. Li and E. J. Cairns, *Journal of the American Chemical Society*, **136** (2014) 4659–4663.
28. X. H. Liu, L. Zhong, S. Huang, S. X. Mao, T. Zhu and J. Y. Huang, *ACS Nano*, **6** (2012) 1522–1531.
29. J. Y. Huang, L. Zhong, C. M. Wang, J. P. Sullivan, W. Xu, L. Q. Zhang, S. X. Mao, N. S. Hudak, X. H. Liu, A. Subramanian, H. Fan, L. Qi, A. Kushima and J. Li, *Science*, **330** (2010) 1515–1520.
30. C. K. Chan, H. Peng, G. Liu, K. McIlwrath, X. F. Zhang, R. A. Huggins and Y. Cui, *Nature Nanotechnology*, **3** (2008) 31–35.
31. L. Su, Y. Jing and Z. Zhou, *Nanoscale*, **3** (2011) 3967–3983.
32. K. S. W. Sing, D. H. Everett, R. A. W. Haul, L. Moscou, R. A. Pierotti, J. Rouquerol and T. Siemieniewska, *Pure and Applied Chemistry*, **57** (1985) 603–619.
33. J. Lee, J. Kim and T. Hyeon, *Advanced Materials*, **18** (2006) 2073–2094.
34. Z. Hu, M. P. Srinivasan and Y. Ni, *Advanced Materials*, **12** (2000) 62–65.
35. A. Ahmadpour and D. D. Do, *Carbon*, **34** (1996) 471–479.
36. H. Marsh and B. Rand, *Carbon*, **9** (1971) 63–77.
37. A. Oya, S. Yoshida, J. Alcaniz-Monge and A. Linares-Solano, *Carbon*, **33** (1995) 1085–1090.
38. N. Patel, K. Okabe and A. Oya, *Carbon*, **40** (2002) 315–320.

39. R. W. Pekala, *Journal of Material Science*, **24** (1989) 3221–3227.
40. T. Kyotani, T. Nagai, S. Inoue and A. Tomita, *Chemistry of Materials*, **9** (1997) 609–615.
41. J. H. Knox, B. Kaur and G. R. Millward, *Journal of Chromatography A*, **352** (1986) 3–25.
42. J. L. Wang, J. Yang, J. Y. Xie, N. X. Xu and Y. Li, *Electrochemistry Communications*, **4** (2002) 499–502.
43. J. Wang, L. Liu, Z. Ling, J. Yang, C. Wan and C. Jiang, *Electrochimica Acta*, **48** (2003) 1861–1867.
44. C. Liang, N. J. Dudney and J. Y. Howe, *Chemistry of Materials*, **21** (2009) 4724–4730.
45. X. Ji, K. T. Lee and L. F. Nazar, *Nature Materials*, **8** (2009) 500–506.
46. B. Zhang, X. Qin, G. R. Li and X. P. Gao, *Energy & Environmental Science*, **3** (2010) 1531–1537.
47. G. He, X. Ji and L. Nazar, *Energy & Environmental Science*, **4** (2011) 2878–2883.
48. M. Rao, W. Li and E. J. Cairns, *Electrochemistry Communications*, **17** (2012) 1–5.
49. G. He, S. Evers, X. Liang, M. Cuisinier, A. Garsuch and L. F. Nazar, *ACS Nano*, **7** (2013) 10920–10930.
50. J. Kim, D.-J. Lee, H.-G. Jung, Y.-K. Sun, J. Hassoun and B. Scrosati, *Advanced Functional Materials*, **23** (2013) 1076–1080.
51. K. Zhang, Q. Zhao, Z. Tao and J. Chen, *Nano Research*, **6** (2013) 38–46.
52. N. Jayaprakash, J. Shen, S. S. Moganty, A. Corona and L. A. Archer, *Angewandte Chemie International Edition*, **50** (2011) 5904–5908.
53. T. Xu, J. Song, M. L. Gordin, H. Sohn, Z. Yu, S. Chen and D. Wang, *ACS Applied Materials & Interfaces*, **5** (2013) 11355–11362.
54. D. S. Jung, T. H. Hwang, J. H. Lee, H. Y. Koo, R. A. Shakoor, R. Kahraman, Y. N. Jo, M.-S. Park and J. W. Choi, *Nano Letters*, **14** (2014) 4418–4425.
55. D. Lv, J. Zheng, Q. Li, X. Xie, S. Ferrara, Z. Nie, L. B. Mehdi, N. D. Browning, J.-G. Zhang, G. L. Graff, J. Liu and J. Xiao, *Advanced Energy Materials*, **5** (2015) 1402290.
56. J. T. Lee, Y. Zhao, S. Thieme, H. Kim, M. Oschatz, L. Borchardt, A. Magasinski, W.-I. Cho, S. Kaskel and G. Yushin, *Advanced Materials*, **25** (2013) 4573–4579.
57. Z. Li, Y. Jiang, L. Yuan, Z. Yi, C. Wu, Y. Liu, P. Strasser and Y. Huang, *ACS Nano*, **8** (2014) 9295–9303.
58. Z. Li, L. Yuan, Z. Yi, Y. Sun, Y. Liu, Y. Jiang, Y. Shen, Y. Xin, Z. Zhang and Y. Huang, *Advanced Energy Materials*, **4** (2014) 1301473.

59. M.-S. Park, B. O. Jeong, T. J. Kim, S. Kim, K. J. Kim, J.-S. Yu, Y. Jung and Y.-J. Kim, *Carbon*, **68** (2014) 265–272.
60. Z. Li and L. Yin, *ACS Applied Materials & Interfaces*, **7** (2015) 4029–4038.
61. D.-W. Wang, G. Zhou, F. Li, K.-H. Wu, G. Q. Lu, H.-M. Cheng and I. R. Gentle, *Physical Chemistry Chemical Physics*, **14** (2012) 8703–8710.
62. C.-P. Yang, Y.-X. Yin, H. Ye, K.-C. Jiang, J. Zhang and Y.-G. Guo, *ACS Applied Materials & Interfaces*, **6** (2014) 8789–8795.
63. J. Song, M. L. Gordin, T. Xu, S. Chen, Z. Yu, H. Sohn, J. Lu, Y. Ren, Y. Duan and D. Wang, *Angewandte Chemie International Edition*, **127** (2015) 4399–4403.
64. R. Demir-Cakan, M. Morcrette, F. Nouar, C. Davoisne, T. Devic, D. Gonbeau, R. Dominko, C. Serre, G. Férey and J.-M. Tarascon, *Journal of the American Chemical Society*, **133** (2011) 16154–16160.
65. H. B. Wu, S. Wei, L. Zhang, R. Xu, H. H. Hng and X. W. Lou, *Chemistry — A European Journal*, **19** (2013) 10804–10808.
66. W. Bao, Z. Zhang, C. Zhou, Y. Lai and J. Li, *Journal of Power Sources*, **248** (2014) 570–576.
67. L. Ji, M. Rao, H. Zheng, L. Zhang, Y. Li, W. Duan, J. Guo, E. J. Cairns and Y. Zhang, *Journal of the American Chemical Society*, **133** (2011) 18522–18525.
68. H. Wang, Y. Yang, Y. Liang, J. T. Robinson, Y. Li, A. Jackson, Y. Cui and H. Dai, *Nano Letters*, **11** (2011) 2644–2647.
69. S. Evers and L. F. Nazar, *Chemical Communications*, **48** (2012) 1233–1235.
70. N. Li, M. Zheng, H. Lu, Z. Hu, C. Shen, X. Chang, G. Ji, J. Cao and Y. Shi, *Chemical Communications*, **48** (2012) 4106–4108.
71. H.-J. Peng, J. Liang, L. Zhu, J.-Q. Huang, X.-B. Cheng, X. Guo, W. Ding, W. Zhu and Q. Zhang, *ACS Nano*, **8** (2014) 11280–11289.
72. H.-J. Peng, J.-Q. Huang, M.-Q. Zhao, Q. Zhang, X.-B. Cheng, X.-Y. Liu, W.-Z. Qian and F. Wei, *Advanced Functional Materials*, **24** (2014) 2772–2781.
73. J.-Z. Wang, L. Lu, M. Choucair, J. A. Stride, X. Xu and H.-K. Liu, *Journal of Power Sources*, **196** (2011) 7030–7034.
74. L. Yin, J. Wang, F. Lin, J. Yang and Y. Nuli, *Energy & Environmental Science*, **5** (2012) 6966–6972.
75. H. Sun, G.-L. Xu, Y.-F. Xu, S.-G. Sun, X. Zhang, Y. Qiu and S. Yang, *Nano Research*, **5** (2012) 726–738.
76. L. Zhang, L. Ji, P.-A. Glans, Y. Zhang, J. Zhu and J. Guo, *Physical Chemistry Chemical Physics*, **14** (2012) 13670–13675.

77. R. Chen, T. Zhao, J. Lu, F. Wu, L. Li, J. Chen, G. Tan, Y. Ye and K. Amine, *Nano Letters*, **13** (2013) 4642–4649.
78. T. Lin, Y. Tang, Y. Wang, H. Bi, Z. Liu, F. Huang, X. Xie and M. Jiang, *Energy & Environmental Science*, **6** (2013) 1283–1290.
79. G. Zhou, L.-C. Yin, D.-W. Wang, L. Li, S. Pei, I. R. Gentle, F. Li and H.-M. Cheng, *ACS Nano*, **7** (2013) 5367–5375.
80. C. Zu and A. Manthiram, *Advanced Energy Materials*, **3** (2013) 1008–1012.
81. Y. Qiu, W. Li, W. Zhao, G. Li, Y. Hou, M. Liu, L. Zhou, F. Ye, H. Li, Z. Wei, S. Yang, W. Duan, Y. Ye, J. Guo and Y. Zhang, *Nano Letters*, **14** (2014) 4821–4827.
82. X. Wang, Z. Zhang, Y. Qu, Y. Lai and J. Li, *Journal of Power Sources*, **256** (2014) 361–368.
83. J. Xu, J. Shui, J. Wang, M. Wang, H.-K. Liu, S. X. Dou, I.-Y. Jeon, J.-M. Seo, J.-B. Baek and L. Dai, *ACS Nano*, **8** (2014) 10920–10930.
84. M.-Q. Zhao, Q. Zhang, J.-Q. Huang, G.-L. Tian, J.-Q. Nie, H.-J. Peng and F. Wei, *Nature Communications*, **5** (2014).
85. L. Zhou, X. Lin, T. Huang and A. Yu, *Journal of Materials Chemistry A*, **2** (2014) 5117–5123.
86. H. Li, X. Yang, X. Wang, M. Liu, F. Ye, J. Wang, Y. Qiu, W. Li and Y. Zhang, *Nano Energy*, **12** (2015) 468–475.
87. C. Wang, X. Wang, Y. Wang, J. Chen, H. Zhou and Y. Huang, *Nano Energy*, **11** (2015) 678–686.
88. M.-K. Song, Y. Zhang and E. J. Cairns, *Nano Letters*, **13** (2013) 5891–5899.
89. Z. Wang, Y. Dong, H. Li, Z. Zhao, H. Bin Wu, C. Hao, S. Liu, J. Qiu and X. W. Lou, **5** (2014) 5002.
90. W. Zhou, H. Chen, Y. Yu, D. Wang, Z. Cui, F. J. DiSalvo and H. D. Abruña, *ACS Nano*, **7** (2013) 8801–8808.
91. B. Wang, S. M. Alhassan and S. T. Pantelides, *Physical Review Applied*, **2** (2014) 034004.
92. Z. R. Ismagilov, A. E. Shalagina, O. Y. Podyacheva, A. V. Ischenko, L. S. Kibis, A. I. Boronin, Y. A. Chesalov, D. I. Kochubey, A. I. Romanenko, O. B. Anikeeva, T. I. Buryakov and E. N. Tkachev, *Carbon*, **47** (2009) 1922–1929.
93. C. Tang, Q. Zhang, M.-Q. Zhao, J.-Q. Huang, X.-B. Cheng, G.-L. Tian, H.-J. Peng and F. Wei, *Advanced Materials*, **26** (2014) 6100–6105.
94. Z. Wang, X. Niu, J. Xiao, C. Wang, J. Liu and F. Gao, *RSC Advances*, **3** (2013) 16775–16780.

95. Z. W. Seh, H. Wang, P.-C. Hsu, Q. Zhang, W. Li, G. Zheng, H. Yao and Y. Cui, *Energy & Environmental Science*, **7** (2014) 672–676.
96. J. Guo, Z. Yang, Y. Yu, H. D. Abruña and L. A. Archer, *Journal of the American Chemical Society*, **135** (2013) 763–767.
97. S.-C. Han, M.-S. Song, H. Lee, H.-S. Kim, H.-J. Ahn and J.-Y. Lee, *Journal of the Electrochemical Society*, **150** (2003) A889–A893.
98. L. Yuan, H. Yuan, X. Qiu, L. Chen and W. Zhu, *Journal of Power Sources*, **189** (2009) 1141–1146.
99. L. Ji, M. Rao, S. Aloni, L. Wang, E. J. Cairns and Y. Zhang, *Energy & Environmental Science*, **4** (2011) 5053–5059.
100. J.-J. Chen, Q. Zhang, Y.-N. Shi, L.-L. Qin, Y. Cao, M.-S. Zheng and Q.-F. Dong, *Physical Chemistry Chemical Physics*, **14** (2012) 5376–5382.
101. Y.-S. Su, Y. Fu and A. Manthiram, *Physical Chemistry Chemical Physics*, **14** (2012) 14495–14499.
102. S. Xin, L. Gu, N.-H. Zhao, Y.-X. Yin, L.-J. Zhou, Y.-G. Guo and L.-J. Wan, *Journal of the American Chemical Society*, **134** (2012) 18510–18513.
103. J.-Q. Huang, H.-J. Peng, X.-Y. Liu, J.-Q. Nie, X.-B. Cheng, Q. Zhang and F. Wei, *Journal of Materials Chemistry A*, **2** (2014) 10869–10875.
104. L. Wang, Y. Zhao, M. L. Thomas and H. R. Byon, *Advanced Functional Materials*, **24** (2014) 2248–2252.
105. J. Guo, Y. Xu and C. Wang, *Nano Letters*, **11** (2011) 4288–4294.
106. G. Zheng, Y. Yang, J. J. Cha, S. S. Hong and Y. Cui, *Nano Letters*, **11** (2011) 4462–4467.
107. S. Moon, Y. H. Jung, W. K. Jung, D. S. Jung, J. W. Choi and D. K. Kim, *Advanced Materials*, **25** (2013) 6547–6553.
108. G. Zheng, Q. Zhang, J. J. Cha, Y. Yang, W. Li, Z. W. Seh and Y. Cui, *Nano Letters*, **13** (2013) 1265–1270.
109. Z. Xiao, Z. Yang, H. Nie, Y. Lu, K. Yang and S. Huang, *Journal of Materials Chemistry A*, **2** (2014) 8683–8689.
110. S. Chen, X. Huang, H. Liu, B. Sun, W. Yeoh, K. Li, J. Zhang and G. Wang, *Advanced Energy Materials*, **4** (2014) 1301761.
111. H.-J. Peng, T.-Z. Hou, Q. Zhang, J.-Q. Huang, X.-B. Cheng, M.-Q. Guo, Z. Yuan, L.-Y. He and F. Wei, *Advanced Energy Materials*, **4** (2014) 1400227.
112. Y. Zhao, W. Wu, J. Li, Z. Xu and L. Guan, *Advanced Materials*, **26** (2014) 5113–5118.
113. F. Wu, J. Chen, L. Li, T. Zhao and R. Chen, *The Journal of Physical Chemistry C*, **115** (2011) 24411–24417.
114. J.-Q. Huang, Q. Zhang, S.-M. Zhang, X.-F. Liu, W. Zhu, W.-Z. Qian and F. Wei, *Carbon*, **58** (2013) 99–106.

115. C. Wang, W. Wan, J.-T. Chen, H.-H. Zhou, X.-X. Zhang, L.-X. Yuan and Y.-H. Huang, *Journal of Materials Chemistry A*, **1** (2013) 1716–1723.
116. L. Xiao, Y. Cao, J. Xiao, B. Schwenzer, M. H. Engelhard, L. V. Saraf, Z. Nie, G. J. Exarhos and J. Liu, *Advanced Materials*, **24** (2012) 1176–1181.
117. W. Zhou, X. Xiao, M. Cai and L. Yang, *Nano Letters*, **14** (2014) 5250–5256.
118. L. Wang, D. Wang, F. Zhang and J. Jin, *Nano Letters*, **13** (2013) 4206–4211.
119. G.-C. Li, G.-R. Li, S.-H. Ye and X.-P. Gao, *Advanced Energy Materials*, **2** (2012) 1238–1245.
120. W. Zhou, Y. Yu, H. Chen, F. J. DiSalvo and H. D. Abruña, *Journal of the American Chemical Society*, **135** (2013) 16736–16743.
121. W. Li, Q. Zhang, G. Zheng, Z. W. Seh, H. Yao and Y. Cui, *Nano Letters*, **13** (2013) 5534–5540.
122. F. Wu, J. Chen, R. Chen, S. Wu, L. Li, S. Chen and T. Zhao, *The Journal of Physical Chemistry C*, **115** (2011) 6057–6063.
123. F. Wu, S. Wu, R. Chen, J. Chen and S. Chen, *Electrochemical and Solid-State Letters*, **13** (2010) A29–A31.
124. Y. Fu and A. Manthiram, *Chemistry of Materials*, **24** (2012) 3081–3087.
125. Y. Fu and A. Manthiram, *The Journal of Physical Chemistry C*, **116** (2012) 8910–8915.
126. F. Wu, J. Chen, L. Li, T. Zhao, Z. Liu and R. Chen, *ChemSusChem*, **6** (2013) 1438–1444.
127. G. Ma, Z. Wen, J. Jin, Y. Lu, K. Rui, X. Wu, M. Wu and J. Zhang, *Journal of Power Sources*, **254** (2014) 353–359.
128. J. Wang, J. Chen, K. Konstantinov, L. Zhao, S. H. Ng, G. X. Wang, Z. P. Guo and H. K. Liu, *Electrochimica Acta*, **51** (2006) 4634–4638.
129. X. Liang, Y. Liu, Z. Wen, L. Huang, X. Wang and H. Zhang, *Journal of Power Sources*, **196** (2011) 6951–6955.
130. Y. Zhang, Z. Bakenov, Y. Zhao, A. Konarov, T. N. L. Doan, M. Malik, T. Paron and P. Chen, *Journal of Power Sources*, **208** (2012) 1–8.
131. M. Sun, S. Zhang, T. Jiang, L. Zhang and J. Yu, *Electrochemistry Communications*, **10** (2008) 1819–1822.
132. H. Chen, W. Dong, J. Ge, C. Wang, X. Wu, W. Lu and L. Chen, *Science Reports*, **3** (2013).
133. Y. Yang, G. Yu, J. J. Cha, H. Wu, M. Vosgueritchian, Y. Yao, Z. Bao and Y. Cui, *ACS Nano*, **5** (2011) 9187–9193.
134. J. Wang, C. Lv, Y. Zhang, L. Deng and Z. Peng, *Electrochimica Acta*, **165** (2015) 136–141.

135. J. Wang, J. Yang, C. Wan, K. Du, J. Xie and N. Xu, *Advanced Functional Materials*, **13** (2003) 487–492.
136. D. M. Ivory, G. G. Miller, J. M. Sowa, L. W. Shacklette, R. R. Chance and R. H. Baughman, *The Journal of Chemical Physics*, **71** (1979) 1506–1507.
137. K. K. Kanazawa, A. F. Diaz, R. H. Geiss, W. D. Gill, J. F. Kwak, J. A. Logan, J. F. Rabolt and G. B. Street, *Journal of the Chemical Society, Chemical Communications*, (1979) 854–855.
138. R. Hernandez, A. F. Diaz, R. Waltman and J. Bargon, *The Journal of Physical Chemistry*, **88** (1984) 3333–3337.
139. M. R. Bryce, A. Chissel, P. Kathirgamanathan, D. Parker and N. R. M. Smith, *Journal of the Chemical Society, Chemical Communications*, (1987) 466–467.
140. J.-C. Chiang and A. G. MacDiarmid, *Synthetic Metals*, **13** (1986) 193–205.
141. V. V. Tat'yana and N. E. Oleg, *Russian Chemical Reviews*, **66** (1997) 443.
142. N. Kohut-Svelko, F. Dinant, S. Magana, G. Clisson, J. François, C. Dagron-Lartigau and S. Reynaud, *Polymer International*, **55** (2006) 1184–1190.
143. G. Ćirić-Marjanović, *Synthetic Metals*, **177** (2013) 1–47.
144. A. Pud, N. Ogurtsov, A. Korzhenko and G. Shapoval, *Progress in Polymer Science*, **28** (2003) 1701–1753.
145. A. Bhattacharya and A. De, *Progress in Solid State Chemistry*, **24** (1996) 141–181.
146. M. C. De Jesus, Y. Fu and R. A. Weiss, *Polymer Engineering & Science*, **37** (1997) 1936–1943.
147. R. Gangopadhyay and A. De, *Chemistry of Materials*, **12** (2000) 608–622.
148. M. Akiba and A. S. Hashim, *Progress in Polymer Science*, **22** (1997) 475–521.
149. W. K. Lewis, L. Squires and R. D. Nutting, *Industrial & Engineering Chemistry*, **29** (1937) 1135–1144.
150. B. A. Dogadkin, *Journal of Polymer Science*, **30** (1958) 351–361.
151. J. Tang, L. Kong, J. Zhang, L. Zhan, H. Zhan, Y. Zhou and C. Zhan, *Reactive and Functional Polymers*, **68** (2008) 1408–1413.
152. F. Sun, J. Wang, D. Long, W. Qiao, L. Ling, C. Lv and R. Cai, *Journal of Materials Chemistry A*, **1** (2013) 13283–13289.
153. K. Dong, S. Wang, H. Zhang and J. Wu, *Materials Research Bulletin*, **48** (2013) 2079–2083.
154. Y. Zhang, Y. Zhao, A. Yermukhambetova, Z. Bakenov and P. Chen, *Journal of Materials Chemistry A*, **1** (2013) 295–301.

155. X. Liang, C. Hart, Q. Pang, A. Garsuch, T. Weiss and L. F. Nazar, *Nature Communications*, **6** (2015) 5682.
156. Q. Pang, D. Kundu, M. Cuisinier and L. F. Nazar, *Nature Communications*, **5** (2014) 4759.
157. X. Tao, J. Wang, Z. Ying, Q. Cai, G. Zheng, Y. Gan, H. Huang, Y. Xia, C. Liang, W. Zhang and Y. Cui, *Nano Letters*, **14** (2014) 5288–5294.
158. S. Evers, T. Yim and L. F. Nazar, *The Journal of Physical Chemistry C*, **116** (2012) 19653–19658.
159. B. Ding, L. Shen, G. Xu, P. Nie and X. Zhang, *Electrochimica Acta*, **107** (2013) 78–84.
160. Z. W. Seh, W. Li, J. J. Cha, G. Zheng, Y. Yang, M. T. McDowell, P.-C. Hsu and Y. Cui, *Nature Communications*, **4** (2013) 1331.
161. Z. Liang, G. Zheng, W. Li, Z. W. Seh, H. Yao, K. Yan, D. Kong and Y. Cui, *ACS Nano*, **8** (2014) 5249–5256.
162. X. Ji, S. Evers, R. Black and L. F. Nazar, *Nature Communications*, **2** (2011) 325.
163. H. Kim, J. T. Lee, D.-C. Lee, A. Magasinski, W.-I. Cho and G. Yushin, *Advanced Energy Materials*, **3** (2013) 1308–1315.
164. M. Yu, W. Yuan, C. Li, J.-D. Hong and G. Shi, *Journal of Materials Chemistry A*, **2** (2014) 7360–7366.
165. K. T. Lee, R. Black, T. Yim, X. Ji and L. F. Nazar, *Advanced Energy Materials*, **2** (2012) 1490–1496.
166. J. Zheng, J. Tian, D. Wu, M. Gu, W. Xu, C. Wang, F. Gao, M. H. Engelhard, J.-G. Zhang, J. Liu and J. Xiao, *Nano Letters*, **14** (2014) 2345–2352.
167. E. J. Cairns, in *Encyclopedia of Electrochemical Power Sources*, (ed). J. Garche, Elsevier, Amsterdam, **5** (2009) 151–154.
168. Y. Liang, Z. Tao and J. Chen, *Advanced Energy Materials*, **2** (2012) 742–769.
169. L. J. Xue, J. X. Li, S. Q. Hu, M. X. Zhang, Y. H. Zhou and C. M. Zhan, *Electrochemistry Communications*, **5** (2003) 903–906.
170. X.-G. Yu, J.-Y. Xie, J. Yang, H.-J. Huang, K. Wang and Z.-S. Wen, *Journal of Electroanalytical Chemistry*, **573** (2004) 121–128.
171. X. Yu, J. Xie, Y. Li, H. Huang, C. Lai and K. Wang, *Journal of Power Sources*, **146** (2005) 335–339.
172. Z. Song, H. Zhan and Y. Zhou, *Chemical Communications*, (2009) 448–450.
173. S.-C. Zhang, L. Zhang, W.-K. Wang and W.-J. Xue, *Synthetic Metals*, **160** (2010) 2041–2044.

174. B. Duan, W. Wang, A. Wang, K. Yuan, Z. Yu, H. Zhao, J. Qiu and Y. Yang, *Journal of Materials Chemistry A*, **1** (2013) 13261–13267.
175. W. J. Chung, J. J. Griebel, E. T. Kim, H. Yoon, A. G. Simmonds, H. J. Ji, P. T. Dirlam, R. S. Glass, J. J. Wie, N. A. Nguyen, B. W. Guralnick, J. Park, Á. Somogyi, P. Theato, M. E. Mackay, Y.-E. Sung, K. Char and J. Pyun, *Nature Chemicals*, **5** (2013) 518–524.
176. S. Wei, L. Ma, K. E. Hendrickson, Z. Tu and L. A. Archer, *Journal of the American Chemical Society*, **137** (2015) 12143–12152.
177. Z. Song, Y. Qian, T. Zhang, M. Otani and H. Zhou, *Advanced Science*, **2** (2015).
178. J. R. Smith, F. C. Walsh and R. L. Clarke, *Journal of Applied Electrochemistry*, **28** (1998) 1021–1033.
179. D. Zhao, J. Feng, Q. Huo, N. Melosh, G. H. Fredrickson, B. F. Chmelka and G. D. Stucky, *Science*, **279** (1998) 548–552.
180. H.-C. J. Zhou and S. Kitagaw, *Chemical Society Reviews*, **43** (2014) 5415–5418.
181. S. J. Visco, C. C. Mailhe, L. C. De Jonghe and M. B. Armand, *Journal of the Electrochemical Society*, **136** (1989) 661–664.
182. X. Zhang, P. N. Ross, R. Kostecki, F. Kong, S. Sloop, J. B. Kerr, K. Striebel, E. J. Cairns and F. McLarnon, *Journal of the Electrochemical Society*, **148** (2001) A463–A470.
183. B. H. Jeon, J. H. Yeon, K. M. Kim and I. J. Chung, *Journal of Power Sources*, **109** (2002) 89–97.
184. Y.-J. Choi, K.-W. Kim, H.-J. Ahn and J.-H. Ahn, *Journal of Alloys and Compounds*, **449** (2008) 313–316.
185. L. A. Montoro and J. M. Rosolen, *Solid State Ionics*, **159** (2003) 233–240.
186. M. J. Lacey, F. Jeschull, K. Edström and D. Brandell, *Journal of Power Sources*, **264** (2014) 8–14.
187. Z. W. Seh, Q. Zhang, W. Li, G. Zheng, H. Yao and Y. Cui, *Chemical Science*, **4** (2013) 3673–3677.
188. Y. Huang, J. Sun, W. Wang, Y. Wang, Z. Yu, H. Zhang, A. Wang and K. Yuan, *Journal of the Electrochemical Society*, **155** (2008) A764–A767.
189. Q. Wang, W. Wang, Y. Huang, F. Wang, H. Zhang, Z. Yu, A. Wang and K. Yuan, *Journal of the Electrochemical Society*, **158** (2011) A775–A779.
190. H. Schneider, A. Garsuch, A. Panchenko, O. Gronwald, N. Janssen and P. Novák, *Journal of Power Sources*, **205** (2012) 420–425.

191. Z. Zhang, W. Bao, H. Lu, M. Jia, K. Xie, Y. Lai and J. Li, *ECS Electrochemistry Letters*, **1** (2012) A34–A37.
192. M. He, L.-X. Yuan, W.-X. Zhang, X.-L. Hu and Y.-H. Huang, *The Journal of Physical Chemistry C*, **115** (2011) 15703–15709.
193. M. Rao, X. Song, H. Liao and E. J. Cairns, *Electrochimica Acta*, **65** (2012) 228–233.
194. J. Wang, Z. Yao, C. W. Monroe, J. Yang and Y. Nuli, *Advanced Functional Materials*, **23** (2013) 1194–1201.
195. A. Jozwiuk, H. Sommer, J. Janek and T. Brezesinski, *Journal of Power Sources*, **296** (2015) 454–461.
196. M. Rao, X. Song and E. J. Cairns, *Journal of Power Sources*, **205** (2012) 474–478.
197. A. H. Whitehead and M. Schreiber, *Journal of the Electrochemical Society*, **152** (2005) A2105–A2113.
198. S.-H. Chung and A. Manthiram, *Journal of Materials Chemistry A*, **1** (2013) 9590–9596.
199. X.-B. Cheng, H.-J. Peng, J.-Q. Huang, L. Zhu, S.-H. Yang, Y. Liu, H.-W. Zhang, W. Zhu, F. Wei and Q. Zhang, *Journal of Power Sources*, **261** (2014) 264–270.
200. G. Zhou, L. Li, C. Ma, S. Wang, Y. Shi, N. Koratkar, W. Ren, F. Li and H.-M. Cheng, *Nano Energy*, **11** (2015) 356–365.
201. Y.-S. Su and A. Manthiram, *Nature Communications*, **3** (2012) 1166.
202. G. Ma, Z. Wen, J. Jin, M. Wu, X. Wu and J. Zhang, *Journal of Power Sources*, **267** (2014) 542–546.
203. X. Wang, Z. Wang and L. Chen, *Journal of Power Sources*, **242** (2013) 65–69.
204. C. Zu, Y.-S. Su, Y. Fu and A. Manthiram, *Physical Chemistry Chemical Physics*, **15** (2013) 2291–2297.
205. T.-G. Jeong, Y. H. Moon, H.-H. Chun, H. S. Kim, B. W. Cho and Y.-T. Kim, *Chemical Communications*, **49** (2013) 11107–11109.
206. G. Zhou, S. Pei, L. Li, D.-W. Wang, S. Wang, K. Huang, L.-C. Yin, F. Li and H.-M. Cheng, *Advanced Materials*, **26** (2014) 625–631.
207. S.-H. Chung and A. Manthiram, *Chemical Communications*, **50** (2014) 4184–4187.
208. H. Yao, K. Yan, W. Li, G. Zheng, D. Kong, Z. W. Seh, V. K. Narasimhan, Z. Liang and Y. Cui, *Energy & Environmental Science*, **7** (2014) 3381–3390.
209. K. Zhang, F. Qin, J. Fang, Q. Li, M. Jia, Y. Lai, Z. Zhang and J. Li, *Journal of Solid State Electrochemistry*, **18** (2014) 1025–1029.

210. J.-Q. Huang, B. Zhang, Z.-L. Xu, S. Abouali, M. Akbari Garakani, J. Huang and J.-K. Kim, *Journal of Power Sources*, **285** (2015) 43–50.
211. J. Song, Z. Yu, T. Xu, S. Chenm H. Sohn, M. Regula and D. Wang, *Journal of Materials Chemistry A*, **2** (2014) 8623.
212. L. Miao, W. Wang, A. Wang, K. Yuan and Y. Yang, *Journal of Materials Chemistry A*, **1** (2013) 11659.

Chapter 3

The Use of Lithium (Poly)sulfide Species in Li–S Batteries

Rezan Demir-Cakan,[*,§] Mathieu Morcrette[†,¶] and Jean-Marie Tarascon[‡,||]

*Department of Chemical Engineering, Gebze Technical University, 41400 Gebze, Turkey
†Laboratoire de Réactivité et Chimie des Solides, Université de Picardie Jules Verne, CNRS UMR 7314, 80039 Amiens, France
‡Collège de France, 11 Place Marcellin Berthelot, 75005 Paris, France
§demir-cakan@gtu.edu.tr
¶mathieu.morcrette@u-picardie.fr
||jean-marie.tarascon@college-de-france.fr

3.1. Introduction: Formation of Lithium Polysulfides and Their Dissolution Controls

Emerging lithium ion (Li-ion) batteries for load levelling and transport is challenging, especially for materials chemistry, and will be a major focus for upcoming years. However, in the longer term, Li-ion batteries (LIBs) cannot deliver high-energy densities and more radical approaches are necessary. There are several options to go beyond this limit and one of the possibilities for achieving longer storage life and high-energy batteries associated with cost and environmental advantages is the lithium–sulfur (Li–S) system which can theoretically offer three to five fold increase in energy density compared with conventional Li-ion cells. Although the Li–S system has interested the battery community for more than five decades,[1] it still faces issues such as poor cycle life, to reach the market place.[2]

Conventional Li–S cell consists of a metallic Li anode and sulfur cathode. Since Li–S cell is assembled at the charge state (oxidation), the cell starts first with discharge (reduction) in which Li polysulfides species (Li_2S_x) are formed throughout the cell operation. During the discharge reaction, which shows a staircase voltage profile (Figure 3.1(a)), Li is oxidized at the anode generating Li ions and electrons. While the Li ions move to cathode internally, the electrons travel to the cathode by the external electrical circuit which creates an electrical current. Consecutively, octasulfur (cyclo-S_8) is reduced to lithium sulfide by accepting the Li-ion and the upcoming electrons at the cathode. The backward reaction occurs during the charging step.

Formation of the Li polysulfides for the discharge step of a Li–S cell can be seen at the Figure 3.1(a). First, molecules of elemental sulfur (S_8) are reduced by accepting electrons which leads to the formation of high-order lithium polysulfides Li_2S_x ($6 < x \leq 8$) at the upper plateau (2.3–2.4 V vs. Li). By stepping down voltage to 2.1 V (vs. Li) further polysulfide reduction takes place, leading progressively to the formation of lower order Li polysulfides Li_2S_x ($2 < x \leq 6$). There are two discharge plateaus

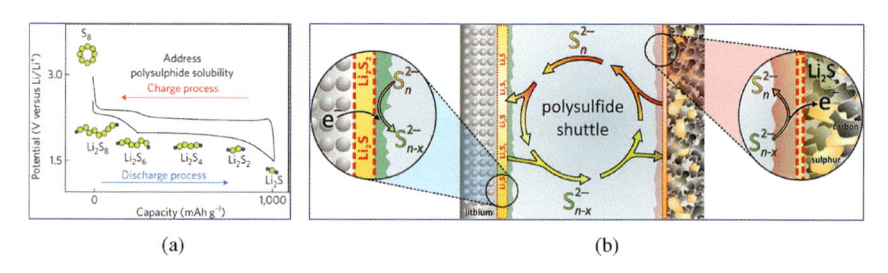

Figure 3.1 (a) Voltage profiles of a Li–S cell. (b) Schematic illustration of the polysulfide shuttle mechanism.

Source: (a) Reproduced with permission from Ref. 4. (b) Reproduced with permission from Ref. 6.

at 2.3 and 2.1 V with ether-based liquid electrolytes, which are associated to the conversion of S_8 to Li_2S_4 and Li_2S_4 to Li_2S, respectively.[3] At the end of the discharge, Li_2S, which is both electronically insulating and insoluble in the electrolyte, is formed as shown in Figure 3.1(a). Owing to the insoluble character of the least polysulfide members, which are poor conductors, the second plateau shows poor kinetics which translate in low discharge efficiency at the current rates lower than C/10[4] (1 C corresponds to charging or discharging of a cell at 1 hour).

Apart from Li_2S, the rest of sulfur reduction species are highly soluble in aprotic solvents. The high solubility of polysulfides, Li_2S_x $x \geq 2$, in organic electrolytes has historically been considered unfavorable in Li–S batteries since they cause a well-known "shuttle" mechanism. The shuttle reaction then leads to both internal self-discharge as well as active material loss due to incidental insoluble Li_2S precipitation within electrochemical cells; either at the metallic Li side or at the sulfur cathode.[5] Schematic illustration of the polysulfide shuttle mechanism is shown in Figure 3.1(b). Long-chain polysulfide diffuse throughout the separator towards the Li anode in which they are further reduced to shorter length polysulfides. Successively, shorter lengths of polysulfides difuse back to the cathode electrode where they are oxidized. Thus, a cyclic process, called shuttle mechanism, can be viewed as a chemical shortcut of the cell.

This chapter mainly focuses on the use of dissolved Li polysulfides species (Li_2S_x), so called catholyte, as electroactive material. Emphasis will be placed on the fundamental understanding of the science linked to polysulfide solubility as well as means to circumvent such an issue in view of practical applications. During the early seventies, even prior to the advent of rechargeable Li batteries, the use of liquid cathode systems were established and have proven to be very successful in yielding high-energy density with excellent discharge characteristics with Li–S dioxide (Li–SO_2) and lithium–thionyl chloride (Li–$SOCl_2$) primary systems. However, even though great advances have been made, safety problems with Li metal in direct contact with a strong oxidizer have persisted until recently. Alternatively, moving to elemental sulfur progressively led to the advent of the rechargeable Li–S battery.

Most of the published research in the field of Li–S batteries focuses on designing tailored cathode configurations with improved performances.

This is without any doubt highly important but not sufficient. One must keep in mind the unavoidable diffusion of Li polysulfides from the confined matrices since the negatively charged polysulfides can easily be dragged toward Li anode side under an electrical field.

Li–S batteries undergo precipitation–dissolution reactions, as discussed earlier, which differ from insertion cathode materials in LIBs.[7] While most of the materials based on intercalation chemistries enlist either solid solution or bi-phasic processes through phase change mechanism, sulfur compounds undergo a transformation from a solid (sulfur) to a liquid (Li polysulfides) during discharging (vice versa during charging). Thus, in order to control polysulfide dissolution many attempts have been suggested, the first one is the well-known confinement strategy by the use of highly ordered porous matrix.[8–10] All aspects of confinement strategy has been discussed at Chapter 2 in detail, which will not be the subject of this chapter.

Li polysulfide migrations are expected to depend highly upon the chemical/physical parameters pertaining to the host structures. By introducing heteroatoms to the carbon matrices or adding metal oxides/sulfides to the sulfur cathodes, long-term cycling stability and good rate performances have been achieved.[11–16] Due to the ineffectiveness of physical interactions of non-polar carbons with polar Li polysulfides, researchers have explored highly modified carbon surfaces in order to better adsorb the dissolved species.[17] For instance, the most studied material includes N-doped carbons.[18–20] Even though they are highly favored, the upper limit for the nitrogen content of N-doped carbons, estimated to be 14 wt.% at the carbonization temperature of 1000°C or 21 wt.% at 900°C,[21] limits the active sites for Li polysulfides adsorption. Moreover, graphene or graphene oxides are also highly used to trap dissolved species.[22–25] Very recently Pang et al.[17] have tested the graphitic carbon nitride to strongly adsorb and entrap Li polysulfides. The authors observed a very low capacity fade; 0.04% per cycle over 1500 cycles at a practical C/2 rate.

Apart from carbonaceous materials, semiconductor metal oxide or metal-organic frameworks (MOFs), because of their highly polar character, were also suggested to adsorb sulfur reduction species. Demir-Cakan *et al.*[26] employed oxygenated porous architecture known as MOFs works and porous SiO_2 as sulfur host matrices. Those structures were found to be more efficient than the mesoporous carbon (MPC) for the proper functioning of sulfur electrodes since they had more polarized surfaces than carbon capable of interacting strongly with charged species. In a similar manner metal oxides, which are inherently polar materials, have also been utilized such as SiO_2,[26,27] TiO_2,[11,28] VO_x[29] or MnO_2.[13,30] Therefore the best results to date were achieved by using highly polar and conducting metal oxides, such as MXene phases[12] or Ti_4O_7[16,31] have been reported to provide a sulfiphilic and conductive surface for enhanced trapping of dissolved Li polysulfides.[17]

Solvent-in-salt, ultrahigh salt concentration, approach was also suggested to better control the dissolution polysulfides. Through this concept the cyclic performance and safety issues of Li–S batteries was enhanced. This positive effect was attributed to (i) the effective suppression of Li dendrite growth, (ii) shape change in the metallic Li anode and (iii) inhibition of Li polysulfide dissolution.[32] Equally, it was also suggested that the polysulfide dissolution from the cathode could easily be controlled by using an electrolyte with a highly concentrated Li salt.[33]

Apart from all aforementioned attempts to control polysulfide dissolution, some reports have proven that those polysulfides are beneficial for the cell life. Xu *et al.*[34] have showed that the self-healing of Li–S batteries occurred in the presence of polysulfide containing electrolyte via the creation of a dynamic equilibrium between the dissolution and precipitation of Li polysulfides at the electrode interfaces. Thus, research in the field of Li–S batteries is slightly moving from those sulfur confinements to the use of dissolved chemically synthesized polysulfides either in static conditions or redox flow configurations.[2,35,36] Alternatively, those polysulfides were even employed as electrolyte additives for improved cycling performances.[34,37,38] However, the Li anode surface chemistry in Li–S batteries still remains a mystery. Thus, at this stage the question that still remains regards the effect of polysulfides, salts and solvents on the Li surface which was not the purpose of this chapter. A deep understanding of the surface growth mechanism and the investigation of the impact of the polysulfide species on Li anode are disused in details in Chapter 5.

3.2. The Use of Dissolved Polysulfides as Electroactive Material in Li–S Batteries

3.2.1. Lithium-dissolved polysulfide batteries under static conduction

Instead of solid sulfur-based electrode, polysulfide-based catholyte can be considered as an alternative to alleviate higher sulfur utilization as well as enhance the reaction kinetics due to the fast ion diffusion in liquid then solid. With the Li-dissolved polysulfide approach up to 10 M concentration can be achieved (e.g. in tetrahydrofuran).[35] First prototype cells of the metallic Li–dissolved polysulfide configuration under static condition was demonstrated by Rauh *et al.* almost 40 years ago.[35] Based on the experiments in which anode was metallic Li while cathode was polysulfide containing carbon electrode, they were able to have a practical energy density of ~300 Wh/kg. However, the cell failed after a couple of cycles, with this failure coming most probably from the non-protected Li anode surface. For the reason of clarity, gravimetric and volumetric energy densities of the catholyte solution with different molar concentration was summarized at the Table 3.1.[39] Note that the concentration is based on sulfur only.

The use of polysulfide approach falls into oblivion for many years until Zhang *et al.*[36] and Demir-Cakan *et al.*[2] concurrently discussed the pros and cons of the use of dissolved polysulfides in comparison with the conventional Li–S cells. Both cell configurations were tested under static condition rather than circulation which will be disused below. Zhang *et al.*[36] assembled and characterized Li/Li$_2$S$_9$ semiliquid full cell configuration with 0.25 M Li$_2$S$_9$ catholyte solution together with a porous carbon current collector that turned to show superior performances to conventional Li–S cells in terms of specific capacity and capacity retention. They have also found that the use of LiNO$_3$ as a co-salt in the Li$_2$S$_9$ catholyte significantly increases the cell's Coulombic efficiency.

Demir-Cakan *et al.*[2] used chemically synthesized Li$_2$S$_5$ liquid cathode which were shown to deliver high-energy densities with large amounts of dissolved species in the electrolyte. Moreover, the authors also note that less viscous (100 mM vs. 300 mM) Li$_2$S$_5$ in sulfone based electrolyte

Table 3.1: Calculation of energy densities of Li–S flow battery with different concentrations of sulfur.

		$S_8^{2-} \to S_4^{2-}$	$S_8^{2-} \to S_2^{2-}$	$S_8^{0-} \to S_4^{2-}$	$S_8^{0-} \to S_2^{2-}$	$S_8^{0-} \to S^{2-}$
Capacity (Ah/kg)	—	209	627	418	836	1672
Voltage (V)	—	2.2	2	2.2	2	2
Gravimetric energy density (Wh/kg)	—	459	1254	919	1672	3344
Volumetric energy density (Wh/L)	0.5 M	7	20	14	26	53
Volumetric energy density (Wh/L)	1 M	14	40	29	53	106
Volumetric energy density (Wh/L)	2 M	29	80	58	107	213
Volumetric energy density (Wh/L)	5 M	73	125	147	214	534
Volumetric Energy density (Wh/L)	10 M	146	250	294	428	1068

Source: Reproduced with permission from Ref. 39.

resulted in an improved discharge capacity which does not come as a surprise since low viscosity favors wettability and enhances the transport of soluble species. Further XPS analysis indicate that the presence of dissolved polysulfide on Li surface leads to an SEI-type protective layer which inhibits supplementary detrimental side effects such as short circuit due to the Li dendrites.

Although a SEI-type of layer is formed on the Li surface, the capacity loss for the cells based on dissolved polysulfide cells still indicate that this layer does not suffice to alleviate the shuttle reaction. Thus, additional protective layer needs to be employed in order to eliminate this detrimental impact. This has motivated the use of Li$^+$ ion-conductive glass-ceramic membranes (i.e. Li$_{1+x+y}$Al$_x$Ti$_{2-x}$Si$_y$P$_{3-y}$O$_{12}$ known as LATP, from Ohara Corporation) to design "two compartment" Li-dissolved polysulfide battery within which the Li surface is protected from the polysulfide poisoning. Even though this sulfide-impermeable membrane is fragile and not cost-effective, Wang et al.[40] used this ceramic membrane with tetrahydrofuran catholyte solution in which they observed a highly stable capacity retention with a good Coulombic efficiency upon prolonged cycling.

Apart from Li$^+$ ion-conductive glass-ceramic membrane, the possibilities of using polysulfide-blocking microporous polymer membranes were also investigated aiming to protect Li anode.[41] As a proof-of-concept demonstration, polymers of intrinsic microporosity (PIMs) membranes were fabricated to control their size- and ion-selective transports. This was done with the help of molecular dynamics simulations taking into account the solvated structures of Li bis(trifluoromethanesulfonyl)imide (LiTFSI) salt and different length of polysulfides (Li$_2$S$_x$, where x = 8, 6, and 4) in different oligomer length glyme-based electrolytes. The simulations suggested that polymer films with pore dimensions of less than 1.2–1.7 nm were able to provide the desired ion-selectivity. Indeed, the polysulfide blocking ability of the PIM membrane (~0.8 nm pores) improved five hundred fold the cell performance compared to mesoporous Celgard separators whose pore size is ~17 nm. As a result, significantly improved battery performance was demonstrated, even in the absence of LiNO$_3$ anode-protecting additives, which is well known as the best electrolyte additives in Li–S batteries.[42]

In order to further protect the Li anode surface from the polysulfide attacks, Li et al.[43] reported a magnetic field-controlled cell using a

Li/metal-polysulfide semiliquid battery. A biphasic magnetic solution of Li polysulfide and oleic acid stabilized γ-Fe$_2$O$_3$ magnetic nanoparticles were used as catholyte, and Li metal was used as anode. Due to the interaction between oleic acid stabilized γ-Fe$_2$O$_3$ and polysulfide through Li atom in Li$_2$S$_8$ and Oxygen atom in oleic acid, adsorption of polysulfide on the γ-Fe$_2$O$_3$ particle surface were observed. Photographs of the biphasic magnetic polysulfide solution can be seen at the Figure 3.2. The polysulfide together with γ-Fe$_2$O$_3$ nanoparticles can be extracted within the

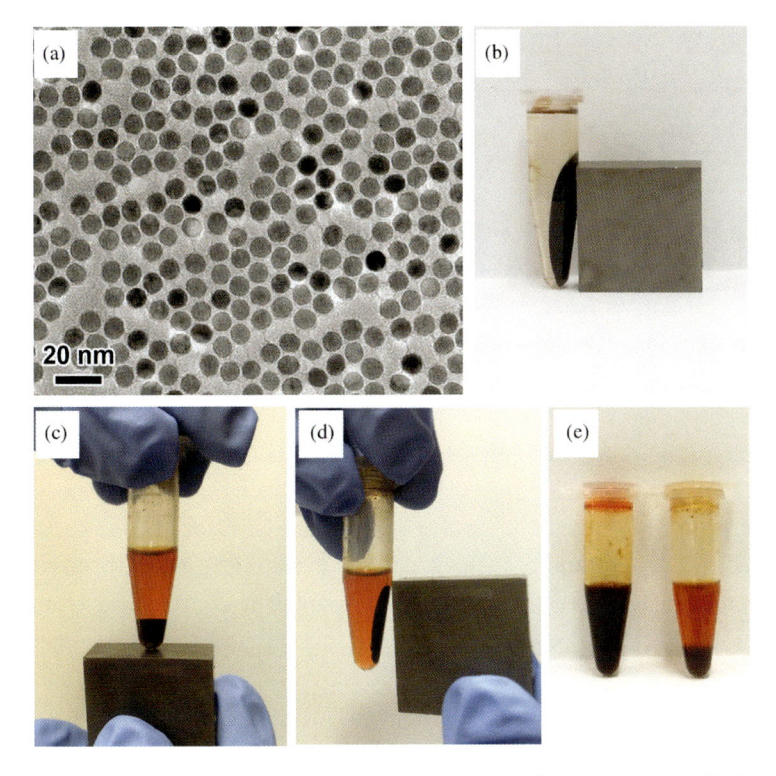

Figure 3.2 (a) TEM image of the superparamagnetic γ-Fe$_2$O$_3$ nanoparticles, (b) ferrofluidic behavior of γ-Fe$_2$O$_3$ nanoparticles in toluene in the presence of an external magnet, (c) biphasic magnetic polysulfide solution, where the polysulfide together with γ-Fe$_2$O$_3$ nanoparticles can be extracted to the same phase that is attracted to the magnetic field, (d) ferrofluidic behavior of the concentrated phase containing polysulfide and the γ-Fe$_2$O$_3$ nanoparticles in the presence of a magnet, (e) photograph showing the color difference of pure polysulfide solution (left) and biphasic magnetic polysulfide solution (right) (with the same amount of polysulfide in the same volume).

Source: Reproduced with permission from Ref. 43.

same phase under a magnetic field. This unique feature helped to maximize the utilization of the polysulfide and minimize the polysulfide shuttle effect, contributing to an enhanced energy density and Coulombic efficiency. Additionally, owing to the effect of the superparamagnetic nanoparticles, the concentrated polysulfide phase showed the behavior of a ferrofluid that was able to shuffle with the control of magnetic field. Although the concept is very elegant, the practical application of the use of dissolved polysulfides with those superparamagnetic of nanoparticles still remains challenging.

In order to accommodate more polysulfides, the group of Manthiram reported 3D current collectors which provide fast ion/electron transport pathways.[44,45] To do so, they first used free-standing multi-walled carbon nanotubes (MWCNTs) paper electrodes[44] and later on binder free carbon fiber electrodes[45] or natural carbon sources such as sucrose-coated eggshell membranes[46,47] or carbonized leafs.[48] As a result, the cycling performances were improved by integrating the interlayers either on the surface of sulfur cathodes or coating of the separators since these interlayers serve as physical barrier to suppress the diffusion of polysulfides. The concept of free-standing electrodes from the same group was also applied to conventional Li–S batteries rather than Li–polysulfide batteries and room temperature Na–S batteries.[49–54]

One of the major challenges in the field of Li–S batteries is to assemble large-sized cells with high sulfur loadings since the performance of the cells are highly dependent on the size of the Li–S cells, unlike LIBs.[55] Challenges of assembling thick sulfur electrodes were discussed in Chapter 1. Instead of starting form carbon/sulfur composites, Qie et al. used polysulfide catholyte as a starting active material to increase the electrolyte uptake of the thick electrodes since liquid media including electrolyte and salt could easily disperse into the carbon matrix as compared to the sulfur powder.[56] They used dual-layer carbon electrode to host dissolved polysulfide in which 18.1 mg/cm^2 sulfur loading was achieved resulting in around 20 mAh/cm^2 ariel capacity. Prior design with carbon nanofiber current collector allowed 5.5 mg/cm^2 sulfur loading.[57] Additionally, same group also tried an effective strategy to obtain Li/polysulfide batteries with high-energy density and long-cycling life using 3D heteroatom codoped nitrogen/sulfur graphene sponge electrodes with strong binding to dissolved polysulfides species resulting a high specific capacity of 1200 mAh/g$_{sulfur}$ at 0.2 C rate.[58]

3.2.2. Lithium-dissolved polysulfide semiliquid redox flow batteries

A rapid and unremitting enervation of the available energy resources has led to serious global energy crises in the past few decades. Thus, researchers are obliged to develop alternative and renewable energy resources to encounter the increasing fuel demand and in the meantime reducing greenhouse gas emissions. However, most renewable energy sources are intermittent, meaning that energy storage is of upmost importance and a critical element in future "smart grid and electric vehicle" applications.

Electrochemical energy storage systems offer the best blend of cost, efficiency and flexibility, with redox flow battery systems.[59] This is because different from a conventional battery cell in which power and energy features are equally rewarding; the redox flow batteries can achieve complete decoupling of power and energy in a single cell. While the size and volume of the cell determine the capacity of redox flow batteries, the power of redox flow batteries depends on the number of cells (stack size) which can be scaled independently for improving system-level energy density.[60]

Although the high temperature (300°C) sodium-sulfur (Na–S) technology has been known since 1970s as the "stellar" batteries for grid applications, they have inherited significant safety issues which has recently resulted in a serious accident in Japan. Thus, the redox flow battery technology functioning at ambient temperature is gaining renewed interest and presently stands as an attractive candidate to Na–S batteries.[61]

Aqueous electrolyte conventional redox flow batteries are of interest for stationary applications owing to their flexible design and cost-effectiveness. However, due to the narrow thermodynamic stability window of water (1.23 V) and limited solubility of the redox species in comparison with the conventional batteries, lesser energy densities are obtained. The electrolysis of water are generated usually above 1.23 V threshold due to the electrical input which should meet the full amount of enthalpy needed for evaluation of oxygen and hydrogen.[59] Additionally a recent report demonstrated a highly concentrated aqueous electrolyte with a stable performance at 2.3 V which was achieved with the formation of electrode-electrolyte interface.[62] Anyhow, the energy densities obtained in aqueous media would be still low in compassion with non-aqueous electrolytes and what matters at the end is the cost/performance ratio.

Moving from aqueous electrolytes to organic electrolytes to circumvent these limitations, Chiang and his coworkers have proposed a new cell design in organic electrolyte which additionally replaced dissolved species with suspended particles.[63–68] The new cell design was called as semisolid Li rechargeable flow batteries. This semisolid redox cell configuration differs from the classical redox flow batteries; Li-intercalation anode and cathode electrodes are circulated in slurries together with conductive carbon additives. Since the dissolution of active materials into the electrolyte limits the energy density, with the use of suspended particles, energy density is no longer limited by the solubility of active species.[7] In the first semisolid Li rechargeable flow batteries proposed by Duduta et al.[68] the cell was circulated with $LiCoO_2$ as the cathode and $Li_4Ti_5O_{12}$ as the anode. Later, Hamelet et al.[69] applied this semisolid concept to a silicon suspension flow battery. Based on Na^+ ion, using $Na_xNi_{0.22}Co_{0.11}Mn_{0.66}O_2$ and $NaTi_2(PO_4)_3$ as the positive and negative electrodes, respectively, has also been demonstrated in non-aqueous electrolytes.[70]

Metallic Li-semiliquid redox flow batteries also represent new trend toward the design of next-generation alkali-ion battery with lower cost than conventional LIBs.[59] In the cell configuration Li^+ ion is used as an energy carrier in which metallic Li is used as anode together with soluble redox molecules with relatively high redox potential at the cathode side. The amount of energy stored is determined by the amount of redox molecules, while power density is determined by redox kinetics of redox molecules and the rate of mass/charge transport inside the cell.[59]

The concept of Li-polysulfide semiflow batteries has been proposed and patented recently.[71–73] Figure 3.3(a) shows the schematic drawing of a Li-polysulfide semiflow battery.[7] Additionally, the proof-of-concept of membrane-free Li–polysulfide semiliquid flow battery was proposed by Yang et al.[74] In this configuration Li polysulfide (Li_2S_8) in ether solvent was used as a catholyte while metallic Li was used as anode. Depending

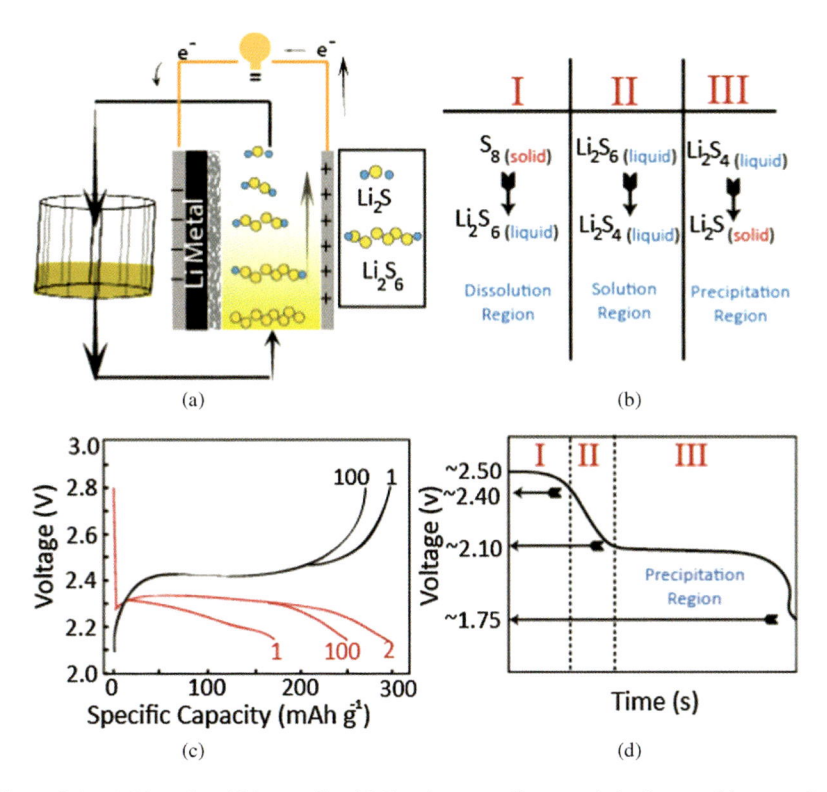

Figure 3.3 A Li–polysulfide semiliquid flow battery. Characteristic flow architecture for a semiliquid polysulfide cathode utilizing sulfur to Li_2S_4 discharge products (a) and summary of different polysulfide products that result during discharge and their active state (solid or liquid), (b) a voltage profile of a 5 M Li_2S_8 catholyte charged and discharged at 0.8 C at different cycle numbers, (c) and breakdown of voltage curve by reaction products (d) reproduced with permission from Ref. 7.

on the catholyte concentration the system provides theoretically 170 Wh/kg or and 190 Wh/L energy densities at the solubility limit (7 M polysulfide catholyte). An energy density of 108 Wh/L$_{catholyte+lithium}$ and 97 Wh/kg$_{catholyte+lithium}$, only based on the mass of the polysulfide catholyte and Li, were demonstrated experimentally with 2000 cycles. During their experiments, consciously the catholyte solution was not discharged to form insoluble Li_2S_2 and Li_2S, thus, the voltage window was limited between 2.15 and 2.8 V vs. Li (Figure 3.3(c)) that is different from the most cases in Li–S batteries whose operating voltage is between 1.5 and 3.0 V vs. Li.

118 *Li–S Batteries: The Challenges, Chemistry, Materials and Future Perspectives*

A simple reaction mechanism of Li-polysulfide batteries can be divided into the three regions; dissolution, solution and precipitation (Figures 3.3(b) and 3.3(d)).[7] Therefore, as was performed by Yang *et al.*[74] although all the side-effects caused by the formation of insulating and insoluble Li$_2$S were avoided, restricting the voltage window only at the upper plateau limits the complete chemical reactions and lowers the capacities and energy densities.

Incontestably, in order to improve the capacity utilization in Li–polysulfide flow batteries one needs to (i) increase the solution concentrations and (ii) enable the cycling of polysulfides deep into the precipitation regime (and yet electronically connected). However, since the polysulfides are not electronically conductive, additional conductive additives (i.e. carbon nanomaterials) need to be dispersed into the slurry to facilitate electrons to overcome huge cell polarization.[5] Fan *et al.*[5] further extended semiredox flow system by incorporating conductive carbon nanoparticles with the polysulfide fluids to form an embedded current collector. Figure 3.4 shows the comparison between the conventional flow cell architecture using stationary carbon fiber current collector and the flowing redox electrodes based on nanoscale percolating networks of conductor particles. The use of conductive suspensions during flow enables a better utilization of the dissolved species by providing a better electrically percolating network, in contrast to the previous studies[74] which avoided the formation of Li$_2$S$_2$ and Li$_2$S. This configuration

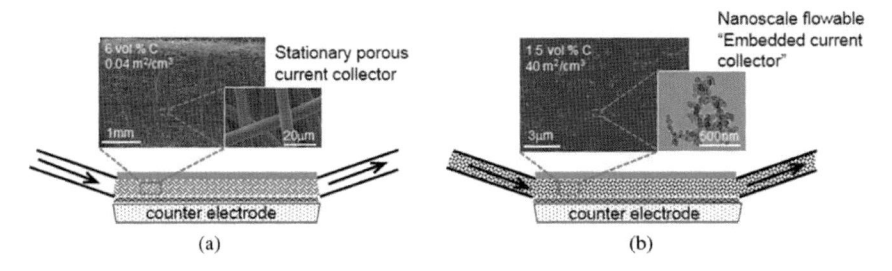

Figure 3.4 (a) Conventional flow cell architecture using stationary carbon fiber current collector with (b) electronically conductive flowing redox electrodes based on nanoscale percolating networks of conductor particles forming an embedded, self-healing current collector. The mixed conducting fluid allows charge transfer reactions throughout volume of flow electrode.

Source: Reproduced with the permission from Ref. 5.

leads reversible cycles even into composition regimes where solid precipitation occurs. They achieved a volumetric capacity of 117 Ah/$L_{catholyte+lithium}$ and energy density of 234 Wh/$L_{catholyte+lithium}$ (where average cell voltage of 2.0 V). The results showed that by addition of 1.5 vol % nanoscale carbons into the Li–polysulfide flow cell, energy density could be increased up to a factor of 5 compared to that for the solution regime.

Li protection is highly important in the field of Li–S batteries especially starting from the dissolved polysulfide species. In order to protect the Li anode surface from the polysulfide species, one of the option is to use $LiNO_3$ additive which was already proven in the ether solvent to help to form a uniform solid electrolyte interphase (SEI) that prevents parasitic reactions between polysulfide and lithium.[42] Certainly, apart from the strategies to protect Li surface by electrolyte additives or the use of ceramic or polymer membranes which were discussed above, another strategy would be the deep investigation of the electrolyte solvents. For instance, it is well known that conventional carbonate-based electrolytes used in LIBs are not suitable for Li–S batteries since polysulfides react with the solvent via nucleophilic addition or substitution to form thiocarbonates and other small molecules.[75,76] On the other hand, it is expected that high polar, electron pair donor solvents such as amides, sulfoxides, nitriles are capable of stabilizing radicals in solution, but the low dielectric constant solvents such as THF or DOL[59,77] are not. For instance, Rauh and coworkers have shown that polysulfides with middle range chain lengths are not formed in imidazolium ionic liquid, while clear trisulfur radical ($S_3^{\cdot-}$) formation is observed in dimethylsulfoxide (DMSO).[78] Consistently, Cuisinier et al. have recently showed that dissociation of the anion precursor, S_6^{2-}, to $S_3^{\cdot-}$ radical in dimethylacetamide (DMA) and DMSO donor solvents let the full utilization of both sulfur and Li_2S.[77] Moreover, Nazar and co-worker also demonstrated that combination of a solvent (acetonitrile) — salt (LiTFSI) complex with a hydrofluoroether (HFE) cosolvent resulted in limited polysulfide intermediates solubility, even though electrochemical data as well as the operando X-ray absorption spectroscopy have proven that Li_2S was detected earlier and its formation proceeded more smoothly along the discharge process compared to the common DOL–DME electrolyte, so that the theoretical capacity could be nearly reached.[79] Since the polysulfide chemistry is indeed rich and complex, Li–polysulfide batteries either in static conditions or in a flow battery configuration might lead to unexpected and prominent results.

Recently, Pan *et al.* examined the factors that govern the solubility of the polysulfide species in the catholyte of Li–S flow cells.[39] For this reason, DMSO has been selected to improve solubility of the least sulfur reduction species (Li_2S_2 and Li_2S). To overcome the incompatibility of DMSO with Li metal, the salt (LiTFSI) concentration was increased from 1 to 3 M and this was shown to help in stabilizing the interface between the electrolyte and the Li surface. Using this trick, a stable capacity of 1200 mAh/g_{sulfur} was obtained at C/5 current density with the capacity retention of 87% under static conditions. Afterwards, Li–polysulfide flow configuration were tested controlling the current densities in which they observed that at relatively small current densities (0.25 mA/cm^2), a specific capacity of 280 mAh/g_{sulfur} could be achieved along with a much lower polarization.

The aforementioned examples witnesses the increasing activities towards the development of high volumetric capacity Li-flow batteries through 2016 via (i) the use of dissolved polysulfide as catholyte, (ii) use of Li-intercalation suspension electrolytes and (iii) sulfur-impregnated carbon composite as a flow cathode[80] (Figure 3.5). Chen *et al.* demonstrated a new flow-cathode concept that could offer higher catholyte volumetric capacity compared with the above approaches. In their concept, instead of chemically synthesized Li polysulfide species, they intentionally applied sulfur-impregnated carbon composite as a circulation cathode to achieve high-energy Li-flow batteries (Figure 3.5(d)). To do so, they were able to achieve 294–192 Ah/$L_{catholyte}$ capacities more than 100 cycles with the highly concentrated active material (12.9 M of sulfur). Later on same group, as a continuation of their C/S circulating concept, used LiI in sulfur-impregnated carbon composite flow cell to improve electrochemical utilization of sulfur as well as to increase the volumetric capacity in which they achieved 500 Ah/L. The demonstrated catholyte volumetric capacity is five times higher than the all-vanadium flow batteries (60 Ah/L) and 3–6 times higher than the demonstrated Li-polysulphide approaches (50–117 Ah/L). Similar strategies were applied to conventional

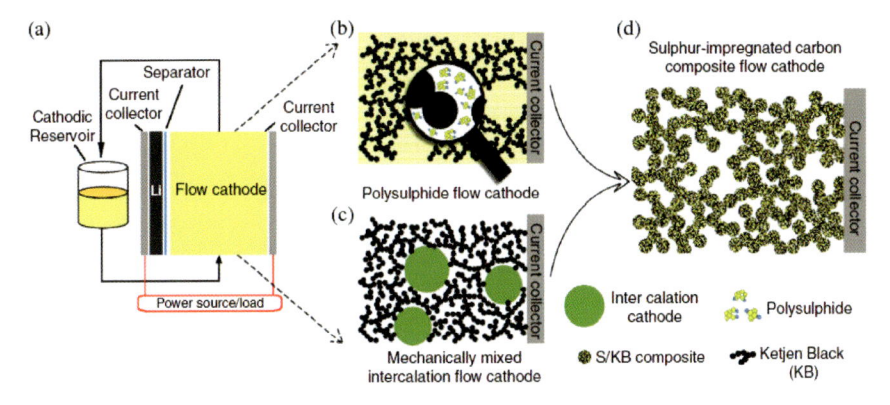

Figure 3.5 Concept of a sulfur-impregnated carbon composite flow cathode. (a) schematic representation of a non-aqueous Li flow batter using, (b) a polysulfide flow cathode[35,74] (c) mechanically mixed intercalation flow cathode[61,68] and (d) a new flow-cathode structure employing sulfur-impregnated Ketjen black composite suspended in the non-aqueous electrolyte.[80]

Source: Reproduced with permission from Ref. 80.

Li–S batteries earlier with LiBr[81] and LiI[82] electrolyte additive in which increased capacities were obtained not only for the cocathode materials, which contribute to the total capacity, but also to the oxidizing agent and internal redox mediator role of the additives.

Supramolecular gel networks of π-stacked redox mediators were also subjected to improve sulfur utilization and rate performance, even deprived of conductive carbon additives. Frischman *et al.* used redox-active perylene bisimide-polysulfide gel that could overcome electronic charge-transport bottlenecks of Li–S hybrid redox flow batteries.[83] A high-throughput computational platform was developed to rapidly screen π-gelators candidate by electron affinity (E_{ea}) and ionization potential (E_i) ensure redox activity at relevant Li–S potentials in which perylene bisimide was identified as a redox mediator. The first demonstration of polysulfide organogel catholyte designed for long-duration grid-scale energy storage application delivered a volumetric energy density of 44 Wh/L$_{sulfur}$ at sulfur loadings of 4 mg/cm². Li *et al.*[60] have also visualized the possibility of applying the redox targeting concept by the use of bis- (pentamethyl-cyclopentadienyl) chromium (CrCp*$_2$) and bis-(pentamethyl-cyclopentadienyl) nickel (NiCp*$_2$) as tandem mediators for the sulfur reduction reaction. In the configuration, sulfur was stored in an energy tank without flowing and the redox mediators were used to extract/inject electrons with the Li⁺ storage compound in the energy tank. Hence, unlike a typical RFB the performance of this redox targeting system is not constrained by the solubility of the redox active species (mediators) in the electrolyte since the mediators are charge transfer agents and charge transporters and not the energy storage compounds. Here the novelty resides in the fact that the containers are becoming electrochemically active and do not any longer remain as dead weight within the system.

3.3. The Use of Dissolved Polysulfides as Electrolyte Additive

The aforementioned beneficial effect of having dissolved polysulfides as a positive electrode in Li–S cells is somewhat counterintuitive in light of early literature reports. Such reports indeed state that those polysulfides are poisoning species for the Li metal electrode and provide a "shuttle" mechanism which causes both internal self-discharge and active material losses upon cycling.

For instance, Xu et al.[34] demonstrated a conceptually new approach underlining the role of polysulfide as self-healing Li–S batteries. They showed that sulfur loss could be mitigated by leveling the concentration gradient of the polysulfide species at the cathode/electrolyte interface which reduces initial dissolution of active materials. Thus, when these species were produced at the cathode, they were not migrated into the electrolyte due to the concentration gradient, instead, they were retained at the cathode. Schematic diagram can be seen at Figure 3.6. By creating a dynamic equilibrium between the dissolution and precipitation of Li polysulfides at the sulfur/electrolyte interface, the Li–S cells were capable of delivering around 1500 mAh/g$_{sulfur}$ capacity over 50 cycles along with the high coulombic efficiency.

Chen et al.[37] also explored the impact of soluble Li polysulfides (Li$_{12}$S$_6$) on the cycling behavior of (Li–S) batteries which was used as cosalts/additives in an ether-based electrolyte. By optimizing the concentration of the polysulfide species and the amount of electrolyte, the Li–S batteries performed high and stable discharge capacity, better rate

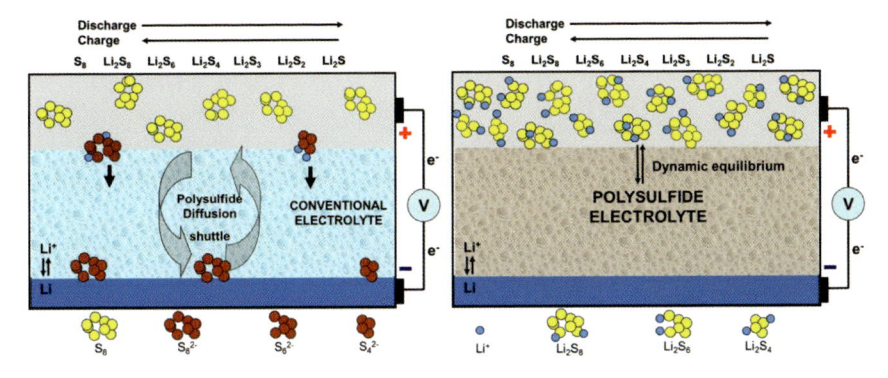

Figure 3.6 Schematic diagram of (left) Li–S battery using conventional electrolyte, such as 1 M LiTFSI in DME/DOL in which polysulfides produced at the cathode during discharge dissolve into the electrolyte; (right) Li–S battery using the polysulfide electrolyte, in which produced polysulfides are retained at the cathode.

Source: Reproduced with the permission from Ref. 34.

capability and cycling performance. The improved performances were attributed to the prevention of insoluble Li_2S formation by the help of dissolved polysulfides.

In order to shed the light on this Li poisoning issue, Demir-Cakan et al.[2] further assembled an atypical cell configuration in which sulfur powder was deposited on the Li anode electrode surface (Figure 3.7). During the cycling progress this new cell configuration performs as good as the conventional Li–S cell setup thanks to the *in situ* created polysulfide species. In order to understand the role of polysulfide species on the Li electrode surface XPS experiments were performed. Results showed that a SEI-type of layer is formed at the surface of Li which prevents the formation of the electronically insulating Li_2S at the surface of Li upon further cycling. A drastic drop in the cell resistance was also observed in the cell which has been already observed by the group of Aurbach as surface film formation with different transport properties.[42]

3.4. Lithium Sulfide Cathodes

One of the fundamental issues in Li–S batteries is the poisoning of Li surface by dissolved polysulfide species. Chapter 5 indicated the

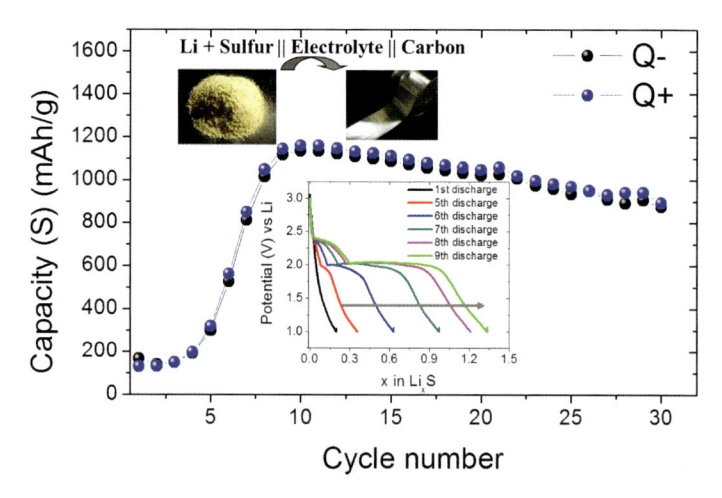

Figure 3.7 (a) Cycling performance of sulfur deposited Li foil which was used as an active material and separated by a Whatman separator from a Ketjen Black conductive carbon. The inset figure shows the *in situ* formation of the polysulfides upon cycling.
Source: Reproduced with permission from Ref. 2.

solutions for the protection of Li metal anodes in rechargeable Li–S batteries. Impending solutions are to use Li-ion conducting glasses or membranes that segregate the active metal from detrimental side reactions.[84–89] Another method is to replace sulfur with its lithiated form, Li_2S. Thus, Li anode can be bypassed with tin, silicon or metal oxide anodes which enable Li-free anodes. The cell is called Li-ion sulfur batteries[90] which was first demonstrated by the group of Scrosati[91,92] and Cui.[93]

Li_2S is the most reduced sulfur discharge species which offers 1166 mAh/g theoretical capacity, that is nearly one order of magnitude higher than traditional metal oxides phosphates based cathode electrodes in LIBs. However, there are several challenges when utilizing Li_2S; low electronic conductivities, sensitivity to moisture and oxygen, limited synthesis routes and high over potential oxidation at the first cycle.[3,90] A number of efforts have been made recently to understand the electrochemical behavior of Li_2S and to develop optimized Li_2S cathodes as discussed next.

3.4.1. Strategies to eliminate high over potential of Li$_2$S

Alike sulfur, Li$_2$S is a semiconductor with a band gap of 3.36 eV as reported by Khachai et al.[94] It was found that the electronic conductivity of Li$_2$S surfaces and nanoparticles is very different from that of the Li$_2$S bulk and depends on the concentration of surface Li atoms (or surface S atoms). The removal of Li atoms from the surfaces of Li$_2$S nanoparticles leads to the metalization of the nanoparticles.[95] In addition, the standard enthalpy of Li$_2$S (−447 kJ/mole) is quite high leading to poor ionic conductivity.[95,96] Overall, Li$_2$S is far from an ideal cathode as compared to Li-ion cathode materials and needs to be activated by applying a higher over potential throughout the initial charging.

Yang et al. demonstrated this potential barrier of 1.0 V at the beginning of the first oxidation of micron-sized Li$_2$S.[38] A schematic of the electrochemical oxidation mechanism of Li$_2$S is shown in Figure 3.8. Initially, Li$_{2-x}$S particles exist as a single phase with a Li-poor shell at their surfaces which continues to become Li deficient during the 2nd step while the core remains nearly at the stoichiometry. Particles through steps 1 and 2 preserve the original Li$_2$S crystal structure. In step 3, polysulfides are

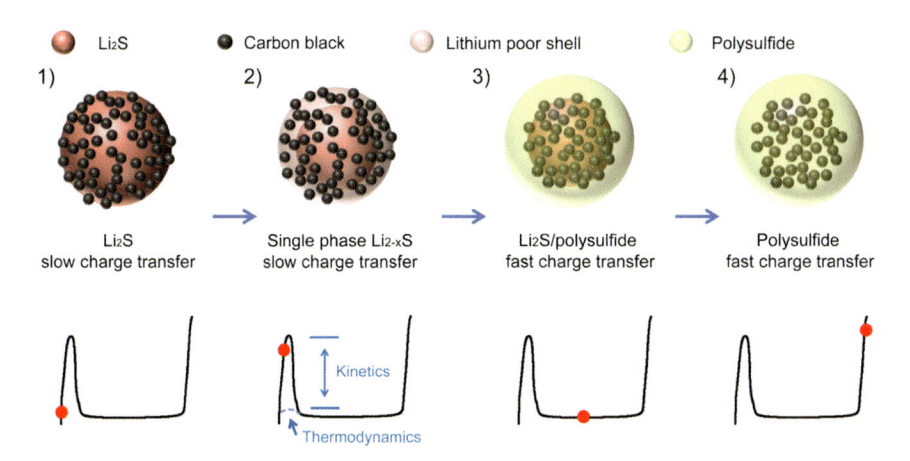

Figure 3.8 A model for an oxidation mechanism of Li$_2$S.
Source: Reproduced with permission from Ref. 38.

formed after overwhelming the initial potential barrier. The coexistence of Li$_2$S and polysulfides facilitates the charge transfer so that in step 4th, Li$_2$S is converted to polysulfides with fast kinetics. After the first nucleation of the polysulfide species into the polysulfide species during the first oxidation step, charge transfer is significantly improved, thus, over potential charging is no longer needed for the following cycles.

The origin of the over potential activation has been the subject of upcoming contributions and the work done by Yang *et al.* provided a practical approach to utilize Li$_2$S as a cathode material. Walus *et al.* has recently showed that different charge voltage profiles could be obtained by using the same system with different type of electrodes hence highlighting the importance of electronic wiring.[97] They also suggested the coexistence of Li$_2$S and long-chain polysulfides due to the high over potential for the electrochemical oxidation of Li$_2$S.

In addition to this phenomenon, Son *et al.*[96] identified another factor for the activation process of Li$_2$S. Since Li$_2$S is highly sensitive to the moisture and oxygen, a stable LiOH layer or unstable S–H layer (due to H$_2$S gas formation) can be easily created on the surface of Li$_2$S, whose standard enthalpy change of formations are 968 and 10.5 kJ/g, respectively. Thus, continuous formation of LiOH and H$_2$S results in the decomposition of Li$_2$S. To prove this finding, the authors sintered Li$_2$S at different temperature to remove the native layer formed on Li$_2$S and noted (Figure 3.9) that the crystalline structure of Li$_2$S remained unaffected up to 850°C. Figure 3.9(b) clearly demonstrates that when Li$_2$S was sintered to higher temperatures (500, 650 or 850°C), the first oxidation capacity values, when the cells were cycled up to 4 V, were improved from 364 (for the pristine one) to 546, 974, and 1017 mAh/g, respectively, with an obvious activation potential barrier.

At this stage, it is worth mentioning that the high over potential of Li$_2$S is beyond the voltage stability of ethereal solvents (e.g. LiTFSI salt in 1–3 dioxolane/dimethoxyethane) which are the common solvents for Li–S batteries. Thus, the widespread application of Li$_2$S as a foremost cathode material for Li-ion sulfur batteries requires additional additives to minimize the high over potentials oxidation of Li$_2$S. To do so, either redox mediators as electrolyte additives or some nanoparticles were applied.[90]

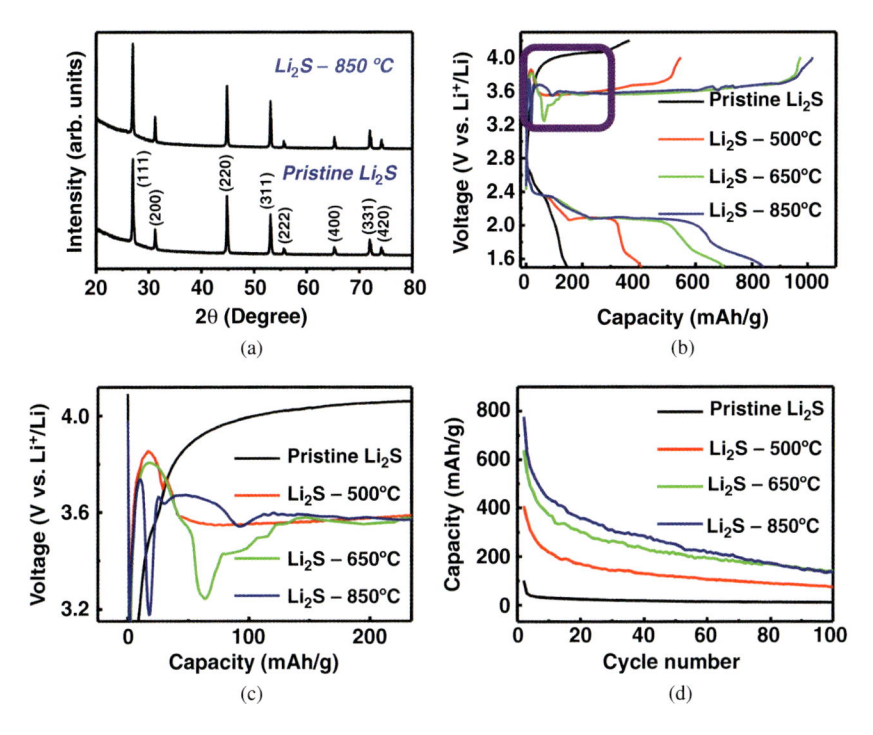

Figure 3.9 (a) XRD pattern of as-prepared Li$_2$S and Li$_2$S heated at 850°C. (b) Voltage profiles of pristine Li$_2$S, Li$_2$S heated at 500, 650, and 850°C between 4 and 1.5 V at a current density of 117 mAh/g during first cycle. (c) Magnified image of voltage profiles within purple boundary line in (b). (d) Cycle performances of discharge capacity of as-prepared Li$_2$S, Li$_2$S heated at 500, 650, and 850°C between 4 and 1.5 V at a current density of 117 mAh/g$_{Li_2S}$. Amount of loading of all cells was more than 2 mg Li$_2$S/cm^2.

Source: Reproduced with permission from Ref. 96.

Aurbach's group has recently demonstrated the possibility of several redox mediators, that are reversible redox couples, capable of undergoing electro-oxidation at the electrode surface, for enhancing the Li$_2$S cathode active materials utilization.[98] Within this study, five different additives were employed whose redox-activity potentials were either lower or higher than that of Li$_2$S. The electrochemical potential of cobaltocene was lower than that of Li$_2$S, while decamethylferrocene, LiI, and ferrocene were chosen to have higher potential. Dibenzenechromium was also tested whose active potential is close to Li$_2$S. Due to the high driving force provided by the several hundreds of millivolts difference compared with the electrochemical potential of Li$_2$S, decamethylferrocene demonstrated an outstanding ability to enable the efficient utilization of Li sulfide at low charging potential of 3.2 V vs. Li. Cells based on decamethylferrocene additive were showing four times higher capacities after prolonged cycling that classical Li$_2$S cells.

Other attempts to lower the over potential of Li$_2$S were tested in the Li–S cell configuration with P$_2$S$_5$,[99] LiBr[81] or LiI.[82,100] Here the additives to the electrolyte were shown to reduce the activation energy through the first oxidation and to decrease the over potential of the first charge as well. For instance, the potential was dropped from 3.3 to 2.75 V in presence of LiI[100] and such a lowering of the voltage was further supported by quantum chemistry calculations bearing in mind that the addition of LiI into the electrolyte forms a protective layer on the surface of Li anode. First, the iodine radicals that are generated at 3 V (vs. Li) reacts with DME solvent forming DME radicals. Then these radical polymerizes and forms a polyether protective layer.

Lastly, heteroatoms on the carbon surfaces were shown to enlist strong chemical interaction with Li$_2$S and Li polysulfides.[101–103] Chen et al.[103] used nitrogen-doped carbon-encapsulated Li$_2$S with 72% active materials loading. Besides improving the conductivity, the nitrogen-doped simultaneously enabled the effective confinement of polysulfides within the conductive shell. Zhou et al.[101] used Li$_2$S/(boron or nitrogen)-doped graphene electrodes to enhanced cyclability while preserving a decent capacity 600 mAh/g$_{Li_2S}$. The improved performances were attributed not only to the porous networks which brings conductivity as well as helps in accommodating dissolved species but also to the strong interaction between N- or B-doped graphene and Li$_2$S.

3.4.2. Li$_2$S composites and their synthesis protocols

Although Li$_2$S is an auspicious cathode material, due to its ability to be combined with Li-free anodes (such as tin and silicon) for minimizing safety issues, Li$_2$S has its drawbacks by being electronically insulated and by forming polysulfides upon cycling, as of sulfur. To address these issues, comprehensive approaches aiming toward the formation of electrically conductive Li$_2$S/C composites were suggested not only to slowdown the dissolution of polysulfides but also to increase the electronic conductivity of the final composites. Metal–Li$_2$S (i.e. Fe–Li$_2$S, Cu–Li$_2$S, Co–Li$_2$S)[104–106] composites have also been suggested to obviate the low electronic conductivity of Li$_2$S. For instance Cu–Li$_2$S[105] system were prepared by mechanical milling and tested with a Li$_2$S–P$_2$S$_5$ glass-ceramic electrolyte, as discussed in Chapter 4 addressing all-solid state Li cells. However, since metal additives are converted to metal sulfides, via well-known conversion reactions,[107] poor electrochemical cyclability as well as increase in polarization upon cycling were obtained independent of all-solid state or liquid electrolyte configuration.

An easiest way of obtaining Li$_2$S/C composites is by mechanochemistry via the use of ball-milling devices capable of intimately mixing commercially available micron size Li$_2$S and carbon powders in an inert atmosphere.[108–110] Cai et al.[110] were the first who prepared Li$_2$S/C composite with 67.5 wt.% Li$_2$S content by use of high-energy ball-milling in which they obtained a specific capacity of 1144 mAh/g$_{Li_2S}$. Aside from the commercially available micron-sized Li$_2$S, other synthesis methods of preparing nanosized Li$_2$S crystals have also been reported.[93,111–115] While Yang et al.[93] used n-butyllithium to lithiate sulfur forming Li$_2$S in MPC, Zheng et al.[114] pursued a different approach to oxidize sulfur. It consists of spraying stabilized lithium metal powder (SLMP) into sulfur containing microporous carbon (MC) so as to form in situ lithiated Li$_2$S carbon composite. As another alternative toward the formation of Li$_2$S in a porous carbon, Fu et al.[115] used prelithiated graphite to reduce chemically synthesized Li$_2$S$_6$ specie with at the end the feasibility of such electrodes to deliver 800 mAh/g$_{Li_2S}$ capacities over 50 cycles.

The insulating properties of Li$_2$S require good electrochemical wiring to enable its conversion into polysulfides and sulfur in the first charge. This calls for the synthesis of carbon-coated Li$_2$S, where the carbon improves the distribution of electrons within the composite electrode and prevents the agglomeration of Li$_2$S particles, hence enabling a better distribution of the electrolyte. Guo et al.[116] obtained Li$_2$S/C composite by mixing PAN polymer with Li$_2$S taking advantage of the strong interaction between the Li ions in Li$_2$S and lone pair electrons pertaining to the nitrile group of PAN. Likewise, Lui et al.[117] obtained amorphous carbon coating on Li$_2$S by poly(vinylpyrorolidone) which were claimed to reduce the shuttle effect. Another strategy to have *in situ* carbon coated Li$_2$S was studied by She *et al.*[118] in which N atoms in polypyrrole (PPy) were found to possess favorable Li–N interaction with Li$_2$S. Due to this Li–N strong bonding, 785 mAh/g capacities over 400 cycles were achieved by the help of constraining intermediate polysulfides. *In situ* synthesis of Li$_2$S/C was also developed by Yang *et al.*[119] based on the carbothermal reduction of lithium sulfate into lithium sulfide in the presence of an excess of a carbon source (Li$_2$SO$_4$ + 2C → Li$_2$S + 2CO$_2$). The process is knowns as Leblanc process and the method is illustrated in Figure 3.10. The improved performances, compared with the physically mixed Li$_2$S and carbon, were ascribed to the uniform dispersion of Li$_2$S in carbon and the ability of the composite to sequester higher dissolved polysulfides generated during electrochemical cycling.

Finally, different carbon sources, i.e. carbon nanotubes, graphene, graphene oxide (GO) or mesophase pitch were also incorporated with

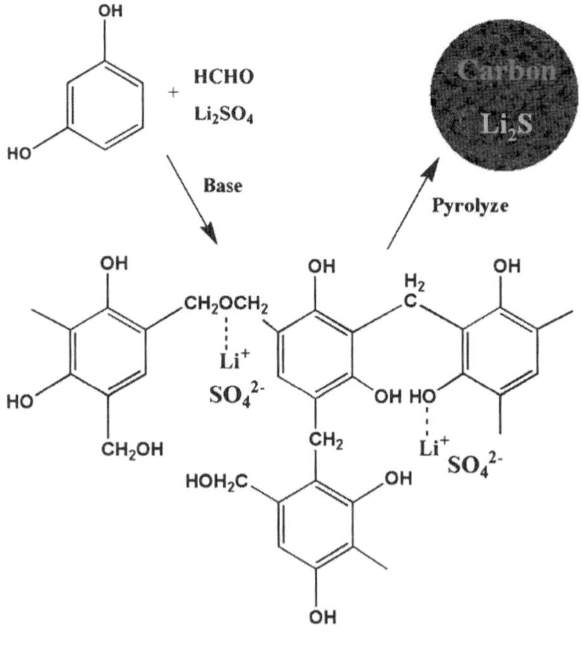

Figure 3.10 *In situ* synthesis scheme for a Li$_2$S@C composite.

Source: Reproduced with permission from Ref. 119.

Li_2S.[101,103,108,113,120–127] Lastly, capitalizing on the higher melting point of Li_2S (938°C), than that of sulfur whose melting point is 115°C, several authors have developed robust, uniform and stable carbon coatings under high temperature treatment conditions. For instance, spark-plasma sintering (SPS)[128] or chemical vapor deposition (CVD)[113] techniques have been used for the preparation of Li_2S/C composites.

3.4.3. Li_2S–Li metal free anode

One of the fundamental issues in Li–S batteries is the poisoning of Li surface by dissolved polysulfide species as well as dendrite formation during stripping/plating of Li ions at the Li anode. Mainly Li-ion conducting glasses or membranes segregate the Li metal from detrimental side reactions.[84–89] However most of the solid ceramics are both too fragile and costly as well as most of them have less ionic conductivity at room temperate.[129] The use of electrolyte additives,[42] coating of the Li foil with protective layer,[130] employing highly concentrated salts,[32] applying pressure onto the Li electrode,[131] use of ionic liquids[132] or polymer electrolytes[91] have been attempted to minimize the dendrite formation on the Li surface improving the performance of Li–S batteries. Some of these solutions are reported in details in Chapter 5. Nevertheless, a method to bypass Li anode with silicon, tin or metal oxide anode together with the deep sulfur discharge product (Li_2S) as cathode could be viewed as ideal. Therefore, such a cell design, called *Li-ion sulfur batteries*,[90] which was first demonstrated by the group of Scrosati[91,92] and Cui,[93] falls short in meeting the expectations.

3.4.3.1. *From lithium powders to carbon-based anodes*

Instead of Li foil, that is commercially used as anode material in primary Li metal or rechargeable solid polymer batteries besides Li–air and Li–S batteries, the use of coated Li (nano-micro) powders is another strategy to suppress dendrite formation.[114,133–136] Heine et al.[133] have compared the performance of coated Li powder (CLiP) pressed on a copper current collector with plain lithium foil. An improved performance was noted. It was explained in terms of better coulombic efficiencies for stripping/plating and lower potentials in the continuous charge–discharge process due to the higher surface area and homogeneous distribution of the CLiP, which leads to an overall reduced current density.

Stabilized Li metal powder (SLMP) together with hard carbon were used as anode in Li–S batteries with the hope to effectively compensate the irreversible capacity and therefore increase the energy density.[137] Using this approach, the first discharge capacity of the battery with 20 wt.% hard carbon could reach a capacity of 1300 mAh/g_{Li_2S} which was therefore attenuated slowly with increasing current densities. This new type of battery anode provides a greater surface area, which facilitates the ion transfer, reduces the shuttle effect and the dendrite growth on the surface of anode.

Graphite, which is used as anode in almost every commercially available Li-ion battery since Sony's first cell back in the early 1990s, has also been used either as an hybrid or primary anode in Li–S batteries.[138,139] Huang et al.[138] demonstrated the use of lithiated graphite which functions as an artificial, self-regulated solid electrolyte interface layer. For proof-of-concept, the lithiated graphite was placed in front of the Li metal so as to actively control the electrochemical reactions and minimize the deleterious side reactions. Jeschull et al.[139] has also described an efficient, reversible Li intercalation into graphite in ether based electrolytes which, in turn, enabled the creation of a stable "lithium-ion–sulfur" cell with an average coulombic efficiency of 99.5%, compared with 95% for Li–S cells.

Full system based on Li_2S cathode and carbon cathode has been demonstrated by Zheng et al.[114] As was discussed above, they showed synthesis of *in situ* lithiated Li_2S carbon composite in a MC through *in situ* lithiation strategy, i.e. spraying commercial SLMP onto a prepared sulfur/MC film cathode prior to the routine compressing process in cell assembly. Later, they coupled those Li_2S-MC together with a Li-free graphite electrode in which they could obtain a stable capacity of around 600 mAh/g_{Li_2S} over 150 cycles. Figure 3.11 shows the cycle performances of individual performances of the sulfur/MC and the as-prepared Li_2S/MC electrode as well as their full cell configuration.

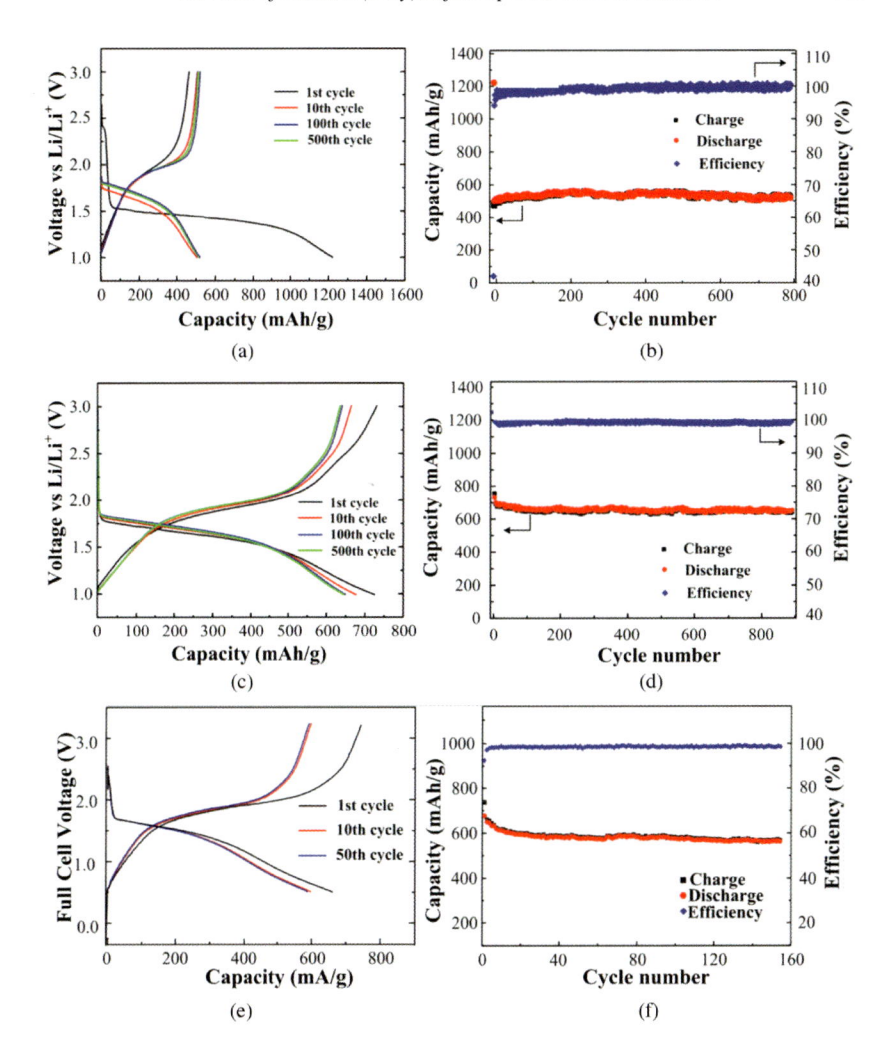

Figure 3.11 Electrochemical cycling properties of the S/MC and the as-prepared Li$_2$S/MC electrode at 0.1 C current density (a) discharge/charge voltage profiles of the S/MC electrode, (b) cycling performance and Coulombic efficiency of the S/MC electrode, (c) discharge/charge voltage profiles of the Li$_2$S/MC electrode, (d) cycling performance and Coulombic efficiency of the Li$_2$S/MC electrode. Electrochemical properties of the full cell with a Li$_2$S/MC cathode and a graphite anode, (e) discharge/charge voltage profiles, (f) cycling performance and Coulombic efficiency curves. The specific capacity values are given with respect to the mass of Li$_2$S.

Source: Reproduced with permission from Ref. 114.

3.4.3.2. Silicon anodes

To date several works have been pursued based on sulfur-lithium-silicon system since silicon offers a very high theoretical capacity of 3580 mAh/g.[93,140–145] Two configurations are mainly pursued regarding the use of sulfur-lithium-silicon system; (i) prelithiated Si together with sulfur powder or (ii) lithiated sulfur (Li$_2$S) together with Si. The group of Cui first used Li$_2$S/MPC composite cathode and a silicon nanowire anode.[93] The Li$_2$S–Si system offered a theoretical specific energy of 1550 Wh/kg, based on only the mass of the active electrode materials, which promises advantages when naively comparing with the practical ~410 Wh/kg offers by some of today's Li-rich NMC/Si Li-ion systems. Although the theoretical values are quite promising, a capacity of only 250 mAh/g that was decaying after 20 cycles was achieved. Such a limited performance could be ascribed (i) volume expansion of Si upon lithiation, (ii) not enough Li content in the silicon nanowires to compensate the cycling, (iii) insulating nature of Li$_2$S which needs to be activated at the first cycle, as discussed above, etc. Later, same group, proposed prelithiated Si nanowires anode together with Li$_2$S cathode[140] with the prelithiation of silicon nanowires done by directly attaching them to a piece of Li metal foil in the presence of electrolyte under pressure (Figure 3.12). After a prelithiation step of 20 minutes, silicon nanowires were paired with Li$_2$S cathode. Even though the concept is promising, the results were still deceiving as witnessed by a fast capacity fade which needs further optimizing the system.

Along the similar path, the group of Scrosati,[141,142] reported a cell based on a sulfur–carbon composite cathode combined with a lithiated, silicon–carbon nanocomposite anode as well as Li-ion sulfur battery based on a carbon-coated Li$_2$S cathode and an electrodeposited silicon-based anode. In the case of Li$_2$S-silicon composite, they obtained a cell having an average voltage of 1.4 V and operating with a very stable capacity of around 300 mAh/g$_{full\ cell}$ over 100 cycles at 1 C current density[141] although the cell was built as anode-limited due to the low loading of the silicon-based electrode. In the case of sulfur-lithiated silicon composite,[142] they have observed a quite stable cycling behavior of the complete cell with the Li triflate, LiCF$_3$SO$_3$, salt in tetraethylene glycol dimethyl ether (TEGDME) solution as electrolyte.

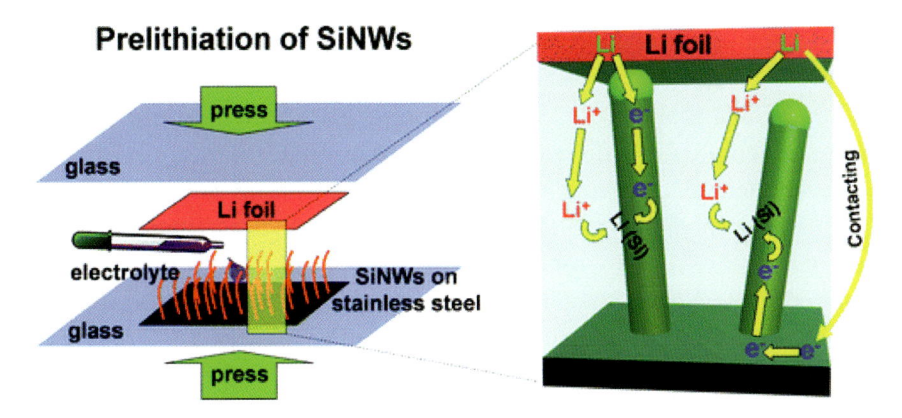

Figure 3.12 Schematic diagrams showing (a) the prelithiation of silicon nanowires on stainless steel foil, and (b) the internal electron and Li⁺ pathways during the prelithiation. Two possible electron pathways are shown in (b) Electron can directly flow into the silicon nanowire if the nanowire tip contacts the Li foil, or the electron can flow across the contacting point of Li foil and stainless steel and enter the silicon nanowire through the bottom.
Source: Reproduced with permission from Ref. 140.

Yan *et al.*[143] have used prelithiated Si/C microspheres as an anode, S/C composites as a cathode and a room temperature ionic liquid of *n*-Methyl-*n*-Allylpyrrolidinium bis(trifluoromethanesulfonyl)imide (RTIL P1A3TFSI) as an electrolyte to address the safety issues of the cell. The reversible capacity was above 900 mAh/g$_{sulfur}$ in the first cycle, and could remain as high as 670 mAh/g$_{sulfur}$ after 50 cycles revealing the favorable compatibility between the RTIL P1A3TFSI electrolyte and the two electrode materials in the Li–S battery.

Elazari *et al.*[144] have also disclosed the use of prelithiated columnar amorphous silicon structures as a promising anode in a Li-ion sulfur system. The characteristics of the sulfur–lithium–silicon cell were similar to the Li–S half cell with slightly reduced cell voltage. Inspired by Aurbach's group studies on Si,[144] besides prelithiated hard carbon electrodes, Brueckner *et al.*[145] tested the performance of prelithiated silicon–carbon coated on a 3D flexible carbon paper. Both configurations gave acceptable capacity retention with more than 400 mAh/g$_{sulfur}$ (roughly half of the initial capacity) after 1300 cycles.

Although elegant and sophisticated for conceptual demonstration, all the aforementioned examples aiming to inject the possible use of Si as replacement of Li in Li–S batteries have been deceiving. This does not come as a surprise since the electrochemical stability of silicon electrodes is still problematic, even for LIBs. Particularly, silicon suffers from low coulombic efficiency due to the decomposition of electrolyte at low potential and huge volume expansion during electrochemical lithiation. Although same electrolyte additives such as fluoroethylene carbonate (FEC)[146] proposed to stabilize the SEI on silicon negative electrode, they would not be suitable for sulfur batteries due to the nucleophilic addition reactions. Si implantation into the Li–S technology will be even more complicated due the presence of dissolved polysulfide and so on. There is a need to go back to fundamentals and study in more detail the effects of solvents, salts and electrolyte additives on both cell compartments.

Along this line, the group of Aurbach[147] has investigated the impact of 3-dioxolane (DOL) based electrolyte solutions (DOL/LiTFSI and DOL/LiTFSI–LiNO$_3$) on the electrochemical performance and surface chemistry of silicon nanowire anodes. Compared with the standard alkyl carbonate solutions (EC–DMC/LiPF$_6$), reduced irreversible capacity losses, enhanced and stable reversible capacities over prolonged cycling lower impedance were demonstrated with DOL solutions, see Figure 3.13. Detailed characterizations (i.e. TEM, XPS, SEM, etc.) indicated a distinctive surface chemistry of silicon nanowires cycled in DOL based electrolyte solutions which was found to be responsible for their enhanced electrochemical performances.

3.4.3.3. Tin anodes

With a theoretical capacity of 990 mAh/g, tin has attracted much attention as an alternative to graphite in LIBs with the first implementation of tin anode together with Li$_2$S cathode being reported by Hassoun et al.[91] By analogy to what has been done in the past within the field of LIBs, they have also replaced the liquid electrolyte by a gel-type polymer membrane which was formed by trapping an ethylene carbonate/dimethylcarbonate Li hexafluorophosphate (EC:DMC LiPF$_6$) solution saturated with Li sulfide in a polyethylene oxide/Li trifluoromethanesulfonate (PEO/LiCF$_3$SO$_3$) polymer matrix. The assembly of the cathode in its discharge

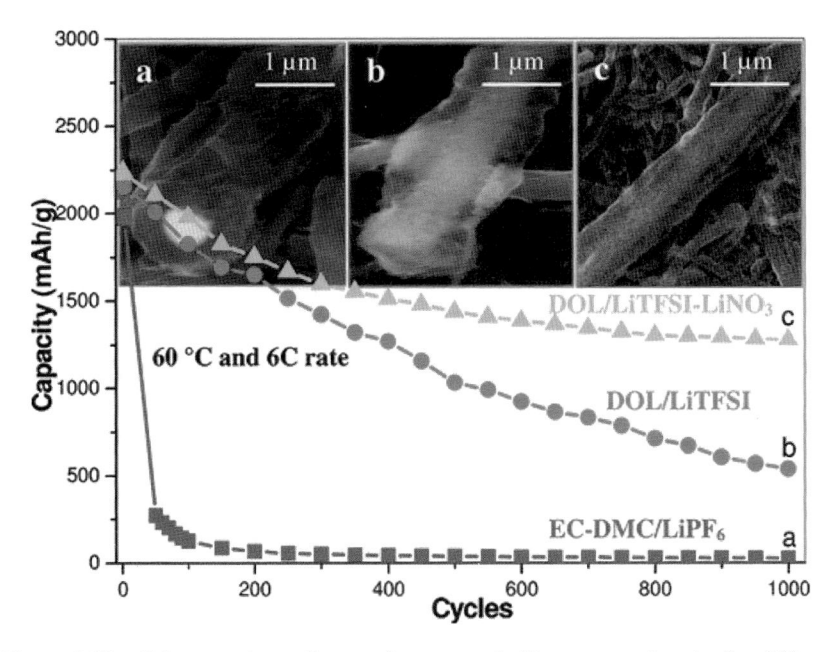

Figure 3.13 Galvanostatic cycling performance of silicon nanowires in the different electrolyte solutions (6 C rate at 60°C): (a) EC-DMC/LiPF$_6$, (b) DOL/LiTFSI, and (c) DOL/LiTFSI–LiNO$_3$.

Source: Reproduced with permission from Ref. 147.

state and combination with a gel electrolyte containing a Li$_2$S-saturated liquid solution was claimed to help in controlling the dissolution of the polysulfide ions. The authors assembled Li$_2$S–Sn cell with specific energy densities of 100 Wh/kg of electrode materials with 100 cycles. Duan et al.[148] also demonstrated the concept of sulfur/Li-ion battery by tin–carbon anode and sulfur cathode which delivered a reversible capacity of 500 mAh/g after 50 cycles at a current density of 200 mA/g.

Like for silicon, tin was also used with its lithiated form as demonstrated by Ikeda et al.[149] Li$_{22}$Sn$_5$ was synthesized by ball milling. It was combined with a carbon–sulfur composite electrode containing 60 wt.% sulfur soaked in ionic liquid electrolyte. Such a cell was able to deliver a capacity of 600 mAh/g$_{Li_2S}$ after 50 cycles, as can be seen from Figure 3.14. Such a limited cycle life was ascribed to the degradation of Li$_{22}$Sn$_5$ electrode since the cells were built as ideally balanced capacity-wise.

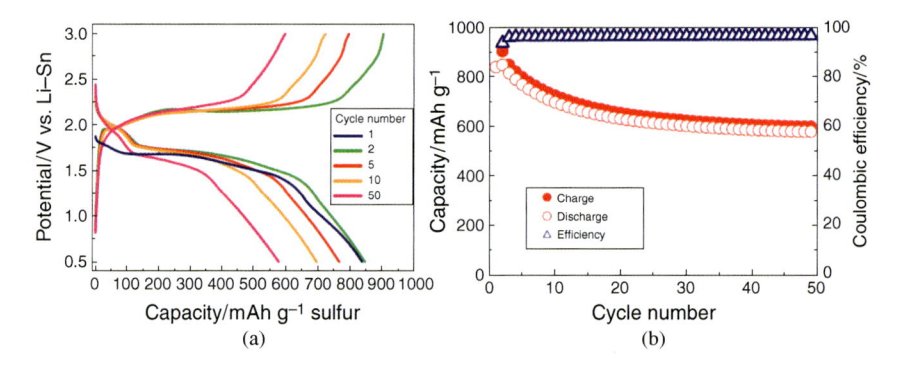

Figure 3.14 (a) Discharge–charge curves and (b) cycle stability of a $Li_{22}Sn_5$/sulfur cell measured at a current density of 85 μA/cm² at 30°C. The loadings of $Li_{22}Sn_5$ and S_8 in the anode and cathode were 5.66 and 0.606 mg/cm², respectively.

Source: Reproduced with permission from Ref. 149.

3.4.3.4. Alloy anodes

Apart from silicon and tin, recently Li–B alloy, which is a two phase material made by coheating Li and crystalline boron, and its structure described as free Li embedded in stable Li_7B_6 framework, have also been suggested to replace metallic Li.[150,151] Duan et al.[150] has demonstrated that the use of Li–B alloy, instead of Li foil, restrained the formation of dendritic Li, reduced the interface impedance of electrode and improved the cycle performance of the cell. The advantages of Li–B alloy over Li anode has been described as two fold. First, the SEI layer forms more quickly on surface of Li–B alloy due to the increased surface area formed by the Li_7B_6 fibrillar network framework in Li–B alloy, which reduces the surface current density during plating, hence enabling an even deposition of Li to some extent. Secondly Li–B alloys provide a better electronic conductivity and a high Li-ion diffusion rate compared to Li metal.

Later, Zhang et al.[151] also obtained similar finding with Li–B alloy anode in sulfur batteries in which the capacity retention for long-term cycling of full cells was improved as compared with that of Li–S batteries. Li–B alloys limit the formation of Li dendrites and cracks, which contributed to its high cycle stability and safety performance.

3.5. Conclusions

Currently, research in the field of Li–S batteries is slightly migrating from the sulfur confinement strategies (i.e. entrapment of polysulfides) to the usage of chemically synthesized dissolved polysulfides. Aware of this fact, new strategies departing from the confinement approaches are being used consisting of dissolved polysulfides. By doing so, the collapse of the porous host matrix due to 80% volume expansion associated to the formation of Li_2S would be also eliminated.

There have been many attempts for the use of dissolved polysulfide in the field of semiredox flow batteries. Hence, in this chapter we have tried to convey most of the ongoing activities dealing with the utilization of polysulfides in Li–S batteries. Many of these studies, while demonstrating elegant approaches and promising results, cannot be relevant to practical applications. Regarding the flow cells for instance, one of the biggest challenges is to ensure a real flow cell configuration considering the entire cell compartments such as external pump, tank and linking pipes. It should be noted that the cycling under flow conditions is much more complicated than the static cells (i.e. coin cells) especially taking into account the viscosity, the concentration of catholyte solutions and the flowing rate that need to be tuned and balanced individually. Additionally different electron donor pair solutions in which sulfur radicals are stabilized should be the subject for better electrochemical utilization of sulfur in flow battery configurations. Therefore, approaches aiming to convert the dead weights of the catholyte and electrolyte containers into electrochemical active reservoirs are quite appealing to pursue as it will be enable to enhance the cost/performance ratio for these systems — one of their main drawbacks.

Alternatively, dissolved polysulfides have even been used as an electrolyte additive for improved cycling performance. Having observed that solubility is not the primary limiting factor for the proper functioning of Li–S batteries, further investigation on the impact of the soluble and non-soluble species on the capacity fading needs to be studied both at the cathode and anode electrodes. Even though those polysulfides are used either as electroactive materials or electrolyte additives, the subject of inhibiting them at the cathode sides are significant measures of this field. However, Li polysulfide migrations are mainly dependent on the nature of the host materials. Thus, by introducing heteroatoms to the conductive matrices, the use of novel materials such as MXenes or graphitic C_3N_4 or adding metal oxides/sulfides to the sulfur cathodes enable better adsorption or entrapping of those polysulfides via strongly interacting with charged species. Such approaches are attractive but time has come to implement them in real cells so as to identify the showstoppers if any.

The Li anode surface chemistry in Li–S batteries still remains surreptitious in the presence of dissolved polysulfides together with salts and solvents. Different from the use of Li anode, another approach is to employ metallic Li-free anode with Li_2S rather than sulfur as the cathode electrode. To do so, high capacity anodes such as silicon, tin, metal oxides or alloys were tested with promising results. Here, progresses are simple and blind implementation of what researchers have tried to do with the LIBs for a decade with the end result being the commercialization of C/Si (90/10 wt.%) anodes by Panasonic. However, as mentioned, the Li–S chemistry is different and more complex than the Li-ion. One must deviate from classical approaches. Gladly, an increasing amount of researchers are considering the use an ionic ceramic membrane. This could be an option, provided we can design more economic and robust membranes. This calls for great research efforts in designing such membranes, whether they are Na or Li-based inorganic compounds.

Bibliography

1. D. Herbert and J. Ulam, US Patent, (1962) **No: 3043896**.
2. R. Demir-Cakan, M. Morcrette, Gangulibabu, A. Gueguen, R. Dedryvere and J.-M. Tarascon, *Energy & Environmental Science*, **6** (2013) 176–182.
3. A. Manthiram, Y. Fu, S.-H. Chung, C. Zu and Y.-S. Su, *Chemical Reviews*, **114** (2014) 11751–11787.
4. P. G. Bruce, S. A. Freunberger, L. J. Hardwick and J.-M. Tarascon, *Nature Materials*, **11** (2012) 19–29.
5. F. Y. Fan, W. H. Woodford, Z. Li, N. Baram, K. C. Smith, A. Helal, G. H. McKinley, W. C. Carter and Y.-M. Chiang, *Nano Letters*, **14** (2014) 2210–2218.
6. M. R. Busche, P. Adelhelm, H. Sommer, H. Schneider, K. Leitner and J. Janek, *Journal of Power Sources*, **259** (2014) 289–299.

7. K. B. Hatzell, M. Boota and Y. Gogotsi, *Chemical Society Reviews*, **44** (2015) 8664–8687.
8. X. Ji, K. T. Lee and L. F. Nazar, *Nature Materials*, **8** (2009) 500–506.
9. N. Jayaprakash, J. Shen, S. S. Moganty, A. Corona and L. A. Archer, *Angewandte Chemie International Edition*, **50** (2011) 5904–5908.
10. H. Wang, Y. Yang, Y. Liang, J. T. Robinson, Y. Li, A. Jackson, Y. Cui and H. Dai, *Nano Letters*, **11** (2011) 2644–2647.
11. S. Evers, T. Yim and L. F. Nazar, *The Journal of Physical Chemistry C*, **116** (2012) 19653–19658.
12. X. Liang, A. Garsuch and L. F. Nazar, *Angewandte Chemie International Edition*, **54** (2015) 3907–3911.
13. X. Liang, C. Hart, Q. Pang, A. Garsuch, T. Weiss and L. F. Nazar, *Nature Communications*, **6** (2015) 5682–5689.
14. X. Liang, C. Y. Kwok, F. Lodi-Marzano, Q. Pang, M. Cuisinier, H. Huang, C. J. Hart, D. Houtarde, K. Kaup, H. Sommer, T. Brezesinski, J. Janek and L. F. Nazar, *Advanced Energy Materials,* **6** (2016) 1501636.
15. X. Liang and L. F. Nazar, *Acs Nano*, **10** (2016) 4192–4198.
16. Q. Pang, D. Kundu, M. Cuisinier and L. F. Nazar, *Nature Communications*, **5** (2014) 4759–4767.
17. Q. Pang and L. F. Nazar, *Acs Nano*, **10** (2016) 4111–4118.
18. J. Song, M. L. Gordin, T. Xu, S. Chen, Z. Yu, H. Sohn, J. Lu, Y. Ren, Y. Duan and D. Wang, *Angewandte Chemie International Edition*, **54** (2015) 4325–4329.
19. Q. Pang, J. Tang, H. Huang, X. Liang, C. Hart, K. C. Tam and L. F. Nazar, *Advanced Materials*, **27** (2015) 6021–6028.
20. J. Song, T. Xu, M. L. Gordin, P. Zhu, D. Lv, Y.-B. Jiang, Y. Chen, Y. Duan and D. Wang, *Advanced Functional Materials*, **24** (2014) 1243–1250.
21. S. Zhang, S. Tsuzuki, K. Ueno, K. Dokko and M. Watanabe, *Angewandte Chemie International Edition*, **54** (2015) 1302–1306.
22. F.-F. Zhang, X.-B. Zhang, Y.-H. Dong and L.-M. Wang, *Journal of Materials Chemistry*, **22** (2012) 11452–11454.
23. G. Zhou, L.-C. Yin, D.-W. Wang, L. Li, S. Pei, I. R. Gentle, F. Li and H.-M. Cheng, *Acs Nano*, **7** (2013) 5367–5375.
24. C. Wang, K. Su, W. Wan, H. Guo, H. Zhou, J. Chen, X. Zhang and Y. Huang, *Journal of Materials Chemistry A*, **2** (2014) 5018–5023.
25. G. Zhou, L. Li, C. Ma, S. Wang, Y. Shi, N. Koratkar, W. Ren, F. Li and H.-M. Cheng, *Nano Energy*, **11** (2015) 356–365.
26. R. Demir-Cakan, M. Morcrette, F. Nouar, C. Davoisne, T. Devic, D. Gonbeau, R. Dominko, C. Serre, G. Ferey and J.-M. Tarascon, *Journal of the American Chemical Society*, **133** (2011) 16154–16160.

27. X. Ji, S. Evers, R. Black and L. F. Nazar, *Nature Communications*, **2** (2011) 325–331.
28. Z. W. Seh, W. Li, J. J. Cha, G. Zheng, Y. Yang, M. T. McDowell, P.-C. Hsu and Y. Cui, *Nature Communications*, **4** (2013) 1331–1336.
29. K. T. Lee, R. Black, T. Yim, X. Ji and L. F. Nazar, *Advanced Energy Materials*, **2** (2012) 1490–1496.
30. Z. Li, J. Zhang and X. W. Lou, *Angewandte Chemie International Edition*, **54** (2015) 12886–12890.
31. X. Tao, J. Wang, Z. Ying, Q. Cai, G. Zheng, Y. Gan, H. Huang, Y. Xia, C. Liang, W. Zhang and Y. Cui, *Nano Letters*, **14** (2014) 5288–5294.
32. L. Suo, Y.-S. Hu, H. Li, M. Armand and L. Chen, *Nature Communications*, **4** (2013) 1481–1490.
33. E. S. Shin, K. Kim, S. H. Oh and W. Il Cho, *Chemical Communications*, **49** (2013) 2004–2006.
34. R. Xu, I. Belharouak, J. C. M. Li, X. Zhang, I. Bloom and J. Bareno, *Advanced Energy Materials*, **3** (2013) 833–838.
35. R. D. Rauh, K. M. Abraham, G. F. Pearson, J. K. Surprenant and S. B. Brummer, *Journal of the Electrochemical Society*, **126** (1979) 523–527.
36. S. S. Zhang and J. A. Read, *Journal of Power Sources*, **200** (2012) 77–82.
37. S. Chen, F. Dai, M. L. Gordin and D. Wang, *RSC Advances*, **3** (2013) 3540–3543.
38. Y. Yang, G. Zheng, S. Misra, J. Nelson, M. F. Toney and Y. Gui, *Journal of the American Chemical Society*, **134** (2012) 15387–15394.
39. H. Pan, X. Wei, W. A. Henderson, Y. Shao, J. Chen, P. Bhattacharya, J. Xiao and J. Liu, *Advanced Energy Materials*, **5** (2015) 1500113.
40. L. Wang, Y. Zhao, M. L. Thomas, A. Dutta and H. R. Byon, *ChemElectroChem*, **3** (2016) 152–157.
41. C. Li, A. L. Ward, S. E. Doris, T. A. Pascal, D. Prendergast and B. A. Helms, *Nano Letters*, **15** (2015) 5724–5729.
42. D. Aurbach, E. Pollak, R. Elazari, G. Salitra, C. S. Kelley and J. Affinito, *Journal of the Electrochemical Society*, **156** (2009) A694–A702.
43. W. Li, Z. Liang, Z. Lu, X. Tao, K. Liu, H. Yao and Y. Cui, *Nano Letters*, **15** (2015) 7394–7399.
44. Y. Fu, Y.-S. Su and A. Manthiram, *Angewandte Chemie International Edition*, **52** (2013) 6930–6935.
45. C. Zu, Y. Fu and A. Manthiram, *Journal of Materials Chemistry A*, **1** (2013) 10362–10367.

46. S.-H. Chung and A. Manthiram, *Acs Sustainable Chemistry & Engineering*, **2** (2014) 2248–2252.
47. S.-H. Chung and A. Manthiram, *Advanced Materials*, **26** (2014) 1360–1365.
48. S.-H. Chung and A. Manthiram, *ChemSusChem*, **7** (2014) 1655–1661.
49. S.-H. Chung and A. Manthiram, *Electrochimica Acta*, **107** (2013) 569–576.
50. S.-H. Chung and A. Manthiram, *Journal of Physical Chemistry Letters*, **5** (2014) 1978–1983.
51. S.-H. Chung and A. Manthiram, *Chemical Communications*, **50** (2014) 4184–4187.
52. A. Manthiram and X. Yu, *Small*, **11** (2015) 2108–2114.
53. R. Singhal, S.-H. Chung, A. Manthiram and V. Kalra, *Journal of Materials Chemistry A*, **3** (2015) 4530–4538.
54. X. Yu and A. Manthiram, *Advanced Energy Materials*, **5** (2015) 1500350.
55. D. Lv, J. Zheng, Q. Li, X. Xie, S. Ferrara, Z. Nie, L. B. Mehdi, N. D. Browning, J.-G. Zhang, G. L. Graff, J. Liu and J. Xiao, *Advanced Energy Materials*, **5** (2015) 1402290.
56. L. Qie, C. Zu and A. Manthiram, *Advanced Energy Materials*, **6** (2016) 1502459.
57. C. Zu and A. Manthiram, *Advanced Energy Materials*, **4** (2014) 1400897.
58. G. Zhou, E. Paek, G. S. Hwang and A. Manthiram, *Nature Communications*, **6** (2015) 7760.
59. Y. Zhao, Y. Ding, Y. Li, L. Peng, H. R. Byon, J. B. Goodenough and G. Yu, *Chemical Society Reviews*, **44** (2015) 7968–7996.
60. J. Li, L. Yang, S. Yang and J. Y. Lee, *Advanced Energy Materials*, **5** (2015) 1501808.
61. S. Hamelet, T. Tzedakis, J. B. Leriche, S. Sailler, D. Larcher, P. L. Taberna, P. Simon and J. M. Tarascona, *Journal of the Electrochemical Society*, **159** (2012) A1360–A1367.
62. L. Suo, O. Borodin, T. Gao, M. Olguin, J. Ho, X. Fan, C. Luo, C. Wang and K. Xu, *Science*, **350** (2015) 938–943.
63. Y.-M. Chiang, W. C. Carter, B. Y. Ho and M. Duduta, World Patent, **WO2009151639**.
64. Y.-M. Chiang and C. W. Carter, US Patent, **US20100047671**.
65. Y.-M. Chiang and C. W. Carter, US Patent, **US2011200848**.
66. Y.-M. Chiang and C. W. Carter, World Patent, **WO2011084649**.
67. Y.-M. Chiang and R. Bazarella, World Patent, **WO2010118060**.
68. M. Duduta, B. Ho, V. C. Wood, P. Limthongkul, V. E. Brunini, W. C. Carter and Y.-M. Chiang, *Advanced Energy Materials*, **1** (2011) 511–516.

69. S. Hamelet, D. Larcher, L. Dupont and J. M. Tarascon, *Journal of the Electrochemical Society*, **160** (2013) A516–A520.
70. E. Ventosa, D. Buchholz, S. Klink, C. Flox, L. G. Chagas, C. Vaalma, W. Schuhmann, S. Passerini and J. Ramon Morante, *Chemical Communications*, **51** (2015) 7298–7301.
71. J. Xiao, J. Zhang, G. L. Graff, J. Liu, W. Wang, J. Zheng, W. Xu, Y. Shao and Z. Yang, US Patent (2012) **US20130260204**.
72. L. C. Jonghe, S. J. Visco, Y. S. Nimon and B. D. Katz, US Patent (2011) **US8758914**.
73. J. Xiao, J. Liu, H. Pan and W. A. Henderson, US Patent (2014) **14/530,442**.
74. Y. Yang, G. Zheng and Y. Cui, *Energy & Environmental Science*, **6** (2013) 1552–1558.
75. J. Gao, M. A. Lowe, Y. Kiya and H. D. Abruna, *The Journal of Physical Chemistry C*, **115** (2011) 25132–25137.
76. T. Yim, M.-S. Park, J.-S. Yu, K. J. Kim, K. Y. Im, J.-H. Kim, G. Jeong, Y. N. Jo, S.-G. Woo, K. S. Kang, I. Lee and Y.-J. Kim, *Electrochimica Acta*, **107** (2013) 454–460.
77. M. Cuisinier, C. Hart, M. Balasubramanian, A. Garsuch and L. F. Nazar, *Advanced Energy Materials*, **5** (2015) 1401801.
78. R. D. Rauh, F. S. Shuker, J. M. Marston and S. B. Brummer, *Journal of Inorganic & Nuclear Chemistry*, **39** (1977) 1761–1766.
79. M. Cuisinier, P. E. Cabelguen, B. D. Adams, A. Garsuch, M. Balasubramanian and L. F. Nazar, *Energy & Environmental Science*, **7** (2014) 2697–2705.
80. H. Chen, Q. Zou, Z. Liang, H. Liu, Q. Li and Y.-C. Lu, *Nature Communications*, **6** (2015) 5877.
81. C. Liang, W. Fu, Z. Lin, Z. Liu, X. Yu, K. Hong, N. Dudney and J. Howe, *220th ECS Meeting*, 2011, **Abstract no 1266**.
82. R. Demir-Cakan, *Journal of Power Sources*, **282** (2015) 437–443.
83. P. D. Frischmann, L. C. H. Gerber, S. E. Doris, E. Y. Tsai, F. Y. Fan, X. Qu, A. Jain, K. A. Persson, Y.-M. Chiang and B. A. Helms, *Chemistry of Materials*, **27** (2015) 6765–6770.
84. M. Buonaiuto, S. Neuhold, D. J. Schroeder, C. M. Lopez and J. T. Vaughey, *ChemPlusChem*, **80** (2015) 363–367.
85. S.-H. Chung and A. Manthiram, *Electrochemistry Communications*, **38** (2014) 91–95.
86. Y. Jung and S. Kim, *Electrochemistry Communications*, **9** (2007) 249–254.
87. D. J. Lee, H. Lee, J. Song, M.-H. Ryou, Y. M. Lee, H.-T. Kim and J.-K. Park, *Electrochemistry Communications*, **40** (2014) 45–48.

88. S. H. Lee, J. R. Harding, D. S. Liu, J. M. D'Arcy, Y. Shao-Horn and P. T. Hammond, *Chemistry of Materials*, **26** (2014) 2579–2585.
89. R. S. Thompson, D. J. Schroeder, C. M. Lopez, S. Neuhold and J. T. Vaughey, *Electrochemistry Communications*, **13** (2011) 1369–1372.
90. A. Rosenman, E. Markevich, G. Salitra, D. Aurbach, A. Garsuch and F. F. Chesneau, *Advanced Energy Materials*, **5** (2015) 1500212.
91. J. Hassoun and B. Scrosati, *Angewandte Chemie International Edition*, **49** (2010) 2371–2374.
92. J. Hassoun, Y.-K. Sun and B. Scrosati, *Journal of Power Sources*, **196** (2011) 343–348.
93. Y. Yang, M. T. McDowell, A. Jackson, J. J. Cha, S. S. Hong and Y. Cui, *Nano Letters*, **10** (2010) 1486–1491.
94. H. Khachai, R. Khenata, A. Bouhemadou, A. Haddou, A. H. Reshak, B. Amrani, D. Rached and B. Soudini, *Journal of Physics: Condensed Matter*, **21** (2009) 095404.
95. Y.-X. Chen and P. Kaghazchi, *Nanoscale*, **6** (2014) 13391–13395.
96. Y. Son, J.-S. Lee, Y. Son, J.-H. Jang and J. Cho, *Advanced Energy Materials*, **5** (2015) 1500110.
97. S. Walus, C. Barchasz, R. Bouchet, J. F. Martin, J. C. Lepretre and F. Alloin, *Electrochimica Acta*, **180** (2015) 178–186.
98. S. Meini, R. Elazari, A. Rosenman, A. Garsuch and D. Aurbach, *Journal of Physical Chemistry Letters*, **5** (2014) 915–918.
99. Z. Lin, Z. Liu, W. Fu, N. J. Dudney and C. Liang, *Advanced Functional Materials*, **23** (2013) 1064–1069.
100. F. Wu, J. T. Lee, N. Nitta, H. Kim, O. Borodin and G. Yushin, *Advanced Materials*, **27** (2015) 101–108.
101. G. Zhou, E. Paek, G. S. Hwang and A. Manthiram, *Advanced Energy Materials*, **6** (2016) 1501355.
102. H.-J. Peng, T.-Z. Hou, Q. Zhang, J.-Q. Huang, X.-B. Cheng, M.-Q. Guo, Z. Yuan, L.-Y. He and F. Wei, *Advanced Materials Interfaces*, **1** (2014) 1400227.
103. L. Chen, Y. Liu, M. Ashuri, C. Liu and L. L. Shaw, *Journal of Materials Chemistry A*, **2** (2014) 18026–18032.
104. M. N. Obrovac and J. R. Dahn, *Electrochemical and Solid State Letters*, **5** (2002) A70–A73.
105. A. Hayashi, R. Ohtsubo, T. Ohtomo, F. Mizuno and M. Tatsumisago, *Journal of Power Sources*, **183** (2008) 422–426.
106. Y. Zhou, C. Wu, H. Zhang, X. Wu and Z. Fu, *Electrochimica Acta*, **52** (2007) 3130–3136.

107. P. Poizot, S. Laruelle, S. Grugeon, L. Dupont and J. M. Tarascon, *Nature*, **407** (2000) 496–499.
108. M. R. Kaiser, X. Liang, H.-K. Liu, S.-X. Dou and J.-Z. Wang, *Carbon*, **103** (2016) 163–171.
109. S. Liang, C. Liang, Y. Xia, H. Xu, H. Huang, X. Tao, Y. Gan and W. Zhang, *Journal of Power Sources*, **306** (2016) 200–207.
110. K. Cai, M.-K. Song, E. J. Cairns and Y. Zhang, *Nano Letters*, **12** (2012) 6474–6479.
111. F. Wu, H. Kim, A. Magasinski, J. T. Lee, H.-T. Lin and G. Yushin, *Advanced Energy Materials*, **4** (2014) 1400196.
112. K. Zhang, L. Wang, Z. Hu, F. Cheng and J. Chen, *Scientific Reports*, **4** (2014) 6467.
113. C. Nan, Z. Lin, H. Liao, M.-K. Song, Y. Li and E. J. Cairns, *Journal of the American Chemical Society*, **136** (2014) 4659–4663.
114. S. Zheng, Y. Chen, Y. Xu, F. Yi, Y. Zhu, Y. Liu, J. Yang and C. Wang, *Acs Nano*, **7** (2013) 10995–11003.
115. Y. Fu, C. Zu and A. Manthiram, *Journal of the American Chemical Society*, **135** (2013) 18044–18047.
116. J. Guo, Z. Yang, Y. Yu, H. D. Abruna and L. A. Archer, *Journal of the American Chemical Society*, **135** (2013) 763–767.
117. J. Liu, H. Nara, T. Yokoshima, T. Momma and T. Osaka, *Chemistry Letters*, **43** (2014) 901–903.
118. Z. W. Seh, H. Wang, P.-C. Hsu, Q. Zhang, W. Li, G. Zheng, H. Yao and Y. Cui, *Energy & Environmental Science*, **7** (2014) 672–676.
119. Z. Yang, J. Guo, S. K. Das, Y. Yu, Z. Zhou, H. D. Abruna and L. A. Archer, *Journal of Materials Chemistry A*, **1** (2013) 1433–1440.
120. M. Zhang, D. Lei, X. Yu, L. Chen, Q. Li, Y. Wang, T. Wang and G. Cao, *Journal of Materials Chemistry*, **22** (2012) 23091–23097.
121. K. Han, J. Shen, C. M. Hayner, H. Ye, M. C. Kung and H. H. Kung, *Journal of Power Sources*, **251** (2014) 331–337.
122. Q. Li, Z. Zhang, K. Zhang, J. Fang, Y. Lai and J. Li, *Journal of Power Sources*, **256** (2014) 137–144.
123. F. Wu, A. Magasinski and G. Yushin, *Journal of Materials Chemistry A*, **2** (2014) 6064–6070.
124. Y. Hwa, J. Zhao and E. J. Cairns, *Nano Letters*, **15** (2015) 3479–3486.
125. Y. Qiu, G. Rong, J. Yang, G. Li, S. Ma, X. Wang, Z. Pan, Y. Hou, M. Liu, F. Ye, W. Li, Z. W. Seh, X. Tao, H. Yao, N. Liu, R. Zhang, G. Zhou, J. Wang, S. Fan, Y. Cui and Y. Zhang, *Advanced Energy Materials*, **5** (2015) 1501369.

126. D. H. Wang, X. H. Xia, D. Xie, X. Q. Niu, X. Ge, C. D. Gu, X. L. Wang and J. P. Tu, *Journal of Power Sources*, **299** (2015) 293–300.
127. F. Ye, Y. Hou, M. Liu, W. Li, X. Yang, Y. Qiu, L. Zhou, H. Li, Y. Xua and Y. Zhang, *Nanoscale*, **7** (2015) 9472–9476.
128. T. Takeuchi, H. Sakaebe, H. Kageyama, H. Senoh, T. Sakai and K. Tatsumi, *Journal of Power Sources*, **195** (2010) 2928–2934.
129. N. Kamaya, K. Homma, Y. Yamakawa, M. Hirayama, R. Kanno, M. Yonemura, T. Kamiyama, Y. Kato, S. Hama, K. Kawamoto and A. Mitsui, *Nature Materials*, **10** (2011) 682–686.
130. G. Ma, Z. Wen, M. Wu, C. Shen, Q. Wang, J. Jin and X. Wu, *Chemical Communications*, **50** (2014) 14209–14212.
131. Y. Mikhaylik, I. Kovalev, R. Schock, K. Kumaresan, J. Xu and J. Affinito, High energy rechargeable Li-S cells for EV application: Status, remaining problems and solutions, In: Z. Ogumi, N. J. Dudney and S. R. Narayanan (Eds.), *Battery/Energy Technology. ECS Transactions*, **25** (2010) pp. 23–34.
132. J. Wang, S. Y. Chew, Z. W. Zhao, S. Ashraf, D. Wexler, J. Chen, S. H. Ng, S. L. Chou and H. K. Liu, *Carbon*, **46** (2008) 229–235.
133. J. Heine, S. Krueger, C. Hartnig, U. Wietelmann, M. Winter and P. Bieker, *Advanced Energy Materials*, **4** (2014) 1300815.
134. M. W. Forney, M. J. Ganter, J. W. Staub, R. D. Ridgley and B. J. Landi, *Nano Letters*, **13** (2013) 4158–4163.
135. Y. Li and B. Fitch, *Electrochemistry Communications*, **13** (2011) 664–667.
136. H. Zhao, Z. Wang, P. Lu, M. Jiang, F. Shi, X. Song, Z. Zheng, X. Zhou, Y. Fu, G. Abdelbast, X. Xiao, Z. Liu, V. S. Battaglia, K. Zaghib and G. Liu, *Nano Letters*, **14** (2014) 6704–6710.
137. K. Fan, Y. Tian, X. Zhang and J. Tan, *Journal of Electroanalytical Chemistry*, **760** (2016) 80–84.
138. C. Huang, J. Xiao, Y. Shao, J. Zheng, W. D. Bennett, D. Lu, L. V. Saraf, M. Engelhard, L. Ji, J. Zhang, X. Li, G. L. Graff and J. Liu, *Nature Communications*, **5** (2014) 3015.
139. F. Jeschull, D. Brandell, K. Edstrom and M. J. Lacey, *Chemical Communications*, **51** (2015) 17100–17103.
140. N. Liu, L. Hu, M. T. McDowell, A. Jackson and Y. Cui, *Acs Nano*, **5** (2011) 6487–6493.
141. M. Agostini, J. Hassoun, J. Liu, M. Jeong, H. Nara, T. Momma, T. Osaka, Y.-K. Sun and B. Scrosati, *ACS Applied Materials & Interfaces*, **6** (2014) 10924–10928.
142. J. Hassoun, J. Kim, D.-J. Lee, H.-G. Jung, S.-M. Lee, Y.-K. Sun and B. Scrosati, *Journal of Power Sources*, **202** (2012) 308–313.

143. Y. Yan, Y.-X. Yin, S. Xin, J. Su, Y.-G. Guo and L.-J. Wan, *Electrochimica Acta*, **91** (2013) 58–61.
144. R. Elazari, G. Salitra, G. Gershinsky, A. Garsuch, A. Panchenko and D. Aurbach, *Electrochemistry Communications*, **14** (2012) 21–24.
145. J. Brueckner, S. Thieme, F. Boettger-Hiller, I. Bauer, H. T. Grossmann, P. Strubel, H. Althues, S. Spange and S. Kaskel, *Advanced Functional Materials*, **24** (2014) 1284–1289.
146. N.-S. Choi, K. H. Yew, K. Y. Lee, M. Sung, H. Kim and S.-S. Kim, *Journal of Power Sources*, **161** (2006) 1254–1259.
147. V. Etacheri, U. Geiger, Y. Gofer, G. A. Roberts, I. C. Stefan, R. Fasching and D. Aurbach, *Langmuir*, **28** (2012) 6175–6184.
148. B. Duan, W. Wang, A. Wang, Z. Yu, H. Zhao and Y. Yang, *Journal of Materials Chemistry A*, **2** (2014) 308–314.
149. K. Ikeda, S. Terada, T. Mandai, K. Ueno, K. Dokko and M. Watanabe, *Electrochemistry*, **83** (2015) 914–917.
150. B. Duan, W. Wang, H. Zhao, A. Wang, M. Wang, K. Yuan, Z. Yu and Y. Yang, *ECS Electrochemistry Letters*, **2** (2013) A47–A51.
151. X. Zhang, W. Wang, A. Wang, Y. Huang, K. Yuan, Z. Yu, J. Qiu and Y. Yang, *Journal of Materials Chemistry A*, **2** (2014) 11660–11665.

Chapter 4

Lithium–Sulfur Battery Electrolytes

Patrik Johansson,[*,†,¶] **Rezan Demir-Cakan,**[‡,‖]
Akitoshi Hayashi[§,**] **and Masahiro Tatsumisago**[§,††]

*Department of Physics, Chalmers University of Technology,
SE-41296 Gothenburg, Sweden
†ALISTORE-ERI European Research Institute,
33 rue Saint Leu, 80039 Amiens, France
‡Department of Chemical Engineering, Gebze Technical
University, 41400 Gebze, Kocaeli, Turkey
§Department of Applied Chemistry, Graduate School of
Engineering, Osaka Prefecture University, 1-1 Gakuencho,
Naka-ku, Sakai, 599-8531 Osaka, Japan
¶patrik.johansson@chalmers.se
‖demir-cakan@gtu.edu.tr
**hayashi@chem.osakafu-u.ac.jp
††tatsu@chem.osakafu-u.ac.jp

4.1. Introduction

The electrolyte is at the very heart of any battery concept physically, but more and more also mentally amongst battery researchers and developers. This is largely due to the growing insight that many problems related to overall efficiency, life-length, and safety often originate in the electrolyte. This is perhaps even more a truth for the Lithium-Sulfur (Li–S) battery technology and hence large efforts are today focused on novel Li–S battery electrolytes — materials as well as concepts. In this chapter we will start by summarizing the similarities and differences in demands and design as compared to the Li-ion battery (LIB) technology, as the latter is more familiar to most readers. We then move to two large sections of liquid and solid electrolytes, respectively, outlining the materials and methods used. In each of the sections we point to a few specific topics and how these are researched today, keeping the comparison with the LIB as a way to more easily understand the unique features/issues/problems that electrolytes for Li–S batteries are facing. The chapter is made at a level and limited to a scope where the open literature is sufficient and plentiful, but of course studying the patent literature and gaining the hidden industry know-how may definitively extend the scope for the interested reader. Overall we hope that after reading this chapter, armed with a basic knowledge of the types of electrolytes and the materials presently in use in Li–S batteries, it will be easier for the reader to understand the needs, limitations, problems, but also the possibilities. This should finally open for suggestions of how to rationally improve the electrolytes with in the end enhanced performance of future Li–S batteries.

4.1.1. General and specific demands

A critical point for any battery technology is the complex, multifaceted role to be filled by the electrolyte. In a generic way it can rather easily be summarized in some basic and general properties needed to be fulfilled to allow for a functional and practical battery:

- ionically conductive,
- electronically insulating,
- thermally stable,
- chemically stable,
- electrochemically (meta-)stable,
- compatible with and wetting the electrodes and the separator,
- non-toxic/green, and
- low cost.

The exact demands on each property and approximate target values will of course differ by the battery technology, but also with the device operation conditions, the user demands, the installation requirements, and the niche market targeted. Nevertheless, for the LIB technology there is more or less a "standard electrolyte". Due to the significant differences in reaction chemistry of Li–S batteries as compared to LIBs it is, however, not possible to do any rational preselection by an adjusted copy-and-paste solution, not even partially, of the LIB electrolyte composition. Yet, many of the basic design parameters and paths of development are the same, as seen by the large overlap of the materials used.

At the same time there are some obvious notable differences in the importance and extent within these general demands, sometimes in favor of the Li–S battery. One such difference is the rather limited electrochemical stability window (ESW) needed: the overall Li–S battery reaction gives a theoretical ESW of no more than *ca.* 2.5 V necessary, to be compared to LIBs which often are at *ca.* 4 V or above. This opens for application of materials previously discarded for LIBs as well as novel materials. The ESW is also more or less fixed; in contrast to LIBs we do not need to worry about any new high-voltage (4.5–5 V) cathodes being launched, making the electrolyte oxidatively unstable.

The specific Li–S battery demands on the electrolyte, as compared to the LIB technology, can be summarized as three basic requirements:

- stability vs. Li metal anodes,
- stability vs. elemental sulfur and polysulfides (PSs) — both anions and radicals, and
- known and controlled solubility of PSs.

Each of these demands will be extensively covered below — as these are the fundamental reasons why we today focus on developing novel Li–S battery electrolytes. The most often mentioned feature of Li–S batteries, mainly related to the last requirement above, is the PS redox shuttle mechanism. The shuttle mechanism in brief means that PSs formed at the sulfur containing cathode diffuse through the electrolyte and are reduced to insoluble Li–PS species and cover the Li metal anode. While indeed acting as an internal protection vs. overcharging, this oxidation of short chain PSs at the anode reduces the coulombic efficiency in the charging stage and the active mass viable for discharge. This can thus be viewed as parasitic reactions causing problems at the Li metal anode. Indeed, the Li metal anode is in itself largely meta-stable with respect to the electrolytes used — connecting to the very first requirement above.

As the liquid and solid electrolytes attack the above requirements very differently the two sections below can be read rather independently from each other, keeping in mind that the problems needed to be solved are the same for both electrolyte concepts.

4.2. Liquid Electrolytes

4.2.1. Non-aqueous liquid electrolytes

All non-aqueous liquid electrolytes investigated for rechargeable intercalation batteries and advanced cell chemistries are based on a matrix of aprotic organic solvent(s), one or more dissolved salts, and some minor amounts of special additives. The additives are basically targeting the shortcomings of the basic salt/solvent composition to create a "functional electrolyte". Liquid Li–S battery electrolytes are no exceptions and in this chapter the components and compositions preferred will be summarized as well as the resulting performance and remaining problems. A major difference and complication for liquid Li–S battery electrolytes as compared to both liquid LIB electrolytes and indeed also to solid Li–S battery electrolytes is that the composition changes as a function of the Li–S battery state-of-charge (SOC). This simple statement indirectly contains many of the problems faced and at a macro-level seen in terms of limited performance, reduced life-length, etc.

4.2.1.1. Basic electrolyte formulations

We here step through the electrolyte components one-by-one with the purpose of providing the basic roles of each for a comprehensive understanding and with only a few representative examples from the literature. For more details there are already a few extensive literature surveys and reviews of Li–S battery electrolytes research and development made since the recent revival of the field i.e. since *ca.* 2010, focusing on the possible choices of salt and solvents as well as the total electrolyte formulation.[1–4]

Starting with the Li salts used, the main design parameter considered is to use weakly coordinating anions to assure a large number of mobile Li$^+$ charge carriers per salt unit and hence per volume unit in the electrolyte. For Li–S battery electrolytes the most common salt choices are LiTf and Li bis(trifluromethanesulfonimidate) (LiTFSI), and less common are Li hexaflurophosphate (LiPF$_6$) and Li perchlorate (LiClO$_4$). Already for LIBs, LiClO$_4$ was early considered inappropriate for commercial cells due to the safety concerns connected to the perchlorate anion, why this salt is mainly used for experimental simplicity (and cost) in academic studies also for Li-S batteries. In contrast, LiTf, a salt less popular for LIBs, together with LiTFSI today totally dominate the Li–S literature. The rationale for this is three fold: (i) the reduced ESW as compared to LIBs as corrosion of the often employed Al current collector by these salts does not start until at *ca.* 2.8 V vs. Li$^+$/Li°, (ii) they are thermally much more stable than LiPF$_6$, and (iii) they are compatible both with elemental sulfur, the PSs, and the ether solvents often used. As for the choice between LiTf and LiTFSI, the latter has a better dissociation ability, while the former has a smaller anion — hence promise of lower viscosity electrolytes and also should be available at a lower cost — even if this is difficult to estimate/verify for larger quantities. This has led to some renewed interest in LiTf, but as when employing LiTFSI the electrolyte conductivity can almost be doubled as compared to using LiTf, LiTFSI has anyhow come to dominate. Looking at the most recent developments, the LiFSI salt, with an anion structurally very similar to LiTFSI, could just as well be employed, and should result in electrolytes with slightly higher conductivities at the expense of some thermal stability. Another candidate could possibly be LiTDI if it can be confirmed to be appropriately stable vs. Li metal anodes. In Figure 4.1 the most common anions and solvents are depicted, while for a more complete overview of Li-salts for Li–S battery electrolytes the review by Younesi *et al.* is recommended.[3]

Turning to the basic choices of solvents, the prime difference as compared to LIBs is that carbonate based solvents are in general not used — they are not sufficiently stable vs. PSs.[5] The detrimental reactions

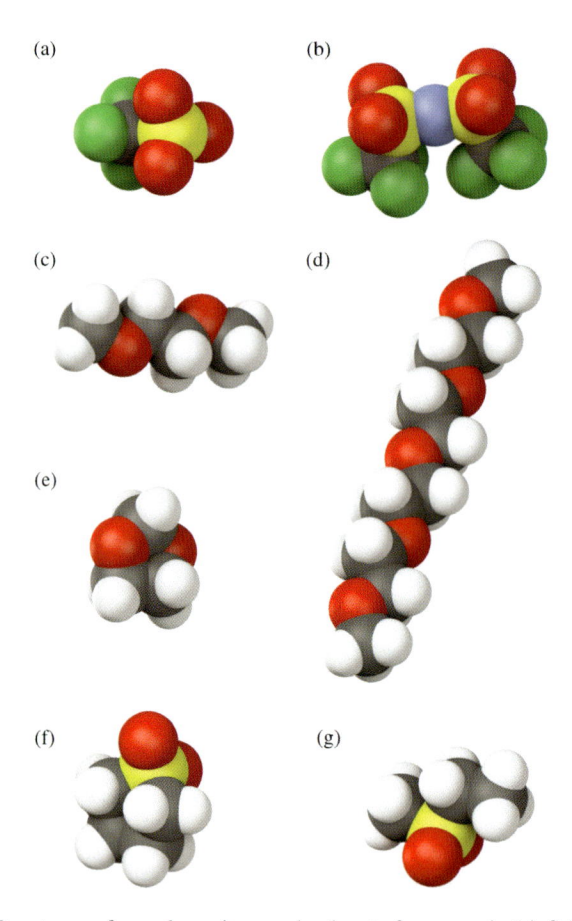

154 *Li–S Batteries: The Challenges, Chemistry, Materials and Future Perspectives*

Figure 4.1 Structures of popular anions and solvents for usage in Li–S battery electrolytes: (a) Tf, (b) TFSI, (c) DME, (d) TEGDME, (e) DOL, (f) TMS, and (g) EMS. Not subject to copyright.

originate in nucleophilic attacks either on the ether or carbonyl oxygen atoms of the carbonates. Yet another problem, albeit much less discussed in the literature, is the risk of formation of highly toxic H_2S if carbonate based solvents is combined with a source of HF, such as hydrolysed PF_6^-. Hence there are several good reasons to avoid both carbonates and $LiPF_6$ i.e. the standard electrolyte choice for LIBs.

In place of the carbonates mainly two other solvent families are employed; ethers — used in *ca.* 75–80% of the Li–S battery studies, and sulfones. They are both stable toward elemental sulfur and PSs including precipitated Li_2S. Most often two or more solvents are combined in order to balance and optimize both general and specific properties for the resulting electrolytes.

Starting with the ethers both linear and cyclic ethers of different sizes are employed. The linear ethers are all methoxy-group end-capped and stretch all the way from the smallest 1,2-dimethoxyethane, known as DME, monoglyme, or G1, to various oligomers of poly(ethylene glycol) dimethyl ether, PEGDME, via the most commonly used intermediate length glymes; diglyme (G2), triglyme (G3), and tetraglyme (G4), the latter perhaps more known as TEGDME. The cyclic ethers are rings containing one or two ether oxygen atoms; tetrahydrofuran, THF, and, 1,3-dioxolane, known under both acronyms DIOX and DOL.

Among all ether based solvents, DME is perhaps the most popular for Li–S battery application. DME is highly polar and has a large donor number (DN) of 18.6, enabling to solvate PSs and is most often used together with DOL, with an even larger DN of 24, which also acts to stabilize the Li metal anode surface by means of creating a solid electrolyte interphase (SEI). Indeed, the archetypical Li–S electrolyte, if any such really exists, is 1 M LiTFSI in 1:1 (v/v) DME:DOL. For example, Sion Power has made use of binary DME:DOL solvent mixtures and especially explored their low temperature (<0°C) operation characteristics.[6] However, at higher temperatures the large vapour pressure and flammability of DME can be a problem, just as for THF, why longer-chain ethers like TEGDME have been employed extensively. In fact, an electrolyte using only TEGDME as solvent can solvate even larger amounts of PSs than DME, due to its higher concentration of ether oxygens, despite a lower DN of 16.6. THF based electrolytes, on the other hand, can solvate up to 10 M(!) of PSs. DOL:TEGDME solvent mixtures have been successfully employed for Li–S battery electrolytes.

We believe that a caveat for the ether solvent family of glymes is highly warranted; while the severe health hazards of diglyme (G2) has been known for long, including damage to the unborn child, tetraglyme (G4, TEGDME) was only very recently reclassified and its usage for production processes and/or consumer products should perhaps be seriously reconsidered.

Sulfone solvents are more limited in scope than ethers in Li–S battery electrolytes, mainly focussing on: the cyclic sulfolane (TMS) and two linear candidates; ethyl–methyl sulfone (EMS) and methyl–isopropyl sulfone (MiPS). The use of sulfones as electrolyte solvents originally arose due to their large ESWs and hence stability in high voltage LIBs. For Li–S battery electrolytes their main promise lies in selectively altering the PS solubility to promote some PSs to be more/less soluble than others.

All sulfones have relatively high viscosities, *ca.* 3–10 times the ethers, with TMS being the most viscous at *ca.* 10 cP. This of course leads to comparatively lower fluidities of their electrolytes and hence lower ionic conductivities, this is why the common resort has been to use them as cosolvents together with ethers. Still most studies employing sulfones are limited to applying rather slow C-rates such as C/20 or C/10.

A recently researched solvent alternative, mainly in order to control the solubility of the PSs, is fluorinated solvents, especially fluorinated ethers. These are used as single solvents, cosolvents, or as additives, all based on a reduced solvation ability of the ether oxygen atoms due to the electron-withdrawing effect of the F-atoms.[7] Furthermore they also offer lower flammabilities, hence safer electrolytes, and lower viscosities, hence higher ionic conductivities. Perhaps the most common fluorinated solvent applied, 1,1,2,2-tetrafluoroethyl 2,2,3,3-tetrafluoropropyl ether, as a representative hydrofluoroether, HFE, solvent, has been shown to improve the Li–S cell cycling efficiency and to result in a capacity fading much slower as compared to the corresponding electrolyte without any fluorinated ether.

Finally, for a complete electrolyte formulation, additives are most often needed. In general additives play important roles in all modern batteries such as the LIBs where additives act to create (better) SEIs, to reduce flammability, to capture products from side reactions, to protect against overcharge, etc. The term additive basically means that they are present only in small amounts, an often quoted limit is <5%, above this level they count as cosolvents or cosalts. For Li–S batteries the additives used are much fewer in numbers as compared to LIBs, but still of significant importance. The totally overwhelming part of the Li–S battery literature quoting any use of additives concerns itself with the salt $LiNO_3$ in order to stabilize the Li metal anode — why we devote a special section to this topic in this chapter. Other approaches are based on other NO_x^- species, on using various PSs directly to the electrolyte to "pre-form" a protective layer on the Li metal anode, or on adding other alkali salts to alter the SEI composition.

4.2.1.2. Role of PS solubility

Having outlined the major formulation components we already have touched on the major importance of the varying PS solubility for various choices and compositions. In more detail, it is the electrode/electrolyte interfaces that can be seen as even more fundamental for Li–S cells as compared to LIBs. Especially, the unique large problem of Li–S cells is to retain the PSs within the cathode architecture to prevent the redox shuttle mechanism. This is a problem that can be attacked both from an electrode, today still the most common mindset, and an electrolyte point of view. Adopting the former, it is often stated that the PS loss and the shuttle mechanism can be limited or even avoided by a properly designed C/S composite, often nanoporous, cathode. However, as the electrolyte eventually will penetrate into the nanopores where the elemental sulfur and PSs are contained, not the least during cell discharge, there will be an inevitable (?) "leakage" and the solubility of PSs in the electrolyte will anyhow play a role and is inherently needed, and to be controlled, for Li–S battery functionality.

As always there are exceptions to the rule, there are for example Li–S cell designs where the sulfur itself is covalently bound to the cathode matrix host, often a polymer, and allowed to be directly transformed to Li_2S. However, for the main track of Li–S cell design and adopting the latter main perspective as above, the goal must be to formulate an electrolyte balancing the electrolyte interplay with both PSs and electrodes. Indeed, the liquid electrolytes employed in general have (very) large solvating abilities for PSs e.g. an electrolyte of 1 M LiTFSI in TEGDME can dissolve up to *ca.* 6 M(!) of Li_2S_8. Hence, the stability of the Li metal anode in a Li–S cell must be assured and the shuttle mechanism prevented by other means. The more common approaches are: to cover the Li metal anode by a thin protective layer e.g. an Al_2O_3 nanometer thin film by ALD or alternatively created by SEI-forming electrolyte additives, apply ion-selective membranes, or add adsorption agents for the PSs within the C/S composite cathode. To truly achieve long-term stability in terms of shelf life (including avoiding self-discharge) and cycle life, one major challenge is in fact the stability of the Li metal anode — proven by many post-mortem analyses of failed lab cells.

An alternative approach to all of the above reasoning is to aim for electrolytes *not* allowing for any large solubility of any sulfur or PS species, while still ensuring Li^+ mobility from anode to cathode. However, due to the very different chemical natures in terms of polarity, hydrophobicity, etc., between sulfur and the various PSs it is difficult to envisage any such liquid electrolyte based on the conventional salt/solvent concept. This approach is therefore more related to a physical hindrance approach by applying semisolid/highly viscous electrolytes such as those based on ionic liquids as solvents or super-concentrated electrolytes. Indeed, the main promise of solid Li–S battery electrolytes such as polymeric, glassy, or crystalline electrolytes lies herein.

Deeply connected to the problem of controlling PS solubility is the ratio between the electrolyte volume and the sulfur loading in the C/S cathode applied in the cells studied.[8] Indeed, the ratios, if reported at all, applied in the scientific literature are often large, the reason why many results cannot easily be transferred to, or even qualitatively interpreted as valid for, final real Li–S cells. Briefly, lab experiments are often made with large quantities of electrolyte and with C/S cathodes having relatively low areal loadings of sulfur. A large ratio for example helps in wetting the C/S cathode and the Li metal anode — perhaps giving promises that cannot be sustained at lower ratios. The major problem is, however, that the excess will serve as a buffer for dissolved PSs and artificially delay any PS concentration increase, that otherwise could lead to low conductivities and very high polarization in the cell. Indeed, cells made with lower, more realistic, electrolyte/sulfur ratios typically display this at the end of the high voltage plateau. At the same time, an excess of electrolyte also often itself leads to more severe PS redox shuttle problems and hence fast capacity fading.[9] As an example; in a cell with the rather "large" ratio of 10 μL/mgS and a 1 M LiTFSI in TEGDME electrolyte, 96% of the elemental sulfur contained in the C/S cathode can in theory be dissolved. The importance of "more relevant" electrolyte/sulfur ratios for almost all evaluations of electrolytes in any cell set-ups is slowly penetrating the scientific literature.

In addition, just as for LIBs, the electrolyte excess itself will negatively affect the final Li–S cell specific capacity obtainable simply by adding non-active material to the cell mass. This can largely be foreseen and subsequently compensated for when moving from lab cells to commercial cells and products. However, in contrast to LIBs, the above problems are connected also to the large variation between different Li–S laboratory cells, not the least between coin-cells, Swagelok cells, and pouch cells, each having different losses of electrolyte. But it also includes the role of the C/S cathode architecture, the separators used, and even the source of the Li metal for the anode — for the very same electrolyte. Yet, there are some estimates made based on various electrolyte/sulfur ratios pointing toward less than 5–6 μL/mgS to be preferred, and maybe even needed to be as low as 3 μL/mgS to achieve the much wanted specific densities >400 Wh/kg at the cell level.[10,11]

4.2.1.3. *Role of salt concentration*

Above Li salts, organic aprotic solvents, and a few additives were briefly covered as basic building blocks for Li–S battery electrolytes, but more or less the 1 M Li salt concentration was assumed to be a constant fact regardless of the materials employed. For LIBs the 1 M salt concentration basically balances the ion conductivity in an optimal way by having enough charge carriers with adequate mobility. However, for Li–S battery electrolytes the Li salt concentration should also affect the solubility of other charged species — the PS anions and radicals — either beneficially by specific interactions or limiting due to the total ion concentration. As the content of PSs will vary with the Li–S cell SOC and depend on the actual cell setup, it is virtually impossible to optimize the salt concentration by means of simply studying ion conductivities or by any other *ex situ* assessment of physical properties of the native electrolyte alone. However, there must be some limiting cases just as for LIBs; a too low Li-ion, i.e. Li salt concentration will not allow cell cycling due to fast depletion of charge carriers in the electrolyte, while for too high concentrations the viscosity will hinder proper wetting of the electrodes and the separator and furthermore not allow any rapid ion transfer (by diffusion). The behavior of this multidimensional system of solvent(s), Li salt and PSs, all of varying types and concentrations is extremely complex. Especially the role of PSs in altering the electrolyte viscosity, and hence the ion conductivity is difficult to both measure and as a resort estimate. As a first approximation advanced modeling can be used to estimate the effect of the dissolved PSs on the electrolyte viscosity by for example, considering their partial molar volumes and applying Einstein's theory of viscosity for a suspension.[12]

Moving to a few descriptive examples, lowering the salt concentration resulted in different maxima for ion conductivities and PS solubility for a model system of $LiClO_4$ in TMS (0.5 and 0.3 m, respectively),[13] while an even lower concentration of 0.1 M LiTFSI in DOL:TEGDME was shown to be quite bad in all aspects; low ionic conductivity, low discharge capacity, and no proper passivation of the Li metal anode.[14] Raising the salt concentration has recently attracted quite some interest — staying at a mere 2 M gave no specific advantages except a slightly higher ionic conductivity, but for 5 M LiTFSI in DME:DOL both reduced PS solubility and diffusion were obtained, which significantly reduced the cell overcharge and improved the coulombic efficiency.[15] Much more interest was, however, stirred by the 2013 Nature Communications paper on the solvent-in-salt concept reaching up to 7 M of LiTFSI in DME:DOL and where the LiTFSI salt becomes the dominant component by both weight and volume.[16] There the optimal properties were all found close to the salt saturation limit; the PS solubility was negligible, the Li metal anode stable, and the ionic conductivity was *ca.* 1 mScm^{-1}. Not the least the latter is impressive as the reported viscosity is very large (72 cP) — which though gives some increased cell polarisation as compared to conventional 1 M electrolytes. The conduction mechanism cannot be solely diffusion based, not the least as the Li$^+$ cation transference number reaches an impressive 0.73. This fact, together with the super-high salt concentration, avoiding anion depletion at the electrode/electrolyte interface,[17] results in a high initial Li–S cell capacity, excellent rate capability — up to 3 C, a high capacity retention, and a high coulombic efficiency. The minor capacity decay observed was attributed to the (in)stability of the C/S cathode rather than the electrolyte or the Li metal anode.

While all this sounds extremely promising and opens for studying completely different salt concentration regimes, the obvious drawback is the cost of the electrolytes, due to the large amount of costly Li salt needed (as compared to inexpensive solvents). As one of the main selling arguments for the Li–S battery technology is low cost this is a bit of a paradox. In addition, there have been very few follow up studies published as to why both the generality of the approach beyond the specific system and the final real Li–S cell promise awaits verification.

4.2.1.4. $LiNO_3$ as additive

With only a bit of exaggeration, anyone aiming at making a Li–S cell using a liquid electrolyte will sooner or later be faced with the question: "Should you not add a few % of $LiNO_3$ to the electrolyte?" As this is such a commonly known aspect of Li–S battery electrolytes we here summarize the basic features of this additive alone and the current state-of-the-art of understanding of its actions. For more extensive coverage of this subject there are some excellent reviews by Zhang.[18,19]

The addition of $LiNO_3$ to the electrolyte is basically made to stabilize the SEI created on the surface of the Li metal anode and hence suppresses the actions of the PSs dissolved in the electrolyte.[20] At the Li–S cell level, this leads to better cycling stabilities and coulombic efficiencies. By quite comprehensive XPS analysis of Li–S cell anodes, combined with electrochemical data and imaging techniques, complex models for the growth of the SEI with and without the presence of $LiNO_3$ in electrolytes and with and without PSs have been created.[21] Starting from the $LiNO_3$ presence it induces the growth of the SEI by its very oxidative nature, resulting in a thick and very stable film of LiN_xO_y, while the combined presence of PSs and $LiNO_3$ leads to a relatively stable SEI film, but now composed of two layers (Figure 4.2).

The layer closest to the Li metal anode is made out of the insoluble Li–S battery discharge products Li_2S and Li_2S_2 coprecipitated with LiN_xO_y. Without any $LiNO_3$ present the former compounds constitute the only "SEI" formed which, however, cannot protect the electrolyte from being decomposed further and the Li metal anode attacked. The layer closest to the bulk electrolyte consists primarily of oxidized sulfides and

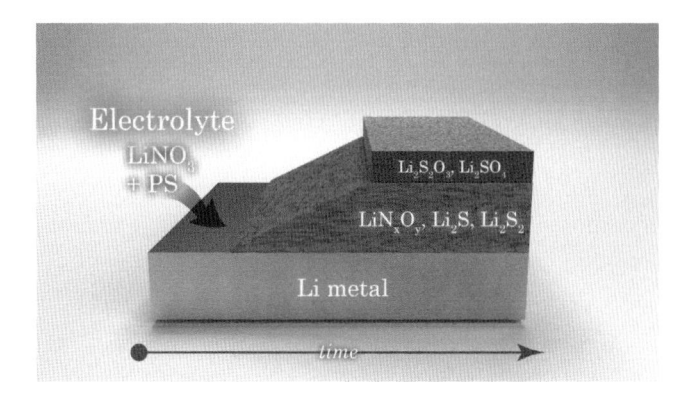

Figure 4.2 An artistic view of the SEI layers created using an electrolyte containing both LiNO$_3$ and PSs. Inspired by Xiong et al.[21]

sulfates, such as Li$_2$S$_2$O$_3$ and Li$_2$SO$_4$. Combined, this SEI blocks any direct contact between the PSs in the electrolyte and the reductive species, both the Li metal and the Li PSs.

Why is then the use of LiNO$_3$ as an additive questioned? There are several reasons. First, due to irreversible reduction of the additive on the C/S cathode when discharged to lower potentials than *ca.* 1.6–1.7 V vs. Li$^+$/Li$^\circ$, with a negative impact on both capacity and reversibility for subsequent cycles, the Li–S battery cells do become limited. Still, however, capacities of *ca.* 500 mAhg^{-1} over 100 cycles can be reached with only 0.2 m LiNO$_3$ in the electrolyte. Second, as some of the LiNO$_3$ indeed is continuously consumed in every cycle at both the anode and the cathode and therefore eventually is depleted in the electrolyte, the concentration needed for any more long-term cycling, is rather at the cosolvent than at the additive level. Hence, a proper optimization of the amount of LiNO$_3$ must be based on the number of cycles aimed for with the cell. The initial concentration must thus be rather high for an extensive number of cycles, but as LiNO$_3$ is such a strong oxidant and a salt, it cannot be added to the electrolyte in very high quantities.

However, as recently highlighted by Zhang, the picture of the actions of LiNO$_3$ at the cathode side of the Li–S cell, often summarized just as less beneficial, should perhaps be revised. LiNO$_3$ actually seems to catalyze the conversion of soluble PSs to elemental sulfur at the end of the charging process via a radical intermediate of the NO_3^- anion. Furthermore, synergies can be created by a combined use of $LiNO_3$ dissolved in the electrolyte and other insoluble (Mg-based) nitrates at the C/S cathode.

Should then finally $LiNO_3$ be added or not to your Li–S electrolyte of choice? It sort of depends… can your electrolyte be made to contain enough of $LiNO_3$ to sustain its beneficial effects throughout the life-length of the Li–S cell at hand? Or are you interested in studying some phenomena that might be masked or delayed by the actions of the $LiNO_3$? Or maybe you find it better to continue to search for intrinsically stable electrolytes avoiding the need of $LiNO_3$?

4.2.1.5. Analysis of PS speciation

As should be clear from the above, the performance of the electrolyte in a working Li–S cell is strongly coupled to the varying PS content, as a function of the cell design, but also due to the cell SOC. The analysis of the PSs present in the electrolytes is thus highly interesting, whether it is made at a macrolevel, i.e. more phenomenological, or at a molecular, identification level, but foremost if it is made under *ex situ*, *in situ*, or *operando* conditions. Indeed, the many coupled reaction steps producing the PSs complicates the analysis and the latter is not helped by the fact that many analysis techniques have difficulties in selectively identifying the various PSs. Furthermore, most *ex situ* approaches in need of reference samples are problematic as only elemental sulfur and Li_2S, the two end-points of the total Li–S cell reactions, are thermodynamically stable. Therefore various *in situ* techniques have gained attention for characterization of both dissolved and precipitated PSs, in particular different spectroscopy techniques like UV/Vis,[22] Raman,[23] NMR,[24] and XAS.[5] Even so only a few have been able to discriminate the exact PSs present, more often they are resorting to resolving only short and long PSs — due to heavy overlapping and/or broad features as in UV/Vis spectra (Figure 4.3).

UV/Vis spectroscopy gains from being a quantitative technique, but for this calibration curves are often needed, and can be considered partially *ex situ*, even if the measurements are made *in situ* or *operando*.[22] Raman spectroscopy on the other hand is more difficult to do

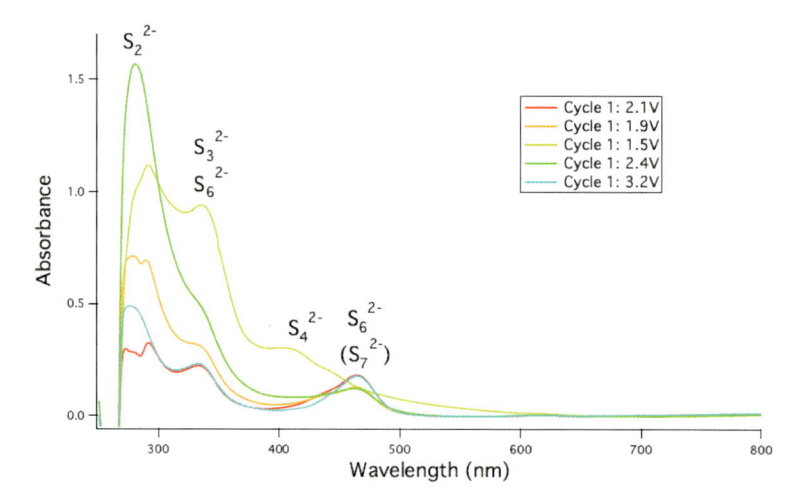

Figure 4.3 UV–Vis analysis of the PSs speciation during a discharge–charge cycle.
Source: Reproduced with permission from the work by Blomberg.[25]

quantitatively, but have other advantages; the features are much narrower and thus it is, at least in principle, easier to discriminate between different PSs. Indeed, with the help from DFT simulations to construct artificial Raman spectra, a confocal Raman *operando* spectroscopy study was able to selectively and unambiguously identify both long-chain PSs, S_8^{2-} and S_6^{2-}, and the by other techniques, except XAS, quite elusive S_3^{*-} radical (Figure 4.4).[23]

Utterly important for all the reasoning above about proper PS detection and evaluation as a function of SOC is to remember the role of the ratio of electrolyte volume to the cathode sulfur loading. This further supports the analysis to be made under *operando* conditions, with proper care taken to construct test cells as close to realistic electrochemical cells as possible — and preferably with setups reproducible in different labs.

4.2.1.6. *Summary and perspectives*

The development of non-aqueous liquid electrolytes for Li–S cells is currently exploring many possible routes to enhanced performance — and there is no consensus on any "standard electrolyte" in sight, not even conceptually. In many ways the electrode/electrolyte interfaces and SEIs

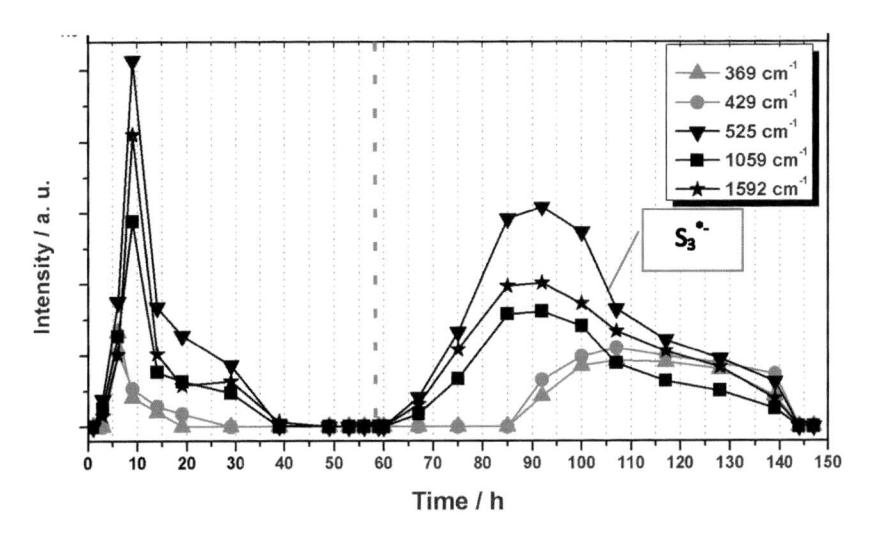

Figure 4.4 Data from the confocal operando Raman analysis by Hannauer et al.[23] highlighting the large contribution from the S_3^{*-} radical.
Source: Partially reproduced with the permission from Ref. 23.

are even more important than for LIBs and can be altered by the native electrolyte and electrode compositions, but will also depend on the cell SOC — especially as the elemental sulfur and various PSs dissolve and precipitate. Both additive and salt concentrations are continuously explored and varied in order to stabilize Li–S cells. A proper analysis of the PSs during cell operation will likely contribute to a better fundamental understanding and in the end be of large use for rational improvement of overall cell performance. However, it is crucial that any studies aiming at such analysis use proper ratios of electrolyte to sulfur to be transferrable to real cells in a straightforward manner. And there are still many openings for disruptive ideas to change the Li–S liquid electrolyte formulation.

4.2.2. Aqueous electrolytes

4.2.2.1. Introduction

The use of earth abundant sulfur provides a promising alternative for high-energy and low-cost battery systems. For instance, classical non-aqueous electrolyte Li–S batteries ensure around three fold increase in gravimetric energy density compared with the present Li-ion technologies.[26,27] Although many attempts have been pursued in this field, these Li–S batteries have still some problematic issues, for instance low utilization of sulfur and electrolyte incompatibility, which constrain their presence at the current battery market.

Aqueous electrolytes offer many advantages over organic electrolytes as both water as well as suitable salts carry a much lower total cost. Additionally, aqueous electrolyte batteries warrant higher rate capabilities due to their high ionic conductivity which is known to be two orders of magnitude higher than that of organic electrolytes.[28] Despite those advantages of aqueous electrolytes, low thermodynamic stability window of water (1.23 V) results in low energy density as compared to the current Li-ion batteries (LIBs). However, the electrolysis of water are generated usually above 1.23 V threshold due to the electrical input which should meet the full amount of enthalpy needed for evaluation of oxygen and hydrogen.[29] Practical applications as well as some recent literature reports depict that this low voltage could be enhanced. For instance, in Pb–acid rechargeable battery technology, aqueous electrolyte stability reaches up to 2.1 V, because of the high hydrogen overpotential of Pb as well as electronically insulating but ionically conductive $PbSO_4$, across which there is a steep potential gradient.[30] Moreover, very recently Suo et al.[31] have presented that low voltage stability of water could be extended with the use of highly concentrated aqueous electrolyte, similar idea was shown earlier with the non-aqueous Li–S batteries counterparts.[32] In their concept, extremely high concentration aqueous solutions containing LiTFSI resulted in a stable performance at 2.3 V. These improvements were explained with the suppressed hydrogen evolution and electrode oxidation.

Aiming to gain the advantages of aqueous electrolyte and Li–S technologies, aqueous electrolyte sulfur based batteries were suggested more than two decades ago by Licht and coworkers[33] and put in practice few years later for PSs based batteries as well as solar cells.[34–37] In 2013, Visco et al. have developed a high performance rechargeable Li–S battery using Li_2S in an aqueous cathode.[38] Since sulfur is not soluble in water, all the cell configurations are made by the reduction of polysulfide compounds.

The electrochemical response and the redox reaction of aqueous PSs are different from the classical organic electrolyte Li–S batteries.

In non-aqueous solvents, the sulfur generates different length of PSs (Li_2S_x, $2 < x < 8$) during reduction leading to insoluble and insulating Li_2S. The reduction phenomenon changes in aqueous electrolyte in which three main chemical species are formed, $H_2S(aq)$, HS^-, and S_x^{2-}, determined by the pH. Thus, an oversimplified reaction mechanism can be written as in (Equation (1))[33,34] where the average voltage lies approximately at −0.5 V vs. SHE corresponding to 2.5 V vs. Li. Additionally, the high capacity and rates result from the anomalously high solubility of PSs in water, contrary to non-aqueous electrolytes.

$$S_2^{2-} + 2H_2O + 2e^- \leftrightarrow 2HS^- + 2OH^- \quad E_o = -0.51 \text{ V vs. SHE.} \quad (1)$$

Recently in the literature, aqueous electrolyte dissolved PS systems were either coupled Li metal anodes or with Li-ion cathode materials.

4.2.2.2. Aqueous electrolyte batteries with lithium metal anodes

After the first report of the use of PSs in aqueous electrolytes more than 20 years ago,[33] the topic almost fell into oblivion until it was revisited by Visco et al.[38] In their patent, the authors demonstrated the feasibility of using aqueous electrolyte Li–S batteries using a Li-ion conductive glass membrane as the Li metal anode protective layer. With the similar approach, later on Li et al.[39] reported an aqueous Li–S battery based on aqueous dissolved Li_2S_4/Li_2S redox couple as the cathode, metallic Li as the anode and $Li_{1.35}Ti_{1.75}Al_{0.25}P_{2.7}Si_{0.3}O_{12}$ (LATP) as the separator. Electrochemical reduction of those aqueous electrolyte PSs resulted in different phenomena than in organic solvents. The authors observed only one distinguishable cathodic peak at 2.53 V vs. Li$^+$/Li, with clearly visualized color change of aqueous electrolyte from yellow to transparent during the discharge (vice versa during charge) (Figure 4.5). A reversible specific capacity of up to 1030 mAh/g was achieved.

Since metallic Li is employed as anode, two compartment cell configurations separated with a ceramic membrane must be used in order to eliminate the contact between water and highly reactive Li. Strategies aiming toward the development of two compartment cell configurations

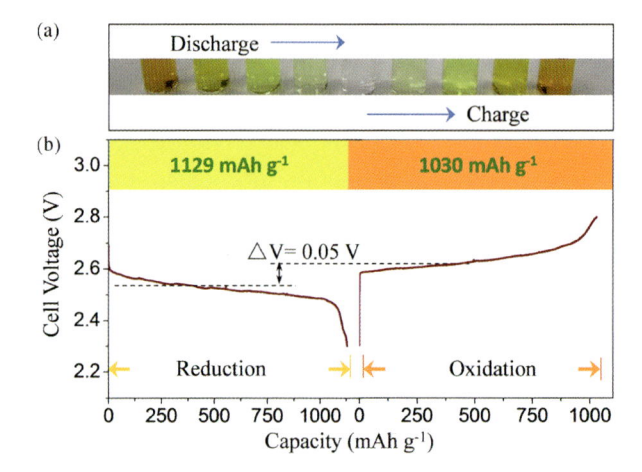

Figure 4.5 (a) An image of color gradients of the Li$_2$S$_4$/Li$_2$S electrolyte, (b) Initial discharge/charge curve of the aqueous lithium–polysulfide battery using 0.1 M Li$_2$S$_4$ solution at 0.2 mA/cm².

Source: Reproduced with permission from Ref. 39.

enabling the use of different electrolytes at the positive and negative electrodes have presently were explored in the literature in order to preserve high voltage utilization. For instance, an acidic–alkaline double electrolyte primary battery operating at a voltage approaching 3.0 V has already been reported.[40] Moreover, water soluble redox couples as cathode (or so called catholyte) together with a metallic Li functioning in a non-aqueous electrolyte as anode were suggested as new concept of alkali-ion batteries.[41] Along that concept Li–iodine batteries using triiodide/iodide couple[42] or Li-bromide batteries using Br$_2$/Br⁻ couples[28] have already been reported. However, all these approaches require utilizing a high cost and fragile Li-ion conducting glass ceramic membrane to separate between aqueous and non-aqueous electrolytes. Although elegant for conceptual demonstration, such a heavy and brittle ionic conducting ceramic is still questionable for practical applications.

4.2.2.3. *Aqueous electrolyte batteries with Li-ion cathodes*

Utilization of metallic Li obliges the use of a two compartment cell configuration with a ceramic conducting membrane to separate the aqueous and non-aqueous electrolyte. In order to eliminate the use of high cost ceramic membranes, an obvious implementation is to replace the ceramic membranes with ion-selective polymer membranes. The concept of which have recently been demonstrated by Demir-Cakan et al.[43] Prior to performing the aqueous PS batteries with an ion-selective polymer membrane, as a prototype example, the authors demonstrated for an aqueous electrolyte Li-ion–PS battery coupling a well-known cathode material (LiMn$_2$O$_4$) with dissolved PSs as anode (so called anolyte).[44] LiMn$_2$O$_4$ together with PS is within the stability window of water, thus, the reduction and oxidation of water does not crucially influence the Li transport kinetics of the electrodes ensuring a safe and reliable working condition. The cell showed an output voltage of ~1.5 V with sustained capacities of ~110 mAh g^{-1} at C/2 rate for more than 100 cycles, resulting almost 80 Wh/kg energy densities which could compete with those of NiMH or Pd-acid aqueous electrolyte rechargeable batteries (Figure 4.6).

Although the Li-ion/PS cell performed well, the configuration was still using a glass ceramic membrane.[44] Thus, the authors reported a Li-ion/PS battery using an ion-selective polymer membrane (Nafion) rather than a ceramic membrane as separator. One of the critical issues of the use of ion-selective polymer membrane is to have their permeability to Li-ions while impermeability to PS species. Prior to assembling a full cell configuration, the authors performed membrane leakage and PS diffusion tests to ensure proper functioning of Nafion membrane (Figure 4.7). Due to the repulsive interaction between the negatively charged membranes

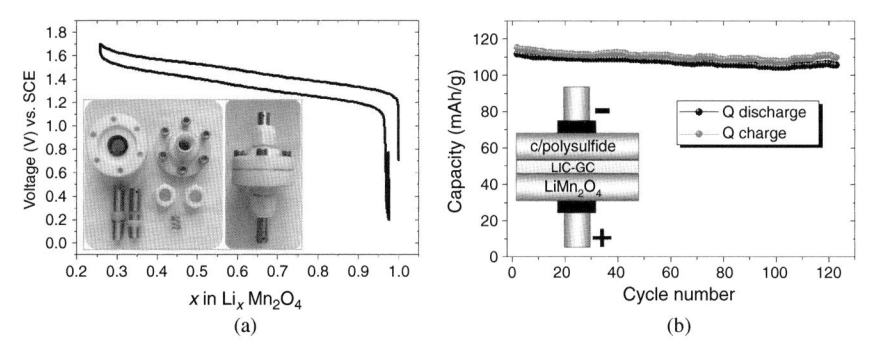

Figure 4.6 (a) First galvanostatic charge–discharge profile of the LiMn$_2$O$_4$/polysulfide full cell (b) cycling performance of the cell in 0.5 M Li$_2$SO$_4$ containing aqueous electrolyte.
Source: Reproduced with permission from Ref. 44.

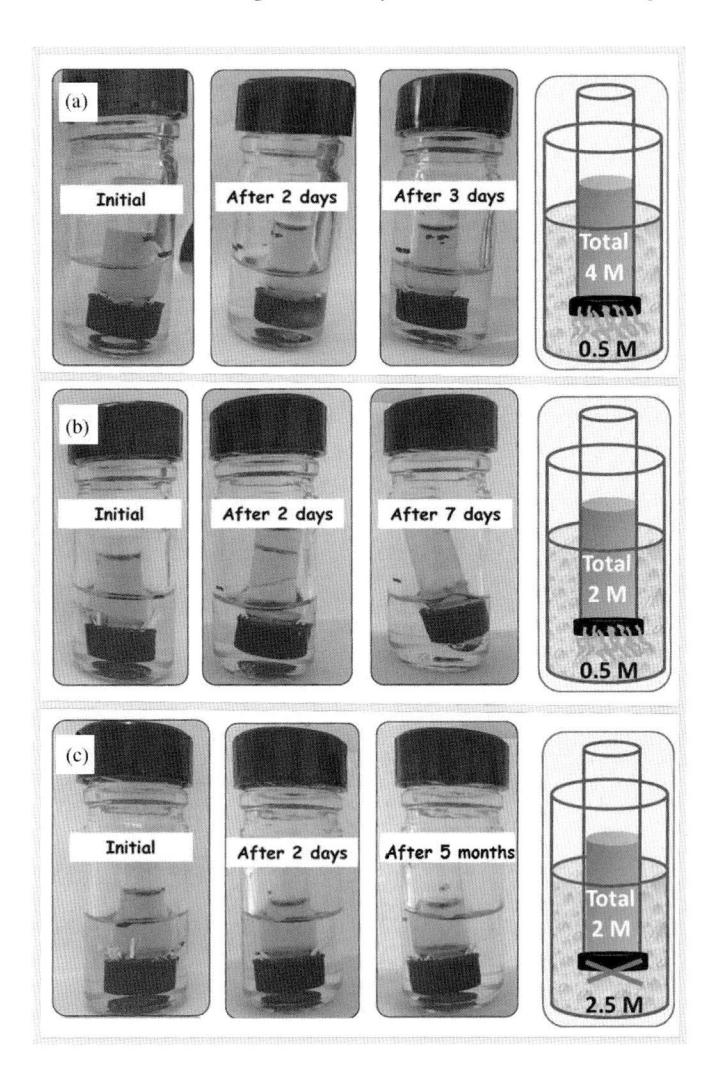

Figure 4.7 Nafion membrane PS leakage test with different salt concentrations after variable resting times (a) and (b) when inside and outside the salt concentrations are not in equilibrium, (c) inside and outside salt concentration is close to equilibrium resulting in no leakage up to five months.

Source: Reproduced with permission from Ref. 43.

(SO_3^- groups) and negatively charged PSs, and after tuning the osmotic movements inside the membrane, dissolved PS leakage from one compartment to another was successfully eliminated, Figure 4.7(c). After all, a 1.5 V battery system, free of costly and brittle ceramic membranes, provides promise to construct a long-lifespan storage device. Sustained reversible capacities of 90 mAh/g at 2 C rate for more than 200 cycles were already achieved. Although the proof-of-concept was established with Nafion, obvious improvements of the use of ion-selective polymer membranes would be exploring other cationic polymer membranes.

4.3. Inorganic Solid Electrolytes for All-Solid-State Li–S Batteries

4.3.1. Introduction

All-solid-state Li–S batteries using inorganic solid electrolytes instead of conventional organic liquid electrolytes have several advantages of high safety with non-flammability of electrolytes and long cycle lives by suppressing dissolution of polysulfides into electrolytes.[45,46] Large-sized solid-state Li–S batteries operating at room temperature are attractive as distributed power sources. Bulk-type all-solid-state batteries which are composed of active material and solid electrolyte powders have been studied. A schematic diagram of bulk-type all-solid-state batteries is shown in Figure 4.8. Electrode layers are composed of active material and solid electrolyte powders; Li$^+$ ions are supplied from solid electrolyte part to active material, and electrons are mobile through active material. The addition of conductive additives such as nanocarbons to electrode layers is needed to form electron conduction paths to active materials such as S or Li$_2$S with poor electronic conductivity. Inorganic solid electrolytes with Li-ion conductivity have been developed in three decades. Solid electrolytes applied to bulk-type solid-state batteries should have high Li$^+$ ion conductivity and good mechanical property to achieve wide electrode-electrolyte solid–solid interfaces facilitating charge-transfer reactions during charge–discharge processes. Electrochemical and chemical stabilities of solid electrolytes are also significant features.

In this section, development of inorganic solid electrolytes, especially sulfide electrolytes (SEs), is reviewed. Several electrolyte properties of conductivity, electrochemical and chemical stability, and mechanical properties of formability and elastic modulus will be demonstrated. All-solid-state

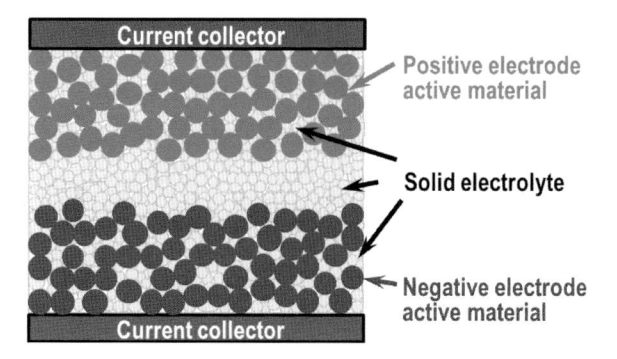

Figure 4.8 Schematic diagram of bulk-type all-solid-state batteries. Not subject to copyright.

Li–S batteries were assembled using sulfide solid electrolytes and their electrochemical performance and reaction mechanism will be reported. Approaches to increase energy density of Li–S batteries will be mentioned.

4.3.2. Development of inorganic solid electrolytes

4.3.2.1. *Conductivity*

Inorganic Li-ion conductors have been widely studied in sulfides and oxides. Conductivities at 25°C for typical sulfide and oxide electrolytes are listed in Table 4.1.[47–63] Sulfides have the highest Li-ion conductivity as inorganic solids; crystalline $Li_{10}GeP_2S_{12}$,[48] $Li_{9.54}Si_{1.74}P_{1.44}S_{11.7}Cl_{0.3}$,[47] and $Li_7P_3S_{11}$[53] show a considerably high conductivity of over 10^{-2} S cm^{-1}, which is as high as that of conventional organic liquid electrolyte. Sulfide solid electrolytes are single ion conductors, where targeted Li-ions are only mobile, while counter anions as well as lithium cations move in liquid electrolyte; Li-ion conductivity of SEs is thus higher than that of liquid electrolytes. $Li_{10}SnP_2S_{12}$[49] and Li_6PS_5Cl[52] have a high conductivity of over 10^{-3} S cm^{-1}.

Champion conductivity is achieved in crystalline electrolyte, but glass (amorphous) solid electrolytes also have several advantages. Glass electrolytes have a relatively high conductivity at the compositions with high

Table 4.1: Room temperature conductivity of typical sulfide and oxide solid electrolytes. Not subject to copyright.

Composition	Conductivity at 25°C (S cm^{-1})	Classification	Reference
$Li_{9.54}Si_{1.74}P_{1.44}S_{11.7}Cl_{0.3}$	2.5×10^{-2}	crystal	Kato[47]
$Li_{10}GeP_2S_{12}$	1.2×10^{-2}	crystal	Kamaya[48]
$Li_{10}SnP_2S_{12}$	4×10^{-3}	crystal	Bron[49]
$Li_{3.25}Ge_{0.25}P_{0.75}S_4$	2.2×10^{-3}	crystal	Kanno[50]
$Li_{3.833}Sn_{0.833}As_{0.166}S_4$	1.4×10^{-3}	crystal	Sahu[51]
Li_6PS_5Cl	1.3×10^{-3}	crystal	Boulineau[52]
$70Li_2S \cdot 30P_2S_5$ ($Li_7P_3S_{11}$)	1.7×10^{-2}	glass-ceramic	Seino[53]
$63Li_2S \cdot 27P_2S_5 \cdot 10LiBr$	8.4×10^{-3}	glass-ceramic	Ujiie[54]
$80Li_2S \cdot 20P_2S_5$	1.3×10^{-3}	glass-ceramic	Mizuno[55]
$50Li_2S \cdot 17P_2S_5 \cdot 33LiBH_4$	1.6×10^{-3}	glass	Yamauchi[56]
$63Li_2S \cdot 36SiS_2 \cdot 1Li_3PO_4$	1.5×10^{-3}	glass	Aotani[57]
$75Li_2S \cdot 25P_2S_5$	1.1×10^{-4}	glass	Hayashi[58]
$La_{0.51}Li_{0.34}TiO_{2.94}$	1.4×10^{-3}	crystal (perovskite)	Ito[59]
$Li_{1.3}Al_{0.3}Ti_{1.7}(PO_4)_3$	7×10^{-4}	crystal (NASICON)	Aono[60]
$Li_7La_3Zr_2O_{12}$	3×10^{-4}	crystal (garnet)	Murugan[61]
$90Li_3BO_3 \cdot 10Li_2SO_4$	1×10^{-5}	glass-ceramic	Tatsumisago[62]
$Li_{2.9}PO_{3.3}N_{0.46}$	3.3×10^{-6}	glass (UPON thin film)	Yu[63]

Li-ion concentration. Crystalline electrolyte has quite high conductivity at a limited composition/structure and drastic conductivity decrease is often observed outside suitable compositions. Glass electrolytes retain high conductivities even with changes in composition. Formation of electrode-electrolyte solid-solid interfaces may change compositions of both materials, and the use of glass electrolyte is useful for application to electrode/electrolyte interfaces because conductivity fluctuation will be suppressed.

Sulfide glasses with Li$_2$S more than 70 mol% have a conductivity of over 10^{-4} S cm^{-1} in the binary system Li$_2$S–P$_2$S$_5$.[58] It is difficult to prepare glass electrolytes with high Li concentration by conventional melt-quenching method because of easy crystallization during cooling process from melting. These glasses are thus prepared by twin-roller rapid-quenching or mechanical milling techniques. The latter preparation technique is done using a high-energy planetary ball mill apparatus; chemical reaction basically proceeds at ordinary temperature and pressure. Electrolyte particles are directly obtained by milling, and can be applied to all-solid-state batteries without additional pulverization of electrolytes. The addition of Li salts is useful for enhancing conductivity of glasses because of the increase in Li concentration and the decrease in activation energy for conduction. The Li salts such as Li halides,[54] Li borohydride (LiBH$_4$)[56] and Li ortho-oxosalts (Li$_3$PO$_4$)[57] were added to sulfide glasses and the conductivity increased from the order of 10^{-4} to 10^{-3} S cm^{-1} at room temperature.

Glasses are also important as a precursor for yielding high temperature phases, which usually exhibit much higher ionic conductivity than low temperature phases. Crystalline Li$_7$P$_3$S$_{11}$ is a high temperature phase at the 70Li$_2$S·30P$_2$S$_5$ composition and the phase is prepared by crystallization of mother glasses and is not obtained by conventional solid phase reaction.[53,64] The crystallization of α-AgI[65,66] and cubic-Na$_3$PS$_4$[67] from mother glasses is also typical example for developing solid electrolytes with high ion conductivity.

Very recently, SEs have also been synthesized by a liquid-phase technique. Crystalline β-Li$_3$PS$_4$ is prepared in tetrahydrofuran[68] or dimethyl carbonate[69] as reaction medium and crystalline Li$_7$P$_3$S$_{11}$ is synthesized in 1,2-dimethoxyethane.[70] A precursor with precipitates is obtained, and compressed pellets of the heat-treated compounds show a conductivity of over 10^{-4} S cm^{-1} at 25°C. Crystalline Li$_3$PS$_4$ is synthesized via a homogeneous liquid, which is prepared from the mixture of Li$_2$S and P$_2$S$_5$ with N-methylformamide (NMF).[71] The Li$_3$PS$_4$ solid electrolytes can also be prepared by a dissolution-reprecipitation process in NMF from 80Li$_2$S·20P$_2$S$_5$ (mol%) glass prepared by mechanical milling in advance.[72] The prepared Li$_3$PS$_4$ electrolyte shows a relatively low conductivity of 10^{-6} S cm^{-1} at the present stage, but conductivity can be enhanced by selecting electrolyte compositions. Argyrodite-type Li_6PS_5Cl is dissolved into ethanol, and the argyrodite phase is reprecipitated by removing ethanol to show higher conductivity than Li_3PS_4.[73]

Oxide solid electrolytes have inherent good chemical stability without generation of harmful H_2S gas like in SEs. However, the highest conductivity in oxides electrolytes is the order of 10^{-3} S cm^{-1}, which is one order of magnitude lower than that of sulfides. Sulfides give higher conductivity than oxides because of higher polarizability of sulfides anions facilitating Li$^+$ ion conduction.

Crystalline oxide electrolytes with NASICON (*Na Super Ionic Conductor*),[60] perovskite,[59] and garnet[61] structures show high conductivity of 10^{-4}–10^{-3} S cm^{-1}. In particular, garnet-type $Li_7La_3Zr_2O_{12}$ (LLZ) has been widely studied as solid electrolyte for all-solid-state lithium batteries because of its high conductivity of 3×10^{-4} S cm^{-1} and high chemical stability against Li negative electrode. Decrease of grain-boundaries by high-temperature sintering is a key to achieve high conductivity because total conductivity is largely affected by grain boundary resistances. To apply crystalline oxide electrolytes to bulk-type all-solid-state batteries, good contacts between electrode and electrolyte should be formed and heat treatment at high temperatures is needed for decreasing grain boundary resistances at solid-solid interfaces. However, side reactions between electrolyte and electrode often proceed at high temperature sintering,[74] and thus formability of oxide electrolytes with decreasing grain boundaries at low temperatures is important.

Increasing Li concentration is a key to enhance conductivity of glass electrolytes. The highest conductivity of 10^{-7}–10^{-6} S cm^{-1} is achieved at the Li orthoborate composition Li_3BO_3 (mol%) in the binary system Li_2O–B_2O_3.[75] LiPON (*Li*thium *P*hosphorous *O*xy*n*itride), which is known as thin-film electrolyte, has the same level of conductivity of 10^{-6} S cm^{-1}.[63] Further enhancement of conductivity is obtained in the pseudobinary Li_3BO_3–Li_4SiO_4 system. Crystallization of the binary glasses increases conductivity up to 10^{-5} S cm^{-1} by precipitation of a high temperature Li_3BO_3 phase: the obtained $90Li_3BO_3 \cdot 10Li_2SO_4$ (mol%) glass-ceramic also has good formability by only cold-press without heat treatment, and thus the electrolyte is useful for having favorable solid-solid interfaces with electrode materials as discussed in detail later.[62]

We focus on sulfide Li-ion conductors as the most promising electrolytes for all-solid-state Li–S batteries in the following section.

4.3.2.2. Electrochemical and chemical stability

Electrochemical and chemical stability are important as solid electrolytes for all-solid-state Li–S batteries. Electrochemical stability of the $Li_7P_3S_{11}$ glass-ceramic electrolyte was examined by cyclic voltammetry (CV). For example, cyclic voltammogram of the glass-ceramic at the first cycle is shown in Figure 4.9. A stainless-steel disk as a working electrode and a Li foil as a counter/reference electrode were attached on each face of a pelletized electrolyte. A cathodic current peak due to Li deposition and an anodic current peak due to Li dissolution are observed reversibly at around 0 V (vs. Li^+/Li). There is no large current peak except these peaks over the whole range from −0.1 to 5.0 V, suggesting that the glass-ceramic electrolyte has a wide electrochemical window of over 5 V and a good compatibility with Li metal.

Sulfides tend to be decomposed by hydrolysis in air atmosphere and generate harmful H_2S. Suppression of hydrolysis of sulfides is a significant task for developing of SEs. Chemical stability of sulfide glass

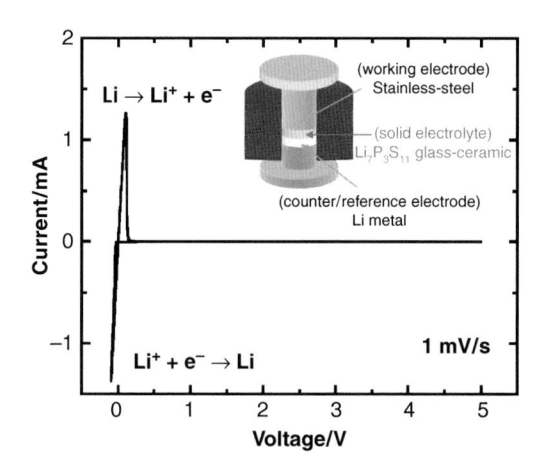

Figure 4.9 Cyclic voltammogram of the $Li_7P_3S_{11}$ glass-ceramic electrolyte at the first cycle. Not subject to copyright.

electrolytes in the binary system $Li_2S-P_2S_5$ was examined by exposing them to air atmosphere and H_2S generation was minimized at the composition $75Li_2S \cdot 25P_2S_5$ (mol%).[76] The glass is composed of Li^+ and PS_4^{3-} ions, and isolated anions PS_4^{3-} without bridging sulfurs are useful for high tolerance for hydrolysis. Li_3PS_4-based solid electrolytes with both good chemical stability and high conductivity have been prepared by combination of oxides (Li_2O or P_2O_5) and iodides (LiI).[77] The addition of metal oxides such as ZnO, which act as an absorbent for H_2S, is also effective in decreasing H_2S.[78] It is noteworthy that the use of a favorable M_xO_y (M_xO_y: Fe_2O_3, ZnO and Bi_2O_3) with a larger negative Gibbs energy change (ΔG) for the reaction with H_2S is effective in improving chemical stability of SEs. Another approach is the use of suitable sulfide compositions based on the hard and soft acids and bases (HSAB) theory. Li tin thiophosphate, Li_4SnS_4 has a better air stability compared to Li_3PS_4, and As-substituted Li_4SnS_4 has good features of both a high conductivity of 10^{-3} S cm^{-1} and high air stability.[51]

4.3.2.3. Formability and elastic modulus

Formability or processability of inorganic solid electrolytes is important to form close and wide contact areas with active material particles and also nanocarbon particles. Adhesion of solid-solid interface is a key to enhancing utilization of electrode active materials in all-solid-state batteries. Densification of sulfide and oxide electrolyte powders by uniaxial pressing at room temperature (coldpress) was examined.[79] Figure 4.10 shows scanning electron microscopy (SEM) images of fracture cross sections of (a) sulfide glass electrolyte ($80Li_2S \cdot 20P_2S_5$) and (b) oxide crystalline electrolyte ($Li_7La_3Zr_2O_{12}$). The samples were prepared from their electrolyte particles by coldpressing at 360 MPa. Densification is induced and grain boundaries among particles almost vanish in the sulfide glass, while voids and grain boundaries appear in the oxide crystal. In general, sintering at high temperatures of over 1000°C is needed to reduce huge grain-boundary resistances of $Li_7La_3Zr_2O_{12}$. It is noteworthy that the sulfide glass prepared by just coldpressing without heat treatment has a high conductivity of over 10^{-4} S cm^{-1} because of low grain boundary resistance via room temperature pressure sintering. Glass electrolytes in

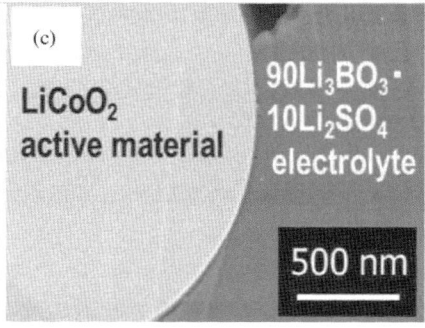

Figure 4.10 SEM images of fracture cross sections of (a) sulfide glass electrolyte (80Li$_2$S·20P$_2$S$_5$) and (b) oxide crystalline electrolyte (Li$_7$La$_3$Zr$_2$O$_{12}$). A cross-sectional SEM image of composite positive electrode layer with LiCoO$_2$ active material and 90Li$_3$BO$_3$·10Li$_2$SO$_4$ glass electrolyte particles is also shown in (c). Not subject to copyright.

the system Li$_3$BO$_3$–Li$_2$SO$_4$ with low glass transition temperatures also have good formability even as oxide materials.[62] Figure 4.10(c) shows cross-sectional SEM image of composite positive electrode layer with LiCoO$_2$ active material and Li$_3$BO$_3$–Li$_2$SO$_4$ glass electrolyte particles. Good adhesion between LiCoO$_2$ and the glass electrolyte particles is formed by only coldpress, suggesting that the oxide glass has deformability as with sulfide glasses. All-solid-state Li cells using the prepared LiCoO$_2$ positive electrode operate as a secondary battery at room temperature.

Retaining solid-solid contacts between active materials and solid electrolytes during charge–discharge processes brings about long cycle lives of all-solid-state batteries. Young's moduli of solid electrolytes are important for keeping favorable contacts even at volume changes of active materials. Young's moduli measured for densified $Li_2S–P_2S_5$ sulfide glasses prepared by hotpressing are 18–25 GPa, and they are gradually increased with an increase in the Li_2S content.[80] These sulfide glasses have an intermediate Young's modulus between oxide glasses and organic polymers. SEs deforming elastically are expected to act as a buffer in response to volume changes of active materials during charge–discharge processes. In fact, most all-solid-state batteries that employ sulfide solid electrolytes exhibit good cycle performance.[81,82]

4.3.3. All-solid-state Li–S batteries with sulfide solid electrolytes

Sulfide glasses are well-balanced solid electrolytes with high conductivity, wide electrochemical window, good formability, appropriate Young's modulus, and moderate chemical stability, and they are thus highly promising solid electrolytes for all-solid-state batteries. Sulfur as a targeted active material in this book has several advantages as a positive electrode such as high theoretical capacity of 1672 mAh g^{-1}, environmental friendliness, and abundant resources. Li polysulfides (Li_2S_x) are formed during discharge process (lithiation process) and easily dissolve in organic liquid electrolytes, leading to lack of a sulfur positive electrode. Dissolution of Li polysulfides is suppressed by absorbing sulfurs in pores of nanocarbons, and this approach has been extensively studied. One essential approach to solve the issue is the use of inorganic solid electrolytes. Because of insulative nature of sulfur, composite positive electrodes, where sulfur is mixed with both nanocarbons (electron conduction additive) and solid electrolytes (SE, Li$^+$ ion conduction additive), should be prepared for utilizing sulfur as an active material in all-solid-state Li–S batteries. Sulfur-carbon nanocomposites are prepared by mixing nanocarbons with sulfurs in solid-state,[83,84] liquid-state[85] and gas-state.[86] Preparation of composite sulfur electrodes by solid-phase milling is mainly reported in this section.

4.3.3.1. Composite positive electrodes with sulfur or Li$_2$S

Composite sulfur electrodes consisting of S, acetylene black (AB), and Li$_2$S–P$_2$S$_5$ SE powders with a weight ratio of 25/25/50 were prepared by high-energy planetary ball milling to make favorable contacts among the three components.[83] Use of Li$_2$S as a discharge product of sulfur active material also has a merit from the viewpoint of a large theoretical capacity of 1167 mAh g^{-1} and versatility of negative electrode materials without Li sources. Composite Li$_2$S electrodes with the same weight ratio of 25/25/50 for Li$_2$S/AB/SE were prepared in the same manner of composite sulfur electrodes.[87] Typical charge–discharge curves for all-solid-state Li–In/S or In/Li$_2$S cells at a constant current density of 0.064 or 0.013 mA cm^{-2} are shown in Figure 4.11. The 80Li$_2$S·20P$_2$S$_5$ glass-ceramic electrolyte was used as a solid electrolyte and Li–In alloy or In metal was used as a counter/reference electrode. Horizontal axis means the utilization of sulfur or Li$_2$S active materials in each composite electrode. A charge–discharge profile of the Li–In/S cell is similar to that of the In/Li$_2$S cell and the utilizations of sulfur and Li$_2$S are respectively *ca.* 88% and *ca.* 84%, suggesting that similar electrochemical reactions proceed in both the cells, and

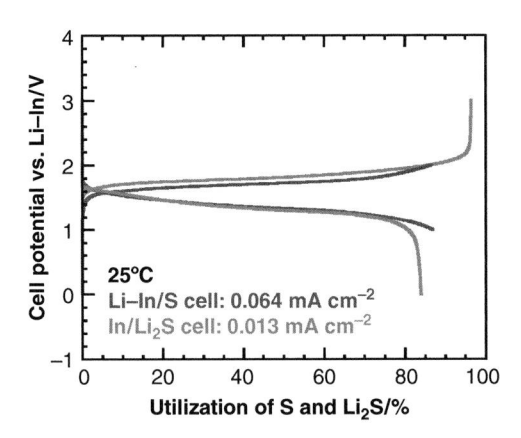

Figure 4.11 Typical charge–discharge curves for all-solid-state Li–In/S or In/Li$_2$S cells at a constant current density of 0.064 or 0.013 mA cm^{-2}. Horizontal axis means the utilization of sulfur or Li$_2$S active materials in each composite electrode. Not subject to copyright.

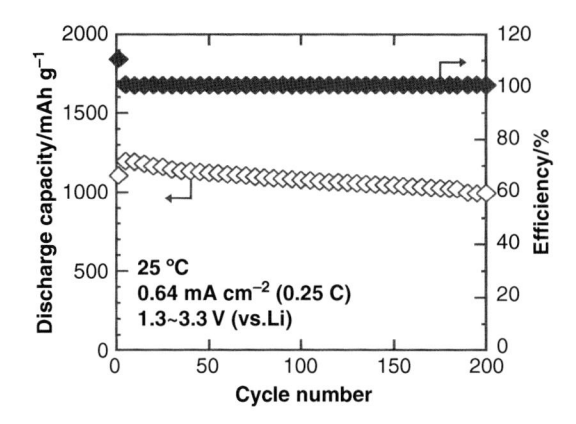

Figure 4.12 Cycle performance of the Li–In/S cell at a current density of 0.64 mA cm^{-2} (0.25 C). Not subject to copyright.

the sulfur cell has a somewhat better rate performance than the Li$_2$S cell. Figure 4.12 shows the cycle performance of the Li–In/S cell at a higher current density of 0.64 mA cm^{-2} (0.25 C). The sulfur cell shows good cyclability and retains about 1000 mAh g^{-1} for 200 cycles, suggesting that the use of solid electrolyte is essentially useful for developing Li–S batteries with long cycle lives.

4.3.3.2. Reaction mechanisms of all-solid-state Li–S batteries

Charge–discharge profiles as shown in Figure 4.11 are different from those of typical Li–S cells using organic liquid electrolytes.[46,88,89] To investigate the difference of the profiles, open circuit voltages for the all-solid-state Li–In/S cell were measured. The cell was discharged at constant current density 0.064 mA cm^{-2} up to a discharge capacity of 50 mAh g^{-1}, and then OCV after the rest for five hours was measured. This measurement was repeated in the same manner to the final discharge capacity of 800 mAh g^{-1} (sulfur utilization of *ca.* 50%). Figure 4.13 shows the OCV profiles for the all-solid-state Li–In/S cell with or without the rest for five hours. The inset is a typical discharge–charge curve of a liquid electrolyte cell.[88] The discharge curve is divided into three regions: (I) conversion of

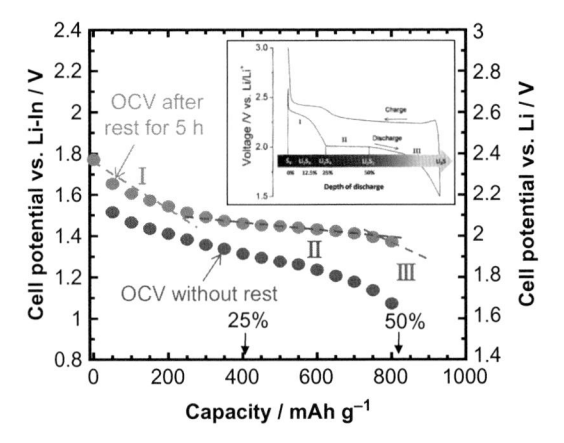

Figure 4.13 OCV profiles for the all-solid-state Li–In/S cell with or without the rest for five hours. The inset is a typical discharge–charge curve of a liquid electrolyte cell.
Source: Reproduced with permission from Ref. 88.

solid sulfur to soluble polysulfides; (II) conversion of polysulfides to solid Li$_2$S$_2$; and (III) conversion of solid Li$_2$S$_2$ to solid Li$_2$S. The OCV profile after the rest for five hours gives a similar profile to that of a liquid-electrolyte cell. A slope from 2.4–2.0 V vs. Li$^+$/Li corresponds to the region (I), while a plateau at 2.0 V is almost the same as that of the region (II). It is noteworthy that the two-stage plateau (the regions (I) and (II)) in all-solid-state Li–S cells is kinetically hindered.

Charge–discharge reaction mechanism for the Li$_2$S composite electrode in an all-solid-state cell was examined by high resolution TEM observation.[90] Figure 4.14 shows the TEM images for the Li$_2$S electrodes (a) before charge–discharge test, (b) after the initial charge, and (c) after the initial discharge. Nanoparticles of *ca.* 5 nm in size with different crystal orientations distribute randomly in the matrix consisting of amorphous SE and AB as shown in the image (a). The nanoparticles are attributable to crystalline Li$_2$S. No lattice fringes due to the crystalline Li$_2$S can be seen and there exists the characteristic contrast due to amorphous structure in the whole region after the initial charge process as shown in image (b). As shown in the image (c), lattice fringes with spacing of about 3.9 Å are clearly seen again after the 1st discharge, suggesting that amorphous sulfur is converted into crystalline nanoparticles during

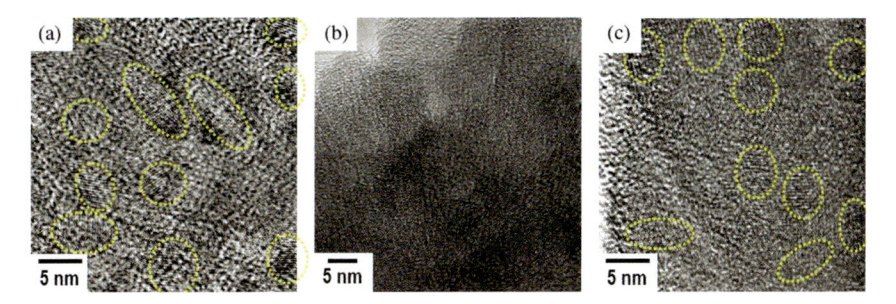

Figure 4.14 TEM images for the Li$_2$S electrodes in all-solid-state cells (a) before charge–discharge test, (b) after the initial charge, and (c) after the initial discharge.
Source: Reproduced with permission from Ref. 90.

discharge reaction. Reversible transformation between crystallization and amorphization of sulfur-based active nanoparticles is responsible for the high capacity and its retention.

Enhancement of conductivity of Li$_2$S is effective in increasing its utilization as an active material. Solid-solutions in the system Li$_2$S–LiI were prepared via mechanochemistry.[91] Lattice constant of Li$_2$S increased with an increase in the LiI content. The addition of 20 mol% LiI increased ionic conductivity by two orders of magnitude. A composite positive electrode with the 80Li$_2$S·20LiI (mol%) solid-solution, vapor grown carbon fiber (VGCF), and Li$_3$PS$_4$ glass electrolyte in the weight ratio of 50/10/40 is applied to all-solid-state cells. An all-solid-state cell (Li–In/Li$_3$PS$_4$/80Li$_2$S·20LiI) operated at a current density of 0.13 mA cm^{-1} (0.07 C) at 25°C and retained a reversible capacity of 930 mAh per gram of Li$_2$S for 50 cycles. The utilization of Li$_2$S considerably increased from *ca.* 50 to 80% by using the Li$_2$S–LiI solid solutions.

4.3.3.3. *Approaches to enhance the energy density*

To increase in energy density of all-solid-state Li–S batteries, the increase in sulfur-based active materials in a composite positive electrode layer is important. We have developed the following three approaches to achieve higher capacities normalized by a total weight of composite positive electrodes: (1) optimization of preparation processes and composite components, (2) use of sulfur-rich transition metal sulfides, and (3) function of sulfide solid electrolytes as active material.

The first approach is preparation of composite sulfur electrode by high temperature ball milling.[85] Composite sulfur electrodes are often prepared by high-energy ball-milling process at room temperature and mixing sulfur and electrolyte solid particles gives a limited contact area between solid-solid interfaces. Ball milling at about 155°C, where sulfur melt has the lowest viscosity, will provide more favorable contacts based on mixing of liquid sulfur and solid electrolyte. Sulfur-AB composites were prepared by high-temperature ball-milling process, and then a Li_2S–P_2S_5 SE was added and ball-milled at ambient condition. The composite S–AB–SE electrode with the 50 wt.% sulfur content was applied to all-solid-state Li–S cells and the cell retained ca. 1000 mAh per gram of sulfur at 0.064 mA cm^{-2} (0.02 C), which is almost twice as high as the capacity of the cell with the 50 wt.% sulfur electrode milled at ambient condition.

Selection of components in composite sulfur electrodes affects battery performance. Nagata et al. have reported the use of $Li_{1.5}PS_{3.3}$ (60Li_2S·40P_2S_5) electrolyte with less Li sources instead of $Li_{4.0}PS_{4.5}$ (80Li_2S·20P_2S_5) in the composite sulfur electrodes significantly enhance a reversible capacity at a high current density of over 1 mA cm^{-2}.[92] The sulfur electrodes consisting of S/Ketjenblack (KB)/$Li_{1.5}PS_{3.3}$ with a weight ratio of 50/10/40 in all-solid-state cells showed a reversible capacity of over 1000 mAh g^{-1}, which is larger than that of the sulfur electrode with the $Li_{4.0}PS_{4.5}$ SE at 6.4 mA cm^{-2}. Although the reason why the cell using the sulfur electrode with the less Li_2S electrolyte with lower conductivity exhibited high-rate capability has not been clarified yet, optimization of components in the sulfur electrode is effective in enhancing performance of all-solid-state Li–S batteries with high-energy density and high rate capability.

The second approach is the use of sulfur-rich transition metal sulfides instead of sulfur active materials.[93–98] For example, titanium trisulfide TiS_3 shows a higher capacity than typical TiS_2, because additional sulfurs in TiS_3 contribute to redox reaction during charge–discharge processes.[93] Amorphous NbS_x (x = 3, 4, 5) are mechanochemically prepared, and electrochemical cells with an organic liquid electrolyte using the amorphous NbS_x (x = 3, 4, 5) show higher discharge capacities with an increase in the sulfur content of NbS_x.[97]

Figure 4.15 shows (a) the initial charge–discharge curves and (b) the cycle performance of all-solid-state cells Li–In/crystalline TiS$_3$ (c-TiS$_3$) or amorphous TiS$_3$ (a-TiS$_3$). A composite positive electrode of TiS$_3$/AB/SE with the weight ratio of 38/5/57 was prepared by hand-grinding. The 80Li$_2$S·20P$_2$S$_5$ glass-ceramic SE was used as a solid electrolyte. The initial discharge capacity of the cell with c-TiS$_3$ at 0.064 mA cm^{-2} (C/40) was about 556 mAh g^{-1}, which was the same as the theoretical capacity of TiS$_3$ (the capacity corresponding to the insertion of 3 mol Li to TiS$_3$). Although

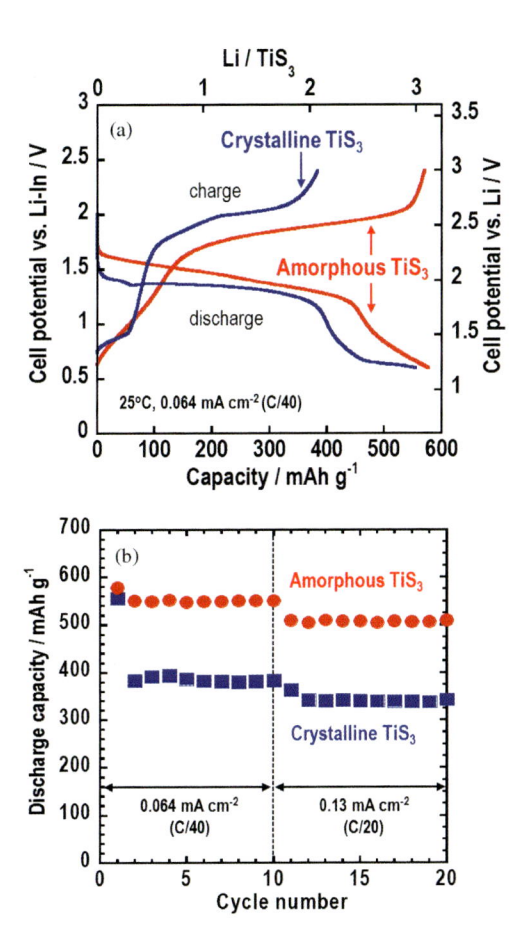

Figure 4.15 (a) The initial charge–discharge curves and (b) the cycle performance of all-solid-state cells Li–In/crystalline TiS$_3$ (c-TiS$_3$) or amorphous TiS$_3$ (a-TiS$_3$).

Source: Reproduced with permission from Ref. 96.

the initial charge capacity for the cell is lower than the discharge capacity, the cell using c-TiS$_3$ shows the reversible capacity of *ca.* 400 mAh g^{-1} from the 2nd to 10th cycle. The reversible capacity decreases with increasing the current density from 0.064 to 0.13 mA cm^{-2}. On the other hand, a-TiS$_3$ shows a reversible capacity of *ca.* 556 mAh g^{-1} for the initial charge–discharge process and retains a higher capacity than c-TiS$_3$ in all-solid-state cells. The cell with a-TiS$_3$ positive electrode including no carbon conductive additives and solid electrolytes also has a reversible capacity for 10 cycles at 0.013 mA cm^{-2} of about 510 mAh gram per weight of a-TiS$_3$, which equals to the weight of total positive electrode.[98] High resolution-TEM observation revealed that crystalline structure of c-TiS$_3$ partially deteriorated and amorphous areas in the c-TiS$_3$ were formed after charge–discharge cycling.[96] These irreversible structural changes of c-TiS$_3$ during charge–discharge tests might be responsible for the irreversible capacity at the initial cycle, but partial amorphization contributes to good cyclability except for the initial cycle. It is noteworthy that sulfur-rich amorphous transition metal sulfides with electrical conductivity are promising positive electrode instead of sulfur active material.

The final approach is the use of sulfide solid electrolytes as active material. Solid electrolytes with Li-ion conductivity function as active material by combining with electronic conductors. A solid electrolyte Li$_3$PS$_4$ was ball-milled with nanocarbons, and the prepared Li$_3$PS$_4$–AB materials are useful as bifunctional electrode acting as not only solid electrolyte but also active material. An all-solid-state cell with the Li$_3$PS$_4$–AB composite as a positive electrode is charged and discharged, and its operation potential of *ca.* 2.6 V vs. Li$^+$/Li is somewhat higher than the sulfur redox potential.[99] Sulfur-rich Li$_3$PS$_{4+n}$ (0 < n < 9) compounds are also reported to show good electrochemical reversibility for all-solid-state cells.[100] The addition of copper is also useful for activating SEs as an active material.[101] Copper metal easily reacts with sulfur to form copper sulfides, and therefore ball milling of sulfur (or Li$_2$S) and copper yielded sulfur-based composite positive electrodes with a high reversible capacity in all-solid-state cells.[102–104] It is noted that the use of the solid electrolytes as an active material in electrode layers is effective in increasing reversible capacity per gram of total mass of positive electrodes.

4.3.3.4. Concluding remarks

Sulfide glass-based materials are superior inorganic solid electrolytes for bulk-type all-solid-state batteries because of their high conductivity of 10^{-4}–10^{-2} S cm^{-1} at room temperature, and favorable mechanical properties of formability and Young's modulus for achieving good electrode/electrolyte interfaces. Very recently, $Li_{9.54}Si_{1.74}P_{1.44}S_{11.7}Cl_{0.3}$ with the highest conductivity of 2.5×10^{-2} S cm^{-1} has been reported and an all-solid-state cell with the electrolyte exhibits superior rate performance.[47] It is noteworthy that the cell can function at 1500 C-rate and the specific power of $LiCoO_2$ in all-solid-state cells is much higher than that of the LIBs and is even higher than those of electrodes used in supercapacitors. Continuous development for well-balanced electrolytes with good conductivity, appropriate mechanical properties, and high electrochemical and chemical stability is desired.

All-solid-state Li–S batteries with sulfide solid electrolytes have inherently good electrochemical performance of high reversible capacity and long cycle lives. The cell with a composite sulfur electrode of S/AB/sulfide SE with a weight ratio of 25/25/50 retained *ca.* 1000 mAh g^{-1} for 200 cycles at 0.25 C. It was revealed from HR–TEM observation that Li_2S active material underwent reversible structural transformation between nanocrystal and amorphous state during charge–discharge cycles. Increasing conductivity of Li_2S by combining with LiI is useful for increasing the utilization of Li_2S. Increasing the sulfur content in composite positive electrodes and improving rate capability are needed to develop the batteries. Milling at high temperatures and optimizing components of composite sulfur electrodes improved cell performance even at a high sulfur content (50 wt.%). The use of superior SEs with higher conductivity will facilitate Li$^+$ ion conduction in composite sulfur electrodes, leading to high rate performance. Sulfur-rich transition metal sulfides with both electronic and ionic conductivity are alternative sulfur-based positive electrodes with high capacity, and amorphous TiS_3 functioned as a positive electrode with high capacity and cyclability even at no addition of nanocarbon and SE. Sulfur or Li_2S active materials combining with a minimum amount of nanocarbons and transition metals will bring about innovative all-solid-state Li–S batteries with high-energy density and rate capability.

Acknowledgments

Patrik Johansson would like to sincerely acknowledge the support by the EUROLIS (Advanced European lithium sulphur cells for automotive application) project (No. 314515) and the HELIS (High-energy lithium sulphur cells and batteries) project (No. 666221) funded by the European Union's FP7 and Horizon 2020 research and innovation programmes, respectively. He also acknowledges the continuous support from several of Chalmers Areas of Advance: Materials Science, Energy, and Transport, and also the many fruitful discussions within the ALISTORE-ERI, especially with Dr. Mathieu Morcrette, UPJV, France, and Dr. Robert Dominko, NIC, Slovenia. Some of the graphic material for the subchapter on non-aqueous liquid electrolytes was made by Dr. Muhammad Abdelhamid, Chalmers, is gratefully acknowledged.

Rezan Demir-Cakan acknowledges the financial support of the Gebze Technical University and The Scientific and Technological Research Council of Turkey (TUBITAK) contract number 214M272 and 213M374. Dr. Mathieu Morcrette and Prof. Jean-Marie Tarascon are acknowledged for their fruitful discussions and continuous supports.

Akitoshi Hayashi and Masahiro Tatsumisago acknowledge the financial support by a Grant-in-Aid for Scientific Research from the Ministry of Education, Culture, Sports, Science and Technology (MEXT) of Japan and the Japan Science and Technology Agency (JST), Advanced Low Carbon Technology Research and Development Program (ALCA), Specially Promoted Research for Innovative Next Generation Batteries (SPRING) Project.

Bibliography

1. J. Scheers, S. Fantini and P. Johansson, *Journal of Power Sources*, **255** (2014) 204–218.
2. M. Barghamadi, A. S. Best, A. I. Bhatt, A. F. Hollenkamp, M. Musameh, R. J. Rees and Thomas Rüther, *Energy & Environmental Science*, **7**(12) (2014) 3902–3920.
3. R. Younesi, G. M. Veith, P. Johansson, K. Edström and T. Vegge, *Energy & Environmental Science*, **8**(7) (2015) 1905–1922.
4. S. S. Zhang, *Journal of Power Sources*, **231** (2013) 153–162.

5. J. Gao, M. A. Lowe, Y. Kiya and H. D. Abruna, *The Journal of Physical Chemistry C*, **115**(50) (2011) 25132–25137.
6. Y. V. Mikhaylik and J. R. Akridge, *Journal of the Electrochemical Society*, **150**(3) (2003) A306–A311.
7. W. Weng, V. G. Pol and K. Amine, *Advanced Materials*, **25**(11) (2013) 1608–1615.
8. J. Brückner, S. Thieme, H. T. Grossmann, S. Dörfler, H. Althues and S. Kaskel, *Journal of Power Sources*, **268** (2014) 82–87.
9. X.-B. Cheng, J.-Q. Huang, H.-J. Peng, J.-Q. Nie, X.-Y. Liu, Q. Zhang and F. Wei, *Journal of Power Sources*, **253** (2014) 263–268.
10. M. Hagen, P. Fanz and J. Tübke, *Journal of Power Sources*, **264** (2014) 30–34.
11. M. Hagen, D. Hanselmann, K. Albrecht, R. Maca, D. Gerber and J. Tübke, *Advanced Energy Materials*, **5** (2015) 1401986.
12. V. Thangavel, K.–H. Xue, Y. Mammeri, M. Quiroga, A. Mastouri, C. Guéry, P. Johansson, M. Morcrette and A. A. Franco, *Journal of the Electrochemical Society*, **163** (2016) A2817–A2829.
13. V. Kolosnitsyn, E. Kuzmina and E. Karaseva, *ECS Transactions*, **19**(25) (2009) 25–30.
14. C. Barchasz, J.-C. Lepetre, S. Patoux and F. Alloin, *Journal of the Electrochemical Society*, **160**(3) (2013) A430–A436.
15. E. S. Shin, K. Kim, S. H. Oh and W. II Cho, *Chemical Communications*, **49**(20) (2013) 2004–2006.
16. L. Suo, Y.-S. Hu, H. Li, M. Armand and L. Chen, *Nature communications*, **4** (2013) 1481.
17. J.-N. Chazalviel, *Physical Review A*, **42**(12) (1990) 7355.
18. S. S. Zhang, *Electrochimica Acta*, **70** (2012) 344–348.
19. S. S. Zhang, *Journal of Power Sources*, **322** (2016) 99–105.
20. D. Aurbach, E. Pollak, R. Elazari, G. Salitra, C. S. Kelley and J. Affinito, *Journal of the Electrochemical Society*, **156**(8) (2009) A694–A702.
21. S. Xiong, K. Xie, Y. Diao and X. Hong, *Journal of Power Sources*, **246** (2014) 840–845.
22. M. U. Patel, R. Demir-Cakan, M. Morcrette, J.-M. Tarascon, M. Gaberscek and R. Dominko, *ChemSusChem*, **6**(7) (2013) 1177–1181.
23. J. Hannauer, J. Scheers, J. Fullenwarth, B. Fraisse, L. Stievano and P. Johansson, *ChemPhysChem*, **16**(13) (2015) 2755–2759.
24. K. A. See, M. Leskes, J. M. Griffin, S. Britto, P. D. Matthews, A. Emly, A. Van der Ven, D. S. Wright, A. J. Morris, C. P. Grey and R. Seshadri, *Journal of the American Chemical Society*, **136**(46) (2014) 16368–16377.

25. E. Blomberg, Redox behavior of Li-S cell with PP14-TFSI ionic liquid electrolyte: Spectroscopic study on speciation of polysulfides during charge/discharge processes, *Master of Science Thesis, Chalmers University of Technology* (2012), http://publications.lib.chalmers.se/records/fulltext/160632.pdf.
26. A. Manthiram, Y. Fu, S.-H. Chung, C. Zu and Y.-S. Su, *Chemical Reviews*, **114** (2014) 11751–11787.
27. A. Rosenman, E. Markevich, G. Salitra, D. Aurbach, A. Garsuch and F. F. Chesneau, *Advanced Energy Materials*, **5** (2015) 1500212.
28. Y. Zhao, Y. Ding, J. Song, L. L. Peng, J. B. Goodenough and G. H. Yu, *Energy & Environmental Science*, **7** (2014) 1990–1995.
29. Y. Zhao, Y. Ding, Y. Li, L. Peng, H. R. Byon, J. B. Goodenough and G. Yu, *Chemical Society Reviews*, **44** (2015) 7968–7996.
30. C. Wessells, R. Ruffo, R. A. Huggins and Y. Cui, *Electrochemical and Solid State Letters*, **13** (2010) A59–A61.
31. L. M. Suo, O. Borodin, T. Gao, M. Olguin, J. Ho, X. L. Fan, C. Luo, C. S. Wang and K. Xu, *Science*, **350** (2015) 938–943.
32. L. Suo, Y.-S. Hu, H. Li, M. Armand and L. Chen, *Nature Communications*, **4** (2013) 1481.
33. D. Peramunage and S. Licht, *Science*, **261** (1993) 1029–1032.
34. S. Licht and D. Peramunage, *Journal of the Electrochemical Society*, **140** (1993) L4–L6.
35. S. Licht, J. Manassen and G. Hodes, *Journal of the Electrochemical Society*, **133** (1986) 272–277.
36. S. Licht and J. Manassen, *Journal of the Electrochemical Society*, **132** (1985) 1076–1081.
37. S. Licht, *Solar Energy Materials and Solar Cells*, **38** (1995) 305–319.
38. Y. S. N. S. J. Visco, B. D. Katz, L. C. De Jonghe, N. Goncharenko and V. Loginova, US Patent, **US 2013/0122334**, 2013.
39. N. Li, Z. Weng, Y. R. Wang, F. Li, H. M. Cheng and H. S. Zhou, *Energy & Environmental Science*, **7** (2014) 3307–3312.
40. L. Chen, Z. Y. Guo, Y. Y. Xia and Y. G. Wang, *Chemical Communications*, **49** (2013) 2204–2206.
41. Y. H. Lu, J. B. Goodenough and Y. Kim, *Journal of the American Chemical Society*, **133** (2011) 5756–5759.
42. Y. Zhao, L. N. Wang and H. R. Byon, *Nature Communications*, **4** (2013) 1896.
43. R. Demir-Cakan, M. Morcrette and J. M. Tarascon, *Journal of Materials Chemistry A*, **3** (2015) 2869–2875.

44. R. Demir-Cakan, M. Morcrette, J. B. Leriche and J. M. Tarascon, *Journal of Materials Chemistry A*, **2** (2014) 9025–9029.
45. M. Tatsumisago, M. Nagao and A. Hayashi, *Journal of Asian Ceramic Societies*, **1**(1) (2013) 17–25.
46. Z. Lin and C. Liang, *Journal of Materials Chemistry A*, **3**(3) (2015) 936–958.
47. Y. Kato, S. Hori, T. Saito, K. Suzuki, M. Hirayama, A. Mitsui, M. Yonemura, H. Iba and R. Kanno, *Nature Energy*, **1** (2016) 16030.
48. N. Kamaya, K. Homma, Y. Yamakawa, M. Hirayama, R. Kanno, M. Yonemura, T. Kamiyama, Y. Kato, S. Hama, K. Kawamoto and A. Mitsui, *Nature Materials*, **10**(9) (2011) 682–686.
49. P. Boron, S. Johansson, K. Zick, J. Gunne, S. Dehnen and B. Roling, *Journal of the American Chemical Society*, **135**(42) (2013) 15694–15697.
50. R. Kanno and M. Murayama, *Journal of the Electrochemical Society*, **148**(7) (2001) A742.
51. G. Sahu, Z. Lin, J. Li, Z. Liu, N. Dudney and C. Liang, *Energy & Environmental Science*, **7** (2014) 1053–1058.
52. S. Boulineau, M. Courty, J. M. Tarascon and V. Viallet, *Solid State Ionics*, **221** (2012) 1–5.
53. Y. Seino, T. Ota, K. Takada, A. Hayashi and M. Tatsumisago, *Energy & Environmental Science*, **7**(2) (2014) 627–631.
54. S. Ujiie, A. Hayashi and M. Tatsumisago, *Materials for Renewable and Sustainable Energy*, **3**(1) (2014) 18.
55. F. Mizuno, A. Hayashi, K. Tadanaga and M. Tatsumisago, *Solid State Ionics*, **177** (2006) 2721.
56. A. Yamauchi, A. Sakuda, A. Hayashi and M. Tatsumisago, *Journal of Power Sources*, **244** (2013) 707–710.
57. N. Aotani, K. Iwamoto, K. Takada and S. Kondo, *Solid State Ionics*, **68**(1–2) (1994) 35–39.
58. A. Hayashi, S. Hama, H. Morimoto, M. Tatsumisago and T. Minami, *Journal of the American Ceramic Society*, **84**(2) (2001) 477–479.
59. M. Ito, Y. Inaguma, W. H. Jung, L. Chen and T. Nakamura, *Solid State Ionics*, **70** (1994) 203–207.
60. H. Aono, E. Sugimono, Y. Sadaoka, N. Imanaka and G. Adachi, *Journal of the Electrochemical Society*, **137**(4) (1990) 1023–1027.
61. R. Murugan, V. Thangadurai and W. Weppner, *Angewandte Chemie International Edition*, **46**(41) (2007) 7778–7781.
62. M. Tatsumisago, R. Takano, K. Tadanaga and A. Hayashi, *Journal of Power Sources*, **270** (2014) 603–607.

63. X. Yu, J. B. Bates, G. E. Jellison and F. X. Hart, *Journal of the Electrochemical Society*, **144**(2) (1997) 524–532.
64. F. Mizuno, A. Hayashi, K. Tadanaga and M. Tatsumisago, *Advanced Materials*, **17**(7) (2005) 918–921.
65. M. Tatsumisago, Y. Shinkuma and T. Minami, *Nature*, **354**(6350) (1991) 217–218.
66. M. Tatsumisago, T. Saito and T. Minami, *Chemistry Letters*, **30**(8) (2001) 790–791.
67. A. Hayashi, K. Noi, A. Sakuda and M. Tatsumisago, *Nature Communications*, **3** (2012) 856.
68. Z. Liu, W. Fu, E. Payzant, X. Yu, Z. Wu, N. Dudney, J. Kiggans, K. Hong, A. Rondinone and C. Liang, *Journal of American Chemical Society*, **135**(3) (2013) 975–978.
69. N. Phuc, K. Morikawa, M. Totani, H. Muto and A. Matsuda, *Solid State Ionics*, **285** (2016) 2–5.
70. S. Ito, M. Nakakita, Y. Aihara, T. Uehara and N. Machida, *Journal of Power Sources*, **271** (2014) 342–345.
71. S. Teragawa, K. Aso, K. Tadanaga, A. Hayashi and M. Tatsumisago, *Journal of Materials Chemistry A*, **2**(14) (2014) 5095–5099.
72. S. Teragawa, K. Aso, K. Tadanaga, A. Hayashi and M. Tatsumisago, *Journal of Power Sources*, **248** (2014) 939–942.
73. S. Yubuchi, S. Teragawa, K. Aso, K. Tadanaga, A. Hayashi and M. Tatsumisago, *Journal of Power Sources*, **293** (2015) 941–945.
74. Y. Kobayashi, T. Takeuchi, M. Tabuchi, K. Ado and H. Kageyama, *Journal of Power Sources*, **81–82** (1999) 853–858.
75. A. Hayashi, D. Furusawa, Y. Takahashi, K. Minami and M. Tatsumisago, *Physics and Chemistry of Glasses — European Journal of Glass Science and Technology Part B*, **54**(3) (2013) 109–114.
76. H. Muramatsu, A. Hayashi, T. Ohtomo, S. Hama and M. Tatsumisago, *Solid State Ionics*, **182**(1) (2011) 116–119.
77. T. Ohtomo, A. Hayashi, M. Tatsumisago and K. Kawamoto, *Electrochemistry*, **81**(6) (2013) 428–431.
78. A. Hayashi, H. Muramatsu, T. Ohtomo, S. Hama and M. Tatsumisago, *Journal of Materials Chemistry A*, **1**(21) (2013) 6320–6326.
79. A. Sakuda, A. Hayashi and M. Tatsumisago, *Scientific Reports*, **3** (2013) 2261.
80. A. Sakuda, A. Hayashi, Y. Takigawa, K. Higashi and M. Tatsumisago, *Journal of the Ceramic Society of Japan*, **121**(1419) (2013) 946–949.

81. M. Tatsumisago and A. Hayashi, *Functional Materials Letters*, **01**(01) (2008) 31–36.
82. M. Tatsumisago and A. Hayashi, *International Journal of Applied Glass Science*, **5**(3) (2014) 226–235.
83. M. Nagao, A. Hayashi and M. Tatsumisago, *Electrochimica Acta*, **56**(17) (2011) 6055–6059.
84. S. Kinoshita, K. Okuda, N. Machida, M. Naito and T. Shigematsu, *Solid State Ionics*, **256** (2014) 97–102.
85. M. Nagao, A. Hayashi and M. Tatsumisago, *Energy Technology*, **1**(2–3) (2013) 186–192.
86. M. Nagao, Y. Imade, H. Narisawa, T. Kobayashi, R. Watanabe, T. Yokoi, T. Tatsumi and R. Kanno, *Journal of Power Sources*, **222** (2013) 237–242.
87. M. Nagao, A. Hayashi and M. Tatsumisago, *Journal of Materials Chemistry*, **22**(19) (2012) 10015–10020.
88. X. Ji and L. F. Nazar, *Journal of Materials Chemistry*, **20**(44) (2010) 9821–9826.
89. D. Bresser, S. Passerini and B. Scrosati, *Chemical Communications*, **49**(90) (2013) 10545–10562.
90. M. Nagao, A. Hayashi, M. Tatsumisago, T. Ichinose, T. Ozaki, Y. Togawa and S. Mori, *Journal of Power Sources*, **274** (2015) 471–476.
91. T. Hakari, A. Hayashi and M. Tatsumisago, *Chemistry Letters*, **44**(12) (2015) 1664–1666.
92. H. Nagata and Y. Chikusa, *Journal of Power Sources*, **263** (2014) 141–144.
93. A. Hayashi, T. Matsuyama, A. Sakuda and M. Tatsumisago, *Chemistry Letters*, **41**(9) (2012) 886–888.
94. T. Matsuyama, A. Hayashi, T. Ozaki, S. Mori and M. Tatsumisago, *Journal of Materials Chemistry A*, **3**(27) (2015) 14142–14147.
95. T. Matsuyama, M. Deguchi, A. Hayashi, M. Tatsumisago, T. Ozaki, Y. Togawa and S. Mori, *Electrochemistry*, **83**(10) (2015) 889–893.
96. T. Matsuyama, A. Hayashi, T. Ozaki, S. Mori and M. Tatsumisago, *Journal of the Ceramic Society of Japan*, **124**(3) (2016) 242–246.
97. A. Sakuda, N. Taguchi, T. Takeuchi, H. Kobayashi, H. Sakaebe, K. Tatsumi and Z. Ogumi, *ECS Electrochemistry Letters*, **3**(7) (2014) A79–A81.
98. T. Matsuyama, M. Deguchi, K. Mitsuhara, T. Ohta, T. Mori, Y. Orikasa, Y. Uchimoto, Y. Kowada, A. Hayashi and M. Tatsumisago, *Journal of Power Sources*, **313** (2016) 104–111.
99. T. Hakari, M. Nagao, A. Hayashi and M. Tatsumisago, *Journal of Power Sources*, **293** (2015) 721–725.

100. Z. Lin, Z. Liu, W. Fu, N. J. Dudney and C. Liang, *Angewandte Chemie International Edition*, **52**(29) (2013) 7460–7463.
101. A. Hayashi, R. Ohtsubo, M. Nagao and M. Tatsumisago, *Journal of Materials Science*, **45**(2) (2009) 377–381.
102. N. Machida, K. Kobayashi, Y. Nishikawa and T. Shigematsu, *Solid State Ionics*, **175**(1–4) (2004) 247–250.
103. A. Hayashi, T. Ohtomo, F. Mizuno, K. Tadanaga and M. Tatsumisago, *Electrochemistry Communications*, **5**(8) (2003) 701–705.
104. A. Hayashi, R. Ohtsubo, T. Ohtomo, F. Mizuno and M. Tatsumisago, *Journal of Power Sources*, **183**(1) (2008) 422–426.

Chapter 5

The Lithium Electrode Revisited through the Prism of Li–S Batteries

Marine Cuisinier[*,‡] and Brian D. Adams[†,§]

*Qatar Environment and Energy Research Institute (QEERI),
HBKU, Qatar Foundation, Doha, Qatar
†Joint Center for Energy Storage Research, Energy &
Environment Directorate, Pacific Northwest
National Laboratory, Richland, WA 99354, USA
‡mcuisinier@qf.org.qa
§brian.adams@pnnl.gov

5.1. Introduction

With a theoretical specific energy of 2500 Wh kg^{-1} and energy density of 2800 Wh L^{-1}, the Li–S system is believed to provide the step-up in energy density necessary for Li-based battery technologies to expand from portable electronics to transportation and grid-storage applications.[1,2] These calculations are based on the one hand, on full sulfur reaction, 16Li$^+$ + 16e$^-$ + S$_8$ ↔ 8Li$_2$S, with a potential of 2.15 V vs. Li$^+$/Li, and on the other hand, the use of a perfectly balanced metallic Li negative electrode. Although the redox potential of sulfur is twice lower than that of the transition metal-based insertion materials used as positive electrodes in Li-ion cells, its specific charge or capacity (1675 mAh g^{-1}) is five to ten times higher, hence the approximate two fold increase in theoretical specific energy. However, not only does sulfur electrochemistry pose several great challenges inherent to the materials and the chemistry at play, but the trade-off made necessary by this low potential positive electrode is not trivial.

Exploring the limitations of the sulfur electrode is beyond the scope of this chapter, yet we note that current commercial Li–S batteries display a maximum specific energy of *ca.* 350 Wh kg^{-1} and energy density of 320 Wh L^{-1}.[3] Whereas tremendous research efforts made on the sulfur positive electrode have permitted to reach these values, the negative electrode (more often made of Li metal) has drawn less attention, possibly owing to the long stagnation in that field deemed unpromising. Only recently have sulfur-based electrodes improved enough to shed light on the limitations arising from the negative electrode side. Future expectations for Li–S batteries are around 500–600 Wh (per kg or L), in which case, the step–up compared to current Li-ion (150 Wh kg^{-1} and 300 Wh L^{-1}) would be effective.[4] This will unarguably necessitate improvements on each and every cell component, together with a better control of their interdependence.

The obvious choice is indeed to pair sulfur with the most electronegative electrode material, i.e. Li itself (−3.04 V vs. SHE).[4,5] However, 40 years of intensive research in rechargeable Li metal batteries have not completely overcome the two critical issues linked to Li electrochemistry in non-aqueous electrolytes which are the growth of dendrites during repeated stripping and plating as well as the low coulombic efficiency (CE) of these processes.[6–11] These are particularly severe under high current densities, which would be the case in the new generation of high-energy battery technologies such as Li/S.[12] While the low CE might be compensated by oversizing the Li electrode, it should be noted that a 300% excess amount of Li would directly result in halving the specific energy of the Li–S cell (to 1300 Wh kg^{-1}). More strikingly, the battery failure related to dendrite growth (via internal short circuit) poses serious safety hazards and, in fact, several dramatic incidents contributed to the massive focus shift of R&D towards safer Li-ion batteries (LIBs) since the early 1990s.

Nonetheless, the characteristics of the sulfur positive electrode (potential, specific capacity) incite to use metallic lithium as the negative electrode; hence the current revival of Li metal batteries (along with Na, Mg, Zn, Al batteries)[13–15] and the avid quest for effective Li metal protection.[16,17] The advantage — so to speak — of working with sulfur when trying to stabilize the Li electrode is that the surface poison (i.e. polysulfides) is present in the electrolyte at the molar concentration, not as a trace contaminant (e.g. Mn dissolution from transition metal oxides), so that the effects of any chemical or physical approach to surface protection are easily and rapidly measured.

The performance of the negative electrode in Li/S batteries is in fact strongly related to the sulfur-based positive electrode and this interdependence is best illustrated through the so-called redox shuttle mechanism of soluble polysulfide intermediates.[18] The traditional reaction pathway for the reduction of sulfur (S_8) which occurs at the positive electrode during discharge is: $Li_2S_8 \rightarrow Li_2S_6 \rightarrow Li_2S_4 \rightarrow Li_2S_2 \rightarrow Li_2S$. The final reduction product (Li_2S) is insoluble in the electrolyte, however, driven by a concentration gradient, the longer chain polysulfide ions (Li_2S_8, Li_2S_6, and/or Li_2S_4) readily diffuse from the positive to the negative electrode, where they may undergo both chemical (Figure 5.1, Equation (1)) and electrochemical reduction (Figure 5.1, Equation (2)), to form either shorter polysulfides or precipitate as Li_2S.

In the former case, if the polysulfides are still soluble, they can diffuse back to the positive electrode. During charge, these short-chain polysulfides are then electrochemically reoxidized into long-chain

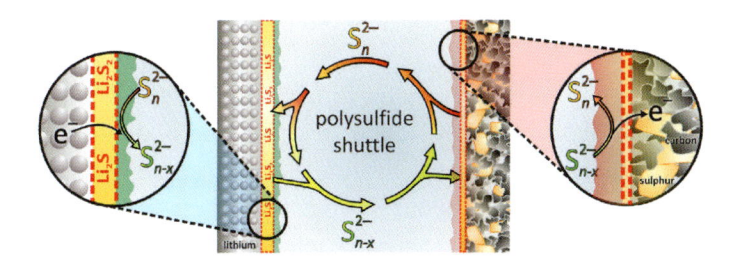

Figure 5.1 Schematic illustration of the parasitic interfacial reactions involving soluble polysulfides in a Li–S cell, reproduced from Ref. 19 (Equation (1)) describes self-discharge, while (Equations (2) and (3)) describe the shuttle effect observed in charge. (Equation (4)) explains the consumption of interphasial Li_2S (formed via Equations (1) and (2)) by chemical equilibrium with soluble polysulfides.

polysulfides (Figure 5.1, Equation (3)), which diffuse again to the negative electrode in a possibly endless process, creating an internal shuttle phenomenon. This redox shuttle has direct consequences, such as a low CE (overcharge) and a severe propensity of Li–S cells to self-discharge (Figure 5.1, Equation (1)). However, the underlying cause of the shuttle is the instability of the negative electrode/electrolyte interface evidenced in both (Figure 5.1, Equation (1)) and (Figure 5.1, Equation (2)): metallic Li is not passivated by a solid electrolyte interphase (SEI) as detailed hereafter.

$$(n-1)Li_2S_n + 2Li^0 \rightarrow nLi_2S_{n-1}, \qquad (1)$$

$$(n-1)Li_2S_n + 2Li^+ + 2e^- \rightarrow nLi_2S_{n-1}, \qquad (2)$$

$$2nLi_2S_{n-1} \rightarrow 2(n-1)Li_2S_n + 4Li^+ + 4e^-, \qquad (3)$$

$$Li_2S + Li_2S_n \leftrightarrow Li_2S_k + Li_2S_{n-k+1}. \qquad (4)$$

In this chapter, we aim to show how the promise of high-energy density sulfur-based batteries (among others) has revitalized the research on Li negative electrodes. Also, and more specifically, how sulfur electrochemistry impacts interfacial processes, fading and failure modes at the negative electrode. After reviewing novel characterization techniques which allow for the examination of air-sensitive Li samples, we carefully examine what has been attempted to stabilize Li electrodes in Li–S cells. The various strategies reported in the literature for protecting Li electrodes are segregated as: (1) electrolyte-based approaches for SEI formation and stabilization, which are often quite specific to the Li/S system, (2) morphology control upon stripping and plating for dendrite inhibition or microstructure stabilization, which can be extrapolated to other Li metal batteries, (3) chemical protection layers on Li to isolate the Li surface from direct contact with the electrolyte (and polysulfides in solution), and finally (4) physical isolation of the Li electrode, which eventually opens the door for aqueous Li–S and countless other promising battery systems.

5.2. Chemistry at the Li/Electrolyte Interface

Li reacts spontaneously upon contact with all polar aprotic solvents, and most of the carrier salt anions to form insoluble Li salts, which precipitate at the Li electrode/electrolyte interface. This reaction is driven by the difference between the Fermi energy of Li metal and the LUMO of the various electrolyte ingredients,[3,5,10,20] and therefore proceeds at vastly different rates so that the surface film formed displays a heterogeneous, multilayered microstructure: the so-called SEI.[6,21]

The concept of SEI was introduced by Peled, who suggested that the first layer that precipitated directly in contact with the negative electrode (the SEI) is thin, dense, and made up of species that are thermodynamically stable with respect to Li.[6] A good SEI is in fact one of the most critical factors allowing the cyclability of Li-based secondary batteries; it stabilizes the negative electrode/electrolyte interface by passivating the surface of Li, hence stopping electrolyte depletion via irreversible reduction reactions. To fulfill this role, the SEI should ideally exhibit high Li^+ conductivity, high density to inhibit electron hopping yet minimal thickness (i.e. requiring minimum amount of active Li to build up) but also high elasticity to prevent the outburst of Li dendrites.

5.2.1. Baseline interfacial processes on metallic Li

Owing to its reactivity towards atmospheric components, the Li electrode is usually covered with a native film comprised of Li_2O, LiOH, and Li_2CO_3.[22] When it comes into contact with most non-aqueous solutions, partial dissolution of the native film often occurs,[23] as well as possible nucleophilic reactions of the Li_2O, LiOH species with electrophilic esters or alkyl carbonates solvent molecules.[24] Consequently, this native film is replaced by surface species originating from reaction of electrolyte species with the alkali metal, in time constants of milliseconds and less.[25] Depending on the electrolyte, the low oxidation state species formed in direct contact with the Li surface include Li_2O, LiF (from all Li salts with perfluorinated anions), or Li_3N;[6,10,26] in principle this SEI is therefore electronically insulating yet ionically conductive.

In practice however, efficient passivation of metallic Li in a non-aqueous electrolyte cannot be achieved, as no SEI is apt to sustain infinite volume changes at the negative electrode and thermodynamic instability of Li cannot be defeated. Upon Li plating and stripping, inner surface films comprised of ionic species especially cannot accommodate the morphological change of the Li surface, and thus are broken down. Not only both Li and solution components are lost upon repeated reactions, this also leads to highly non-uniform current distribution so that Li deposition tends to form dendrites, as illustrated in Figure 5.2(a).

Besides the obvious safety problems (shorting, formation of high area reactive Li, etc.), the exact chemical composition and morphology of the SEI evolve with the history of the Li cell (storage and cycling conditions), resulting in the degradation of its conduction properties leading upto cell failure. A summary of SEI components from exposure of Li surfaces to various solvent/salt combinations is displayed in Table 5.1.

Further reduction of electrolyte species takes place under more selective conditions via electron transfer through the inhomogeneous

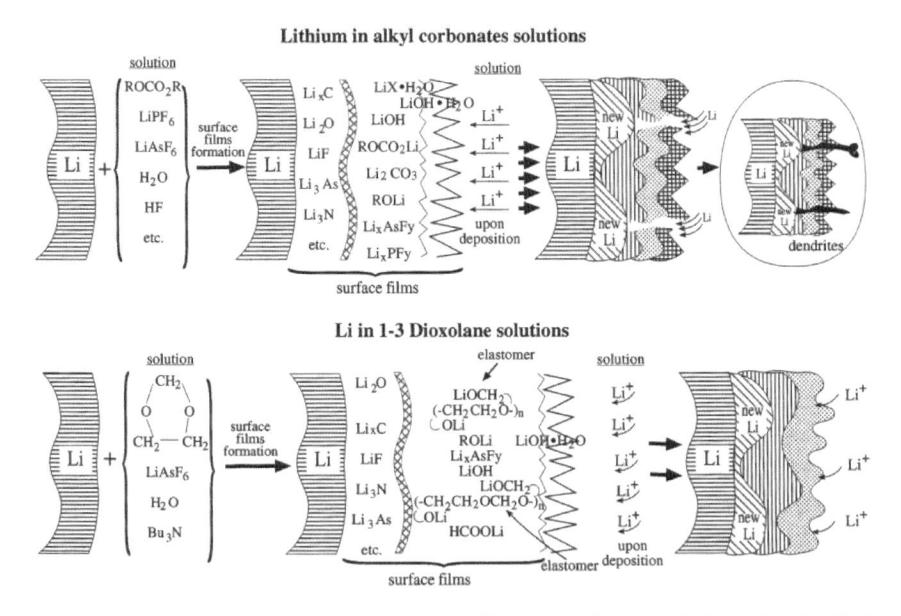

Figure 5.2 Schematic illustration of surface film formation on Li electrodes in alkyl carbonates (a) and in 1–3 dioxolane (DOL) solutions (b).

Source: Reproduced from Ref. 26.

Table 5.1: Chemical components of passivation films on Li metal.

Electrolyte	Passivation film
Without electrolyte (native film)	LiOH, Li$_2$CO$_3$, Li$_2$O
General PC systems	CH$_3$CH(OCO$_2$Li)CH$_2$OCO$_2$Li
PC/LiPF$_6$	LiF, Li$_2$O, LiOH
PC/LiClO$_4$	Li$_2$CO$_3$, LiOH, Li$_2$O, LiCl, ROCO$_2$Li, LiCHClCHCl, LiCH$_2$CHClCH$_2$Cl
General EC systems	(CH$_2$OCO$_2$Li)$_2$
General DMC systems	CH$_3$OCO$_2$Li
General DEC systems	CH$_3$CH$_2$OCO$_2$Li, CH$_3$CH$_2$OLi
SO$_2$/LiAlCl$_4$	Li$_2$S$_2$O$_4$, Li$_2$SO$_3$, LiSnO$_6$, Li$_2$S$_2$O$_5$
General THF systems	BuOLi
THF/LiAsF$_6$	BuOLi, RLi, –As–O–As–, ROLi, –O(CH$_2$)$_4$-THF$^+$, F$_2$–As–O–As–F$_2$
General 2MeTHF systems	Li pentoxides
2MeTHF/LiAsF$_6$	(–As–O–)$_n$, LiAs(OR)$_n$F$_{6-n}$, AsO$_n$F$_{5-n}$, As(OR)$_n$F$_{3-n}$
General DME systems	CH$_3$OLi
DME/LiAsF$_6$	CH$_3$OLi, LiF
General BL systems	C$_3$H$_7$OCO$_2$Li, cyclic β-keto ester
General DOL systems	CH$_3$OLi, C$_2$H$_5$OLi, LiOC$_2$H$_4$(OCH$_2$)$_n$OX (X = OLi, H, OR)

Source: Reproduced from Ref. 27.

SEI. In this upper layer, there is dynamic precipitation and dissolution of surface species with higher oxidation states, forming a porous polymeric matrix made of solvent reduction products such as Li alkyl carbonates (ROCO$_2$Li, ROLi, RCOO$_2$Li, etc.) in the case of alkyl carbonates,[26] with inclusions of inorganic Li salts (LiOH, Li$_x$PF$_y$...). As the surface films become thicker, the driving force of the solution–lithium reactions (the difference in Fermi energy) decreases, until passivation is finally reached. The resulting porous structure governs the overall mechanical and conduction properties of the interphase.[10] By extension, this whole negative electrode/electrolyte interphase is often referred to as the SEI.

This complex topic is well documented in the context of Li metal batteries incorporating a classical insertion-type positive electrode,[16] and explains the persistent dominance of Li-ion technologies based on low energy density graphite negative electrodes instead.

5.2.2. Specificities of the Li-S system

5.2.2.1. *Interphase formed on Li in DME:DOL solutions*

In the Li–S system, the strong nucleophilic character of polysulfide dianions and radicals precludes the use of carbonate solvents commonly used in LIBs.[28–31] Among the different solvents explored in the early Li–S works, ethers have since become dominant.[32] Despite their oxidative breakdown at relatively low potentials (4.0 V vs. Li$^+$/Li), they can safely be operated within the Li–S window (0–3 V vs. Li$^+$/Li) and possess favorable physico-chemical properties. In their recent statistical overview of the Li/S literature, Hagen *et al.* observed in particular that a 1,3-DOL and 1,2-dimethoxyethane (DME) combination was used in 43% of all 274 publications over the last 12 years.[33] The binary mixture DOL:DME was historically developed by Sion Power for its good rate capability at low temperatures:[34] DME exhibits a relatively high dielectric constant and low viscosity, as well as high polysulfide solubility while DOL generates a flexible SEI on the surface of metallic Li by reductive cleavage.[35]

In fact, as reported by Youngman *et al.* DOL polymerizes via an anionic mechanism to form oligomers of polydioxolane.[36] These polyDOL oligomers are insoluble and adhere to the Li surface via alkoxy edge groups (-OLi), while their elasticity makes the SEI much more flexible than the above described fully ionic surface layers formed on Li in other electrolyte solutions. Hence, when Li is deposited or dissolved in these systems, the flexible surface films can better accommodate the volume changes of the active metal, thus providing good passivation to the Li deposits. This concept of flexible passivation film depicted in Figure 5.2(b) provides an explanation for the improved reversibility of Li electrodes in DOL solutions. Indeed, cycling efficiency is very high,[37] and Li deposition morphology very smooth[38] which eventually allowed the use of DOL-containing electrolytes in commercial rechargeable Li batteries.[39,40]

It should be noted that DOL may also polymerize through a variety of processes: cationic (in presence of acidic traces), and/or radical when exposed to sulfur intermediates during cycling.[30,41] Therefore, to prevent the bulk electrolyte from polymerizing (increasing viscosity and degradation of transport properties), binary mixtures are always used in Li–S electrolytes.[42–44] The second solvent must exhibit high solvation properties and is typically a glyme, DME or sometimes tetraethylene glycol dimethyl ether (TEGDME). Aurbach and Granot demonstrated that 1 M electrolytes with shorter chain glyme solvents (DME, 1,2-diethoxyethane and diglyme (G2)) and various salts (LiAsF$_6$, LiBF$_4$, LiClO$_4$, LiCF$_3$SO$_3$, LiTFSI, LiI, and LiBr) resulted in poor Li cycling efficiencies due to the formation of Li alkoxy species (ROLi), which covered the Li surface.[45] These glyme solvents were, however, noted to be less reactive than cyclic ethers, esters and alkyl carbonates,[45,46] and the only product formed from the surface reaction of Li and the shortest chain glyme, DME, is Li methoxide (CH$_3$OLi, Table 5.1).

5.2.2.2. Influence of LiTFSI salt

Regarding the improved performance of Li electrodes in DOL solutions reported in the late 1990s, it should be noted that the Li salt was also playing an important role. Comparing SEM micrographs obtained from Li electrodes after being cycled several times (Li dissolution, deposition) in DOL solutions of LiAsF$_6$ and Li bis(trifluoromethane)sulfonyl imide (LiN(SO$_2$CF$_3$)$_2$ or LiTFSI), Aurbach et al. observed much rougher Li morphology in the latter case. These results correlated well with the Li cycling efficiency measurements (98 and 90% for LiAsF$_6$ and LiTFSI, respectively).[47] Despite the better performance using LiAsF$_6$, its use has been inhibited today owing to the release of highly toxic arsenic compounds and because it is relatively expensive.[8]

According to Hagen et al. a consensus was reached around LiTFSI (57% of all electrolytes) salt in Li–S literature works.[33] This can be explained by its non-coordinating, high dissociation ability[48] and good compatibility with ether solvents. Solutions of 1 M concentrations are commonly used, however, this concentration might be due to convention only. While the high thermal stability of LiTFSI seems to be of limited relevance when combined with volatile DOL:DME (1:1), its well-documented decomposition on the surface of both Li and non-active metals — with an onset reduction potential between 2.0 and 1.5 V vs. Li$^+$/Li — makes this widespread use in low voltage Li/S batteries questionable.[47]

Like any other Li salt with a perfluorinated anion (Table 5.1), LiTFSI reduction results in LiF deposition at the surface of Li. Historically, LiF-rich SEI on Li have been desired for their grain, lumpy surface leading to high cycling efficiency.[26] The N–SO$_2$ groups are the most reactive centers of these anions toward electron transfer or nucleophilic attack and therefore LiTFSI also produces Li$_3$N, Li$_x$CF$_y$, Li$_2$NSO$_2$CF$_3$, LiSO$_2$CF$_3$, and Li$_x$SO$_y$ interphasial species.[8] However, their precipitation in the surface films formed on Li may be too inhomogeneous to achieve the passivation of Li needed to avoid any electron transfer to polysulfides in solution (Figure 5.1, Equation (2)).[35] Besides, these salt reduction processes actually compete with DOL reductive polymerization, so the more reactive the salt, the rougher the Li-morphology upon cycling, and the lower the Li cycling efficiency.[47]

As a matter of fact, salts that are most typically used in Li-ion cells are not really suitable either: DOL is prone to ring-opening polymerization in the presence of acidic impurities which are inherent to LiPF$_6$ and LiBF$_4$, whereas LiClO$_4$ may pose safety issues owing to its oxidative properties.[49] However, other Li salts such as Li bis(fluorosulfonyl)imide (LiFSI) or Li triflate (LiCF$_3$SO$_3$ or LiOTf) could be employed just as well as the LiTFSI, with differences in terms of ionic conductivity, viscosity, and cost.[50] There are very few reports assessing the impact of the Li salt on the performance of Li–S cells.[29,51] In fact, the influence of the salt can be twofold: regarding sulfur electrochemistry in solution first, the salt plays a role in terms of reactivity towards polysulfides (e.g. LiBF$_4$, LiFSI) and in terms of donor ability (e.g. LiNO$_3$ vs. LiTFSI).[52] In the present chapter however, we focus on the negative electrode side, where as mentioned in the introduction, the morphology and the chemistry of the interphase (and therefore its conduction properties) are strongly dependent on the salt used, based on their reduction potential and the solid products formed.

5.2.2.3. Interphase formed in an "uncontrolled" Li–S cell

The chemical and electrochemical reactions shown in (Figure 5.1, Equation (1)) and (Figure 5.1, Equation (2)) between polysulfides in solution and the negative electrode can precipitate insulating Li$_2$S onto its surface, leading to both active material loss and impedance increase.[10,53,54] Specifically in the case of Li metal, the self-discharge reaction (Figure 5.1, Equation (1)) not only oxidizes but also corrodes the negative electrode, thus altering its microstructure towards a mossy/powder-like roughness which increases significantly its surface area.[3,55] This in turn eventually leads to cell swelling and electrolyte redistribution to fill the newly created porosity (thus drying the positive electrode and separator),[3] but it also aggravates the thermal and chemical instability of Li metal towards the electrolyte.[10,20]

The extent of the self-discharge reaction in DOL based solutions was investigated in 1989 by Peled *et al.* who immersed Li metal in various solvent mixtures and electrolytes at 70°C and calculated corrosion rates.[42] In the absence of soluble polysulfides, no Li was consumed even after 90 days (1%), whereas up to 19% loss of the Li mass was observed in the presence of 0.1 M Li$_2$S$_6$ or Li$_2$S$_8$ after two months via (Figure 5.1, Equation (1)), with only minor effect of the volume ratio between tetrahydrofuran (THF), toluene and DOL in the electrolyte. The addition of sulfur powder in contact with the polysulfide containing electrolytes was found to dramatically increase the Li corrosion rates initially (above 80% loss after three weeks), although the values decreased with storage time, indicating the eventual equilibration between the solid and solution phases. It was already established that Li$_2$S precipitated within the SEI, as the final product of (Figure 5.1, Equations (1) and (2)), so that the accelerated corrosion observed in the presence of sulfur can be explained by the complex interplay of chemical (anti)dismutations consuming the interphasial Li$_2$S (Figure 5.1, Equation (4)), and therefore exposing the Li underneath which can get further corroded. This experiment was in fact quite illustrative of a Li–S cell at rest, and correlates well with the severe self-discharge observed prior cycling in the absence of sulfur confinement in the positive electrode or electrolyte additives.[56]

Recently, Aurbach et al. showed that the addition of Li_2S_6 to a DOL/LiTFSI solution remarkably decreased the impedance at the Li electrode/electrolyte interface.[35] Using state-of-the-art FTIR and XPS analyses, this pronounced effect of Li_2S_6 on the Li surface chemistry was interpreted as leading to the formation of a different interphase, with improved transport properties for Li^+ compared to that formed in DOL/LiTFSI. Based on the FTIR spectra, it appeared that the presence of polysulfides in solution attenuated some of the Li interfacial reactivity related to the reduction of $TFSI^-$ anions, while the S_{2p} core spectrum of Li stored in the Li_2S_6-containing solution revealed the presence of Li_2S, and maybe Li_2S_2 in the interphase.[35] In fact, comparing the reduction potential of LiTFSI (1.5–2.0 V vs. Li/Li^+, see above) to that of Li_2S_n, polysulfides are expected to get reduced preferentially, (Figure 5.1, Equation (2)), therefore minimizing $TFSI^-$ decomposition at the interface. Nevertheless, this study confirmed Peled's observations;[42] because several polysulfides are soluble in solution (e.g. Li_2S_n with $2 < n \leq 8$), chemical reactivity (Figure 5.1, Equation (1)) and/or electron-transfer processes through the interphase (Figure 5.1, Equation (2)) do not effectively passivate the Li electrode through reductive precipitation.[35] In the end, the decreased impedance observed at the Li/electrolyte interface may not be attributed to a better interphase, but instead to its continuous detachment/dissolution upon Li_2S oxidation by polysulfides in solution (Figure 5.1, Equation (4)).

This dynamic precipitation/dissolution of the SEI on the Li electrode is maintained upon cycling. Xiong et al. observed that it maintained a stable impedance and potential change in symmetrical Li|catholyte|Li cells as shown later in Figure 5.4(b).[57] This suggested that the interphase structure came to equilibrium after some time. Yet again, this does not mean that passivation of the Li surface is ever reached, as proven by the redox shuttle observed during charge. Instead, electrolyte solvents and salts are continuously consumed at the Li/electrolyte interface,[10] which eventually leads to electrolyte depletion (the cell dries up). Depending on the extent of these parasitic reactions, solvent degradation products with -OLi edge groups are generated and if soluble, can become a source of oxygen in Li/S cells. This could explain the formation of Li_xSO_y species in the positive electrode by irreversible oxidation of the sulfur active material.[54]

In this sense, the absence of a stable SEI on the Li electrode has irreversible detrimental impact on each and every active component of a Li–S cell *in spite of* the equilibria between Li$_2$S and polysulfides which prevent sulfur accumulation onto the negative electrode. Li itself gets corroded, leading to a mossy microstructure with increasing surface reactivity, electrolyte salts and solvents are consumed, and their decomposition products oxidize sulfur into inactive sulfate species.

5.3. Development of Characterization Techniques to Examine Li Metal

Several characterization techniques have been developed to visualize and/or analyze the complex composition and multilayer structure of electrode/electrolyte interphases in batteries. The most useful tools to study surface chemistry are fourier transform infrared (FTIR) and photoelectron X-ray (XPS) spectroscopies, while the morphology and microstructure evolution upon aging/cycling are typically monitored by scanning electron microscopy (SEM). The high reactivity of Li metal, its low molecular weight and the dynamic nature of the interphase each pose specific challenges that are being addressed more and more rigorously and successfully along with the recent advances in *in situ* and *operando* technologies.

5.3.1. Technical challenges of Li characterization

5.3.1.1. *Handling precautions*

Most of the historical work of Aurbach *et al.* in the late 1980s was performed with the FTIR spectrometer isolated from the room atmosphere and purged with dry air or nitrogen.[58] Nowadays, it is commonly admitted that further precautions are needed and FTIR instruments are often placed in Argon-filled glove boxes, to avoid any contamination from H$_2$O and CO$_2$ during the measurements. This enables experimenters to increase the acquisition time, and therefore, improve the signal to noise ratio.[35] Sample transfer is also a major concern, and while instrument manufacturers increasingly offer commercial devices, many research groups have successfully developed home-made transfer systems such as hermetically closed FTIR cells with KBr windows or articulated manipulators equipped with gate valves and vacuum flanges to fit the entry ports of SEM[58,59] or XPS instruments.[35] For XPS measurements in particular, not only is the depth probed limited to *ca.* 10 nm, but also the extreme surface species are overrepresented in the quantification so that absolutely no air exposure should be tolerated. In the following, we therefore focus on literature in which strict handling precautions were taken.

5.3.1.2. Beam damages and sulfur sublimation

Another challenge faced when characterizing Li is its low molecular weight, and therefore low cross section. High beam doses easily damage the surface films, and very cautious working conditions such as ultra-low acceleration voltage are required during SEM, transmission electron microscopy (TEM), or XPS analyses. It should also be noted in the specific case of Li/S batteries, elemental sulfur readily sublimes under strong vacuum conditions, which was found to create porosity in sulfur-based positive electrodes under the SEM for instance.[60] Yet, Wenzel *et al.* showed that cooling down samples to −80°C enabled proper XPS analyses with a chamber pressure in the range of 10^{-7} Pa.[61] Based on the interfacial reactions detailed in Section 1, one would not expect to find elemental sulfur at the surface of the negative electrode so that sublimation should not be an issue, yet polysulfide intermediates might be disproportionate owing to beam radiation to form species which are more stable in the solid state such as elemental sulfur and Li_2S, or even oxidize to sulfates.[62] Therefore, a commonly applied strategy consists of repeating short measurements of each sample several times rather than increasing acquisition time.[35] In the absence of significant changes in the spectra, those can still be averaged to gain resolution, while validating the operating condition chosen.

5.3.1.3. Sensitivity towards light elements and chemical contrast

Last, X-ray and electron-based techniques intrinsically show little to no sensitivity towards Li, and poor contrast is achieved in electron microscopy between the light elements that constitute the SEI. To circumvent this limitation, Zier *et al.* exposed a cycled graphite negative electrode with its SEI to osmium tetroxide (OsO$_4$) vapor.[63] The OsO$_4$ oxidant was found to react differently with the different SEI components, for example, components like LiF or Li$_2$O — showing little or no reaction — appeared as transparent phases (i.e. unstained). At the opposite end, osmium tetroxide preferably reacted with Li dendrites since metallic Li is a strong reducing agent, which led to considerably higher osmium concentrations. Li dendrites therefore appeared in high-angle annular dark-field imaging (HAADF) as very bright regions, while by EDX, osmium content was eighteen fold higher compared to the rest of the SEI. In addition to tremendously improving contrast within the interphase in electron microscopy, staining by OsO$_4$ vapor minimized the damage inflicted on the sample by the electron beam in the SEM and TEM, and also stabilized lithiated graphite and Li samples under ambient air. The main downside of this approach is of course that osmium tetroxide is highly toxic, corrosive and volatile, and also the exact reaction mechanism remains ambiguous. In the case of Li–S batteries, an alternative idea to enhance poor elemental contrast at the Li surface when investigating failure of the negative electrode could be to replace sulfur by selenium,[64] in order to exploit its higher molecular weight and higher boiling point.

5.3.2. Characterization tools moving towards *in situ* and *operando*

5.3.2.1. *Imaging by electron microscopies*

As the interphase at the negative electrode evolves continuously upon stripping and plating of Li or reaction with soluble redox species, it is essential to characterize the negative electrode and its surface in a working cell, either *in situ* (static mode, i.e. under OCV) or *operando* (dynamic, i.e. during cycling). Imaging the electrode/electrolyte interface by electron microscopy is complicated by the presence of the liquid, which is also prone to degradation (formation of bubbles and/or precipitates). After carefully calibrating the electron beam dose to be below the damage threshold, it is now possible to make truly quantitative *operando* observations and couple high spatial resolution measurement with real time quantitative electrochemistry using modern liquid stages in an electrochemical scanning transmission electron microscopy (ec-STEM).[65,66] Sacci et al. demonstrated the use of chemically sensitive annular dark-field (ADF) STEM imaging to determine density changes in the constituents electrodeposited onto a glassy carbon working electrode, to reveal that the SEI is approximately twice as dense as the liquid electrolyte.[67] Using a similar approach, Mehdi et al. monitored SEI formation and Li dendrite evolution during Li plating onto a Pt electrode.[68] From STEM images formed by mass–thickness contrast, they managed to quantify the mass of Li involved in electrodeposition and building of the SEI layer and to correlate these values as a function of the total current passed through the cell.[68] It should be noted that these recent studies were conducted in typical non-aqueous electrolytes such as $LiPF_6$ in EC:DMC[67] or PC,[68] rather than solid or ionic liquid electrolytes as were used just a couple of years ago. The implications of *operando* STEM liquid-cell imaging for Li-battery applications are tremendous, and as we see these experiments being performed under more and more realistic conditions, we will be able to move from a "demonstration stage" to answering real questions. *In situ* observations can also help to clarify the electrochemical Li deposition and dissolution mechanism in Li metal batteries with a solid electrolyte. For example, Nagao et al. observed by SEM that when using Li_2S–P_2S_5 glass-ceramic electrolyte, electrodeposited Li grew along grain boundaries inside the electrolyte layer, and then inside cracks generated locally, to finally cause short circuit.[69] In fact, we expect mechanical failure and dendritic growth to be the major limitation in Li–S cells incorporating a protected Li electrode (PLE, see Section 6.3).

5.3.2.2. *Spectroscopies*

Not only microscopy capabilities are improving drastically, but essentially all characterization methods can now be performed under more realistic cycling conditions. For instance, Cheng et al. had designed a proper *in situ* spectro-electrochemical FTIR instrument and demonstrated its feasibility for the study of the interface between Li and a solid polymer electrolyte (SPE) in 2007,[70] while the *ex situ* work of Aurbach et al. on Li electrodes in contact with polysulfide containing electrolyte clearly shows that FTIR is sensitive to sulfur-containing interphasial species.[35] Even more exciting, we anticipate that the development of hard X-ray photoelectron spectroscopy (XPS) under high pressure (HP-HAXPES) will eventually yield to the *operando* characterization of solid–liquid interfaces in Li batteries,[71] to provide insight to intermediate (electro)chemical processes.

In situ nuclear magnetic resonance (NMR) spectroscopy is a useful and non-invasive tool to study electrochemically induced structural changes, and despite the number of technical challenges, it has now been implemented by a number of groups.[72] Resonances from diamagnetic Li (in the electrolyte, in the interphases but also in Li_2S and Li_2S_n) appear at approximately ±10 ppm, while metallic Li is shifted to approximately 250 ppm, by a Knight shift mechanism.[72] It is therefore easy to discriminate the negative electrode from everything else in Li/S cells.[73] Bhattacharyya et al. first applied *in situ*[7] Li NMR spectroscopy to the Li negative electrode by collecting time-resolved, quantitative information about the growth and stripping of Li microstructures (moss, dendrites) during electrochemical cycling.[74] To understand how this was made possible and to explore the morphology of the Li electrodes, it should be noted that the bulk-metal signal is proportional to the area of the metal, and not to its volume (the radiofrequency only penetrates a finite thickness, the so-called "skin depth" calculated at 15 μm under the given experimental conditions). In contrast, mossy or dendritic Li microstructures with thickness around 1–2 μm are fully penetrated, so that unlike the bulk, their NMR signal from these Li microstructures is directly proportional to their volume or mass, hence the ability to separate it from that of bulk Li. And, in fact, dendrites were seen by NMR before they became apparent in the electrochemistry. Moreover, noting that the growth of Li dendrites was associated with the observation of an additional resonance, the authors demonstrated the strong effect of bulk magnetic susceptibility (BMS), which results in dependence of the chemical shift (with [7]Li peak ranging between 250 and 270 ppm) on the orientation between the metal and the direction of the static magnetic field. The additional resonance appearing during Li electrodeposition was therefore assigned to dendritic or mossy Li growth perpendicular to the surface of the electrode.[74]

212 *Li–S Batteries: The Challenges, Chemistry, Materials and Future Perspectives*

More recently, Xiao *et al.* applied *in situ*[7] Li NMR spectroscopy to a functioning Li–S cell held within an inhouse cylindrical microbattery.[75] In Figure 5.3, the evolution of Li microstructures can be followed with the variation of integrated intensities under peaks 5–8. These show periodic decrease and increase in good agreement with stripping and plating, respectively, yet the maxima are not reached for fully charged states, but rather on 2.3 V plateau. The microstructural evolution of the Li metal electrode is in fact entangled with parasitic reactions involving sulfur intermediates (see Figure 5.1): the surface area of metallic Li increases during charge because the plated Li deposit is not compact, especially at the beginning of the high-voltage plateau where the concentration of long-chain polysulfides is maximum. The slight decrease in the surface developed by Li microstructure over the high-voltage plateau was interpreted by the authors as the densification of the negative electrode: dendrites are

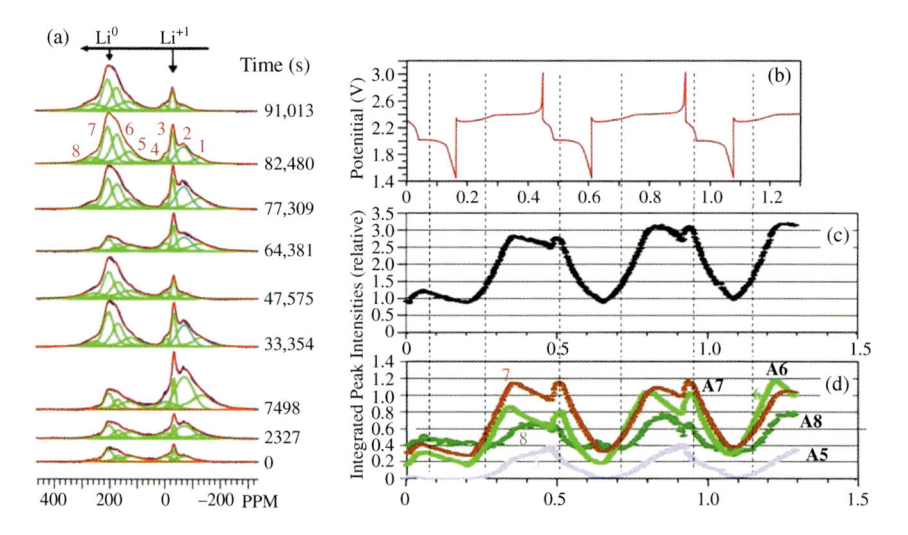

Figure 5.3 (a) Representative *in situ* ^{7}Li NMR spectra as a function of time during discharge/charge of a Li–S cell. Major peaks of interest within the range of −260–300 ppm are distributed between two broad peaks from −260 to 100 ppm (diamagnetic Li, peaks 1–4) and from 100 to 300 ppm (metallic Li, peaks 5–8). (b) Voltage profile of the Li/S cell used for the *in situ* NMR measurement. (c) Evolution of the overall integrated area under the broad corresponding to Li with a Knight shift and (d) individual integrated areas for peaks 5–8.

Source: Reprinted and adapted with permission from Ref. 75. Copyright (2015) American Chemical Society.

consumed via interphasial reactions (Figure 5.1, Equations (1) and (2)), even though the pristine density is never fully recovered. Overall, the total of ^{7}Li signals corresponding to electronegative Li accumulates over repeated cycles, indicating the buildup of porous microstructures at the negative electrode (Figure 5.3(c)), which is supported by the observation of severe Li corrosion by SEM.[55,75,76] In future development, we anticipate the application of ^{7}Li magnetic resonance imaging (MRI)[77] to truly indicate the location and change of microstructural Li morphology in Li–S cells, without any of the beam-induced complications met with *in situ* TEM (see above).

5.3.3. Electrochemical methods

With the development of sophisticated and highly impressive *operando* microscopy techniques, we are observing the decline in the battery literature of traditional electrochemical characterization, wrongly thought to be too obscure for the average reader. However, some of these experiments are not only accessible to anybody owing a potentio/galvanostat, but also unrivaled in terms of experimental conditions reproducing that of "real" or "practical" battery operation.

5.3.3.1. *Li metal plating/stripping*

The cycling stability of plating/stripping is the most relevant phenomenon to monitor when optimizing the Li electrode. To isolate the relevant aspects of aging, symmetrical cells that use two electrodes of Li foil that are fundamentally infinite Li sources are typically assembled (see Figures 5.4 and 5.13(a) for examples). Cycling capacity is limited for both charging and discharging processes, and figures of merit consist in voltage hysteresis and cycle life.[46,57] In this type of cell, observations of short-circuiting due to dendrite growth can be followed by erratic voltage behavior as current pathways are broken and reformed.[78] Alternatively, a short-circuit may cause a significant *decrease* in the polarization if direct solid-to-solid contact is made internally between the two electrodes. In this scenario, the direct passage of electronic current between the metallic electrodes results in a voltage which is strictly related to the resistance of the circuit (i.e. the

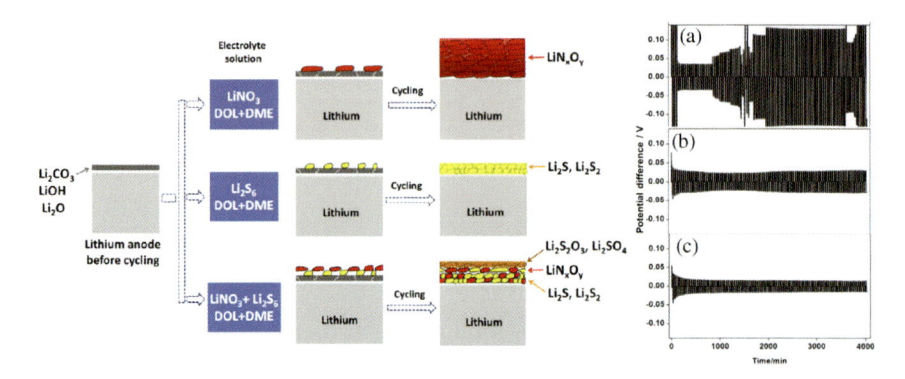

214 Li–S Batteries: The Challenges, Chemistry, Materials and Future Perspectives

Figure 5.4 Illustration of the surface film behavior on Li electrode upon contact and cycling in different electrolyte solutions. Cycling behavior of symmetrical cells with the electrolytes (a) DOL:DME/LiNO$_3$ (0.2 M), (b) DOL:DME/Li$_2$S$_6$ (0.2 M), and (c) DOL:DME/Li$_2$S$_6$ (0.1 M) + LiNO$_3$ (0.1 M).
Source: Reproduced from Ref. 57.

IR drop). This phenomenon is usually observed when more dense deposition morphologies are formed, rather than dendrites, as in the work of Harry *et al.* who observed globular Li structures.[79]

Asymmetric cells, using copper or stainless steel as working electrodes for the deposition and stripping of Li, are also extensively used. The efficiency of Li plating/stripping can be directly obtained from the CE which is the ratio of the total charge for each process. This simple coulometry technique allows the amount of Li loss during cycling to be accurately determined. Here, the amount of Li removed from the working electrode (Q_{strip}) is recorded relative to that deposited (Q_{plate}) during the same cycle.[46,80] A fade in the CE indicates that less and less Li can be stripped from the working electrode, meaning that some of the Li deposited previously reacted with the electrolyte,[10] and therefore cannot be recovered during the given stripping process. Cell failure can be observed as the same as with symmetric cells. Voltage hysteresis (or cell polarization) typically increases with the growth of an insulating SEI, and complete cell death is characterized by erratic voltage spikes or the observation of overcharging during the stripping process (i.e. internal electronic current pathways).

When adding 0.18 M Li$_2$S$_8$ to a LiNO$_3$-containing DOL:DME electrolyte, Li *et al.* reported that the CE improved from 95.6 to 98.1%.[80] The ultimate target to enable Li metal batteries in general is to exceed 99.8%. This poses a novel technical challenge since the coulometer used must certify this high level of accuracy[81,82] and there are very few commercial systems which are capable of high CE precision measurements.[81] With Li–S specificities, high precision CE measurements in the presence of polysulfides (as in the work of Li et al.[80]) will become extremely important and relevant. In this aspect, along with the development of ultra-low current potentio/galvanostat equipment for *operando* microscopic techniques, the use of ultra-high precision coulometers should be expanded.

5.3.3.2. Impedance spectroscopy

Electrochemical impedance spectroscopy (EIS) is a common technique used to characterize SEI. EIS results have become widely published since it is easy to carry out these experiments and is routine in electrochemistry labs. For example, the ionic conductance of SEI layer can be obtained by the EIS experiments.[83] However, the most troublesome step of EIS experiments is the interpretation of the original spectroscopy, and this is a challenge to even the most experienced electrochemists for complex systems. Historically, this technique was used on planar electrodes in three-electrode cells with model electrolytes containing minimal contaminant species. Typically, impedance spectra are fit with an equivalent electrical circuit, however, very few models can clearly describe and obtain a theoretical fitting between the calculated and experimental impedances for porous electrodes coupled with mixed electrolytes (e.g. containing polysulfides) in two-electrode configurations.[84] Although routine EIS measurements can be easily acquired on assembled Li–S batteries, the results should not be over-interpreted. Without the use of a reference electrode, the impedance-forming layers (i.e. SEI) at the Li electrode cannot be decoupled from processes occurring at the positive electrode, unless symmetrical cells are used.[85] Woo et al. observed three semicircles in their Nyquist plots, suggesting that at least three interfaces coexist in the Li/Li cell. Modeling the EIS results with an equivalent electrical circuit consisting of an ohmic resistance in series with three R//CPE elements (R and CPE connected in parallel), they proposed distinct Li structures interface with the liquid electrolyte namely (i) bulk Li (at 10^4 Hz), (ii) cycled Li underneath the SEI (most prominent phenomenon, at

10^2 Hz), and (iii) cycled Li in the outer part of the SEI (at 10^0 Hz).[85] In contrast, in their two-electrode Li/S cell, Cañas *et al.* only attributed one semicircle to the Li/electrolyte interface (at 10^4 Hz) while interfacial phenomena at lower frequencies (e.g. at 10^1 and 10^{-1} Hz) were assigned to sulfur species at the positive electrode/electrolyte interface.[84] This is just one example of how interfacial phenomena get entangled without the proper experimental setup. Fundamental EIS experiments which carefully describe one specific interface along with computational modeling would also be a step forward in describing the underlying chemistry at Li electrodes.

The specificity of the Li/S system — with the presence of sulfur intermediates in solution — makes it irrelevant to investigate and/or characterize the negative electrode by itself. In addition to the challenges inherent to metallic Li, one must also consider the volatility of sulfur and the coexistence of species with various oxidation states and different physical states. Direct observation methods such as *operando* studies must demonstrate control experiments and clearly state the limitations, otherwise research groups will keep contradicting (as seen for x-ray absorption spectroscopy[62,86,87] and x-ray transmission microscopy[88,89]) rather than building upon each other's knowledge and knowhow. There is a surprising lack of indirect observations methods developed for the Li/S system, one could imagine that chemical derivation[90] or even freezing samples[61] could widen the possibilities for the characterization of intermediate states. Last, isolating interfacial processes at the Li electrode is of primary importance, and this will not be achieved by electrochemical methods unless two-electrode Li–S configurations are abandoned for a more rigorous cell setup. We believe, for instance, that partially lithiated $Li_{4+x}Ti_5O_{12}$ would make an excellent reference electrode, minimizing the cold finger effect and corrosion by polysulfides compared to a metallic Li reference. To the best of our knowledge, such experiments have never been reported.

5.4. Electrolyte-Based Approaches Towards Passivation Film Formation

Electrolyte composition has a major effect on the performance (CE and morphology of deposits) of the Li electrode. The electrolyte can easily be controlled and formulated without physically modifying the Li electrode in any way. For this reason, additives and selection of appropriate solvent/salt combinations have been favored approaches to study these factors on Li deposition/stripping processes. Many of the types of additives which have been tested in the past are summarized in Table 5.2. An additional advantage of this approach is that electrolyte modifications are, for the most part, quite reproducible and search groups have conducted analytical studies aimed at better understanding.

5.4.1. LiNO$_3$, the ultimate additive for Li–S?

The use of Li nitrate (LiNO$_3$) as an additive or a co-salt has become so widespread as a shuttle-inhibitor in Li–S batteries that its mechanism of action is now under intense scrutiny.

5.4.1.1. *LiNO$_3$-induced surface film*

In terms of interfacial chemistry, the NO$_3^-$ ions have a double-edge impact: At the negative electrode/electrolyte interface, first, they are spontaneously

Table 5.2: Additives used to enhance the cycling efficiency of Li metal.

Group	Additives	Major acting mechanism
Inorganic	HF	Formation of LiF layer
	AlI$_3$, MgI$_2$, SnI$_2$	Formation of Li-alloy layer
	S$_x^{2-}$	Formation of protecting film
Organic	2Me-furan, 2Me-THF, Pyridine derivatives, Dipyridyl derivatives	Surface adsorption, formation of organic protecting film, solvation of Li$^+$
	Cetyltrimethylammonium chloride	Electrode surface adsorption
	Non-ionic surfactants, Crown ethers	Electrode surface adsorption, solvation of Li$^+$
	Benzene	Electrode surface adsorption
Gas	CO$_2$	Formation of Li$_2$CO$_3$ layer
	N$_2$O, CO	Formation of protecting film

Source: Reproduced from Ref. 27.

reduced on the surface of Li metal, hence forming a Li_xNO_y and Li_2O rich interphase.[35] Moreover, NO_3^- ions can also be reduced at the negative electrode by low-order polysulfides and Li_2S, resulting in Li_xSO_y surface species as depicted in Figure 5.4. Owing to surface passivation by both Li_xNO_y and Li_xSO_y species, further internal redox reaction of soluble S_n^{2-} species onto the negative electrode is impeded, so that both resistance to self-discharge and CE are significantly improved.[35,57] Several groups reported that passivation could be realized by just pretreating (i.e. soaking) Li electrodes in $LiNO_3$ solution.[91,92] Using a $LiNO_3$ pretreated Li negative electrode, Han et al. managed to cycle Li–S cells with relatively high sulfur loading (3.25 mg cm^{-2}) with improved CE (>85 vs. 50% without pretreatment), however a continuous decrease of the CE was observed from 100 to 85% over 100 cycles,[92] suggesting that the passivation layer was damaged over repeated cycling. In fact, comparing symmetrical Li|catholyte|Li cells, Xiong et al. observed that the one cycled in the polysulfide-free $LiNO_3$ electrolyte solution had higher charge-transfer resistance. Also, the much higher and irregular potential difference during galvanostatic cycling compared to Li_2S_6-containing cells (see Figure 5.4(a)), indicated that the properties of this SEI film was changing during cycling, suggesting that it was repeatedly broken during cycling.[57] Therefore, efficient Li protection by $LiNO_3$ seems to require both NO_3^- and S_n^{2-} in solution (see below). Another experiment consisted of cycling a Li–S cell in a $LiNO_3$-containing electrolyte and then transferring the Li electrode into a new cell, without the additive. As reported by Xu et al. for instance, the new cell with the precycled Li exhibited overcharge (indicative of shuttle reactions) very similar to a cell with no Li protection.[93] In fact, later in this chapter we will show that chemical pretreatments of the Li electrode have shown limited effectiveness in Li–S cells. Eventually, studies reporting long-term cycling[94] or varying $LiNO_3$ concentration in the electrolyte[93,95] demonstrated its irreversible consumption, so that the benefits of this additive on the cycling performance of Li–S cells are limited to its availability in solution. As mentioned in the introduction, the Li negative electrode undergoes severe volume changes upon cycling (stripping/plating), so that $LiNO_3$ is consumed each and every time new surfaces are created or exposed, explaining why pretreatment cannot be effective in the long term.

Several parasitic reactions can explain NO_3^- depletion, mostly owing to its narrow potential stability window. In particular, reduction of NO_3^- to NO_2^- occurs at the positive electrode between 1.6 and 1.9 V vs. Li/Li$^+$, depending on the current rate as well as on the nature and electrochemical surface of the sulfur host. As reported recently by Rosenman *et al.* not only is this reduction irreversible — thus consuming $LiNO_3$ and resulting in poor negative electrode passivation — but it also generates strong nucleophiles (such as O^{2-}), to which the authors attribute the decomposition of their electrolyte solvent (DOL) and salt (LiTFSI).[96] A surface film containing resistive LiF as well as carbonyl and carbonates from the polymeric reduction of the electrolyte is formed on the surface on the sulfur positive electrode below 1.9 V. In the light of these findings, it seems that increasing the initial concentration of $LiNO_3$ to compensate for its irreversible consumption may not be an appropriate solution; hence conservative voltage cut-off should be employed and novel shuttle inhibitors explored.

5.4.1.2. Synergy between $LiNO_3$ and Li_2S_n at the negative electrode interface

The addition of long-chain polysulfides in electrolyte solutions has long been shown to have a positive impact on Li–S cell performance. As mentioned above, polysulfide addition affects the nature and composition of the SEI on the Li negative electrode in ethereal solutions, which decreases its impedance and stabilizes it more readily. It appears that the coprecipitation of reduction products from $LiNO_3$ and Li_2S_n results in a smooth and compact SEI.[35,91,57] It was also shown that the interphase formed on Li in electrolytes containing both $LiNO_3$ and Li_2S_n retained a stable chemical composition and structure in both the charged and discharged state of the Li/S cell.[97] All these observations suggest that the interphase formed in the presence of polysulfides is mechanically more resilient than the one formed from $LiNO_3$ alone. Besides, it must exhibit a higher ionic conductivity, since the charge-transfer resistance is significantly decreased.[57] This in turn favors the homogeneous distribution of current density upon stripping and plating of Li which results in improved cycling efficiency and cycle life of the Li metal electrode.

Furthermore, the oxidation of sulfur species by nitrates in the outer interphase produces relatively stable sulfate compounds. This top layer (see Figure 5.4(c)) plays a critical role to prevent the contact between the S_n^{2-} in solution and reductive Li^0 (Figure 5.1, Equation (1)) or Li_2S (Figure 5.1, Equation (4)) at the negative electrode.[57] In this sense, the presence of some soluble S_n^{2-} in the electrolyte before the initial reduction of sulfur is a further protection against self-discharge: proper negative electrode passivation can be achieved upon cell assembly, without any contribution from active sulfur mass in the positive electrode.

5.4.2. Bulk electrolyte

5.4.2.1. *Fluorinated solvents (ethers)*

Fluorinated ethers represent an illustrative case of non-polar solvents which exhibit chemical stability against polysulfides as well as low donor ability. Owing to fluorine stronger electronegativity and increased steric hindrance of the chelating oxygen compared to hydrogen, partially fluorinated ethers exhibit very distinct properties such as a lower melting point, higher oxidation potential, and lower flammability. Fluorination also results in severely diminished ionic solvation ability. For instance, 1,1,2,2-tetrafluoroethyl 2,2,3,3-tetrafluoropropyl ether (TTE) was shown incapable of Li^+ solvation; *de facto* preventing Li_2S_n and even LiTFSI solubility.[98,99] Moreover, TTE seems to be electrochemically reduced at 1.8 V, to form a stable thin film on the Li metal, which protects the negative electrode from reacting with polysulfides in the electrolyte. It could therefore substitute and/or complement DOL or $LiNO_3$ additive in the building of an efficient SEI.[100] In a follow up work, Gordin et al. reported a synergistic effect between another partially fluorinated ether (bis(2,2,2-trifluoroethyl) ether, or BTFE) and $LiNO_3$ to decrease self-discharge of Li/S cells in DOL:DME-based electrolytes.[101] The relatively high reduction potential of these partially fluorinated ethers, similar to $LiNO_3$, suggests that the improvements in CE and resistance to self-discharge may be due to surface passivation at the positive electrode even more so than the Li negative electrode. This raises the question, once again, of the stability and the transport properties of such interphase, with possible detrimental long-term impact of Li^+ consumption and polarization increase.

5.4.2.2. Superconcentrated solutions

In their study of the polysulfide shuttle phenomenon, Mikhaylik and Aldridge had already observed that electrolytes with higher LiTFSI salt concentration showed lower rates of Li corrosion in the presence of polysulfides and lower overcharge.[18] They suggested that increasing the salt concentration could play a role by increasing electrolyte viscosity (hence reducing polysulfide mobility) but also by decreasing polysulfide solubility via common ion effect (or approaching saturation).[102,103] Even more interestingly, higher salt concentration could have a beneficial influence on the negative electrode/electrolyte interface.

In this respect, Yamada *et al.* reported ultra-fast, reversible, Li intercalation into graphite from super-concentrated LiTFSI or Li bis(fluorosulfonyl)imide (LiFSI) solutions in DME,[104] and even acetonitrile (ACN),[105] in which all solvent molecules are coordinated to Li$^+$ to form a solvate ionic liquid.[104] One of the factors explaining the exceptional electrochemical behavior of the graphite electrode in (ACN)$_2$LiTFSI was the specificity of the interphasial film formed. Indeed, they noted the formation of TFSI-derived film during the first discharge, which worked as a protective layer to kinetically suppress further reductive decompositions of TFSI anions as well as ACN solvent molecules. Based on DFT calculation, they attributed this enhanced reductive stability to the unique solvate structure of the (ACN)$_2$LiTFSI solvate, in which the LUMO, dominating the behavior of its reduction reaction, shifts from ACN to TFSI$^-$. These anions are therefore predominantly reduced, instead of ACN, to form a TFSI-derived surface film (rich in S and F species) on a graphite electrode during the first Li intercalation, which suppresses further decompositions of the electrolyte.[105] One of the highest coulombic efficiencies (up to 99.1%) for cycling of a Li electrode was reported by Qian *et al.* using a highly concentrated 4 M solution of LiFSI in DME.[46] Li/Li cells were able to achieve more than 6000 cycles at 10 mA cm^{-2} and copper|Li cells were cycled for more than 1000 cycles at 4 mA cm^{-2} without dendrite growth. The outstanding performances were attributed to increased solvent coordination and increased Li-ion concentration at the Li surface. There are complex relationships between solvation and anion decomposition which are dependent on concentration, anion type, solvent type, and the presence or absence of polysulfides. Anions which are coordinated to Li⁺ cations are expected to be more susceptible to reduction than their uncoordinated counterparts.[106,107] This is referred to as the sacrificial anion reduction mechanism and is thought to result in the formation of a greater amount of low resistance inorganic components in the SEI (than for 1 M electrolytes). Recent density functional theory and *ab initio* molecular dynamics calculations show that LiTFSI reacts extremely fast when in contact with a Li surface and this anion decomposition precedes salt dissociation.[108] Differences in stability between FSI⁻ and TFSI⁻ anions are also expected to influence the nature of the SEI and its components.[106]

Such superconcentrated electrolytes have proven extremely efficient in Li–S cells, yet it is difficult to isolate the role of such interphasial film on the Li negative on the improved CE and capacity retention from the effect of reduced polysulfide solubility/mobility in the electrolyte.[103,109] As illustrated in Figure 5.5, the absence of coloration in the ACN:HFE-containing cells supports the conclusion of negligible solubility/mobility of Li polysulfides in such superconcentrated and fluorinated electrolytes.[109] The lithium/electrolyte interface can therefore be considered exempt from sulfur redox species and the benefit of such superconcentrated electrolytes is two fold. At the Li negative electrode, excellent CE for Li plating/stripping occurs[46] along with smooth, dendrite-free, shiny, globular deposits (Figure 5.5(d)). This is in sharp contrast to other electrolytes, where Li is deposited with dendritic or mossy morphologies (Figure 5.5(c)). Nearly 100% CE is observed for Li/S batteries using superconcentrated electrolytes[103,109] since the polysulfide shuttle is effectively inhibited, with all sulfur species confined at the positive electrode, and a stable SEI is formed on Li to prevent its corrosion and consumption of the electrolyte. One drawback of the use of highly concentrated electrolytes in Li/S cells is the poor rate capabilities which arise from the immobilization of all sulfur species. This can clearly be observed when comparing the voltage profiles in Figures 5.5(a) and 5.5(b). The standard DOL:DME electrolyte (Figure 5.5(a)) possesses low overpotentials for sulfur reduction and its reoxidation where these reactions occur in the solution-phase. On the other hand, the (ACN)$_2$LiTFSI–HFE electrolyte (Figure 5.5(b)) portrays substantial overpotentials since the electrochemical reactions are

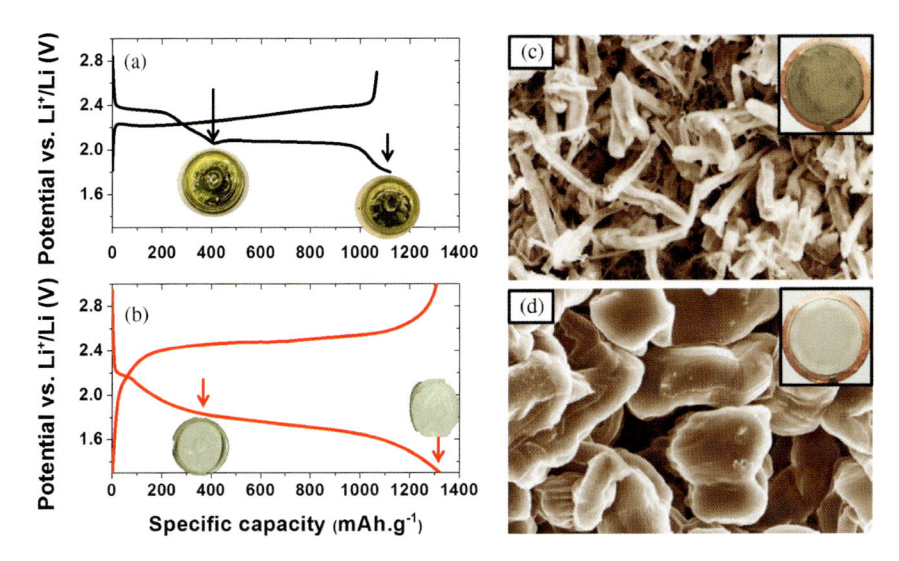

Figure 5.5 Photographs of separators from Li–S cells employing a DOL:DME (a) (black) or an ACN:HFE. (b) (red) ACN:HFE electrolyte; arrows indicate the state of discharge where the cells were opened. Reproduced from Ref. 109, with permission of The Royal Society of Chemistry. SEM images of the morphologies of Li metal after plating onto Cu substrates in different electrolytes. (c) 1 M LiPF$_6$ in PC and (d) 4 M LiFSI in DME. The current density was 1.0 mA cm^{-2} and the deposition time was 1.5 hours reproduced from Ref. 46, distributed under a Creative Commons CC-BY license.

mainly solid-state conversions. It should be noted, however, that this challenge is exclusive to the positive electrode, since the polarization is extremely low for Li plating/stripping.[46,73] In the work of Cuisinier *et al.* the polarization was attributed almost entirely to the sulfur/electrolyte interface (0.23 V, i.e. 95%) and the contribution of the Li/electrolyte interface was minimal (0.01 V, i.e. 5%).[109]

5.4.3. Electrolyte additives

5.4.3.1. *Sulfur-based additives*

Similar to Li polysulfides, phosphorus pentasulfide (P$_2$S$_5$) can be reduced at the surface of metallic Li. Using it as an electrolyte additive in Li/S cells, Lin *et al.* attributed the significantly improved CE to the presence of a dense and smooth film, mainly composed of Li_3PS_4 instead of Li_2S.[110] Li_3PS_4 is a Li ionic conductor,[111] so that the passivation film truly functioned as a SEI which allowed efficient Li-ion transport to and from the negative electrode surface while still preventing electron-transfer to and from polysulfides in solution (i.e. inhibiting the redox shuttle). This translated into a low voltage hysteresis, hence improved energy efficiency for the Li/S cell. One possible concern with this approach arose from a recent study by Hakari *et al.* showing that Li_3PS_4 glass could function not only as a Li^+-conductor but also as an active material in itself when ball-milled with acetylene black.[112] Although they could not elucidate the electrochemical reaction associated with the delithiation of Li_3PS_4, this suggests that the solid electrolyte might not be rigorously stable in contact with metallic Li and could therefore deteriorate upon prolonged cycling.

Departing from the strict notion of electrolyte additive, Demir-Cakan *et al.* found that the performance of Li–S cells could be further enhanced compared to a polysulfide additive when elemental sulfur powder was directly deposited onto the Li negative electrode.[53] They observed by XPS that this simple approach resulted in a unique interphasial chemistry, with lower Li_2S content and a longer average polysulfide chain length, which was maintained for at least eight cycles. The conduction properties at the Li/electrolyte interface were investigated by impedance spectroscopy, and it was suggested that the SEI formed by direct contact of sulfur with the Li electrode may act as a viscous ionic-blocking barrier in solution, preventing further chemical reactions (see Figure 5.1, Equation (1)). Similar to Li_3PS_4,[110] it is not clear however if such sulfur-rich interphase on the Li electrode can maintain its initial properties over hundreds of cycles to achieve sustainable performance in a practical Li/S cell.

5.4.3.2. *Other reduction-type additives*

Many other electrolyte additives have been investigated in Li metal batteries with an aim to stabilize the interface between electrolyte and the Li electrode. Here, we report some of these works, when relevant to the Li–S system. In particular, we saw that interfacial films composed mainly of inorganic species, such as the one formed from $LiNO_3$, tend to be brittle and therefore lack mechanical stability upon volume changes during cycling.

Li bis(oxalato)borate (LiBOB) has been used extensively in LIBs as an additive or sole electrolyte salt due to its dense SEI-forming nature, particularly on graphite negative electrodes.[113,114] In the work of Xiong et al. varying amounts of LiBOB were added to a 1 M LiTFSI DOL:DME electrolyte.[115] SEM observations shows that Li negative electrode with LiBOB has a smoother and denser surface morphology than the electrode without LiBOB. Although no CE was reported, the inhibition of overcharge in Li/S cells strongly suggest that the parasitic reaction between higher polysulfides and Li metal is prevented after adding 1–4 wt.% LiBOB in the electrolyte (at least on the first cycle).[115]

Song et al. studied the effect which fluoroethylene carbonate (FEC) had on the Li SEI and its impact in Li–S batteries.[116] The protective layer formed on Li metal was based on a semiinterpenetrating polymer network (IPN) structure generated by UV-curing. The UV-curable formulation consisted of a curable monomer, P(VdF–co–HFP), liquid electrolyte (1 M LiPF$_6$ in EC/EMC/FEC), and a photoinitiator. The protective layer with FEC drastically reduced the impedance of Li vs. Li symmetrical cells after cycling when compared to a protective layer formed using 1 M LiPF$_6$ in TEGDME as the electrolyte instead. FEC also effectively suppressed the shuttle-induced overcharge and improved cycling performance of Li/S cells when compared to both the TEGDME-based UV-protection layer and unprotected Li.[116] Although this work was complicated by the UV-curing polymerization technique, it is suspected that the use of FEC as a simple electrolyte additive (similar to its use in LIBs), could improve the SEI and aid in the suppression of the shuttle process in Li–S batteries, via its reduction to vinylene carbonate (VC) and subsequent polymerization into Li-rich oligomers.[117] It is unclear however, if FEC and VC could withstand the presence of nucleophilic polysulfide intermediates during the formation cycle.

Some metal iodides such as tin iodide (SnI$_2$) and aluminum iodide (AlI$_3$) can enhance Li cycling performance by making a Li-alloy layer at the Li anode interface (see Table 5.2).[118,119] The alloying layer can limit the growth of Li dendrites, while the I$^-$ anions also improves cycling efficiency by the reaction of I⁻ anions with Li-ions on the surface of Li to form a conductive LiI layer.[120] Organic additives, such as 2-methylthiophene and 2-methylfuran are known to form an SEI whose conductance is higher than that formed in electrolytes without the additives (Table 5.2).[119] The synergistic effect of AlI$_3$ with 2-methylfuran on the cycling efficiency was reported by Matsuda *et al.*[118,119]; and was also observed by Wu *et al.* who used a combination of AlCl$_3$ and pyrrole.[120] When Li deposition/stripping cycling experiments were performed on Cu substrates, the CE was only improved (relative to the electrolyte with no additives) when both AlCl$_3$ and pyrrole were present. To the best of our knowledge, this combination approach of an alloy-forming metal and organic additive where both can be chemically or electrochemically reduced to form a conductive and mechanically flexible SEI has not yet been applied to Li–S batteries.

Zu and Manthiram showed that copper acetate can be used as an electrolyte additive to stabilize the surface of Li metal in a polysulfide-rich environment through the chemical formation of a passivation film consisting of mainly Li$_2$S/Li$_2$S$_2$/CuS/Cu$_2$S along with electrolyte decomposition products.[121] This passivation film was reported to control the Li deposition sites, leading to a stabilized Li surface characterized by a dendrite-free morphology and improved surface chemistry as shown in Figure 5.6. The SEM images in Figures 5.6(a) and 5.6(c) show the rough and inhomogeneous Li surface from the control cell (no Cu acetate additive) after the first and 100th charge of Li–S cells. In contrast, however, when copper acetate (0.03 M) is added to the electrolyte, the surface of the Li remains smooth even after the 100th cycle as displayed in Figure 5.6(d). Although the organic component of the passivation film derived from Cu acetate is not clearly characterized, it likely contributes to improving the mechanical flexibility of the SEI similarly to the above mentioned additives.

Li nitrate has incontestably dominated Li protection in Li–S batteries, owing to its demonstration as an effective shuttle inhibitor. However, its film-forming properties are not ideal, and its reduction potential is slightly too high. Hence, it gets progressively consumed upon prolonged cycling. A possible research avenue would be to explore combinations of multiple additives — as commonly reported for Li-ion systems — to tailor the chemistry and mechanical properties of Li SEI. More indepth discussion on

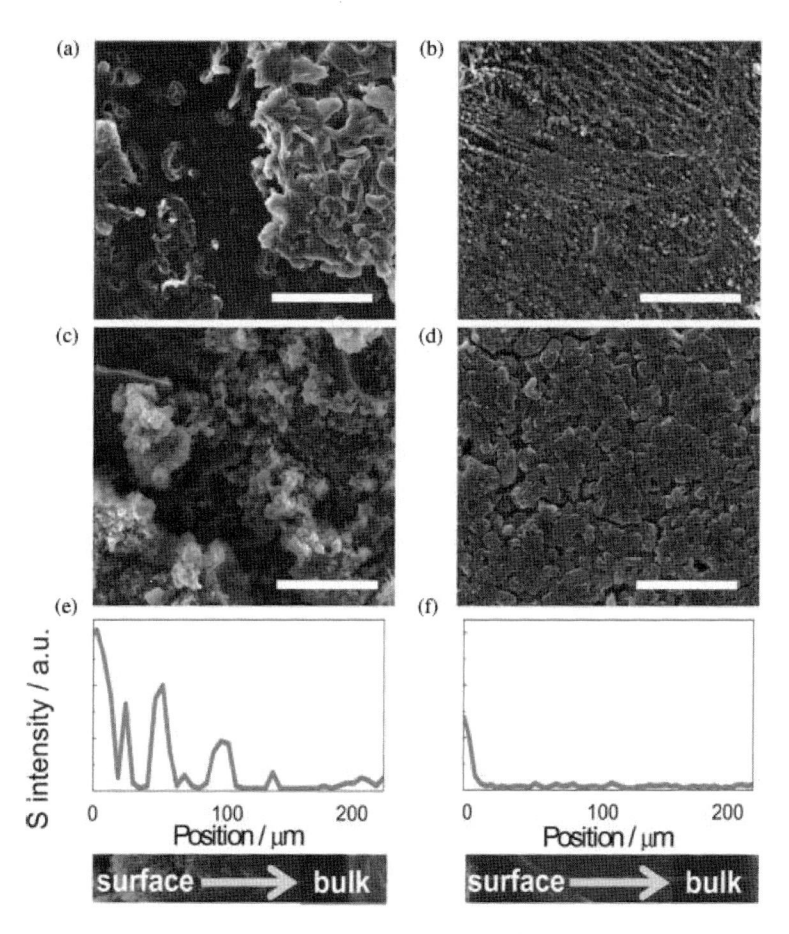

Figure 5.6 SEM characterizations of the Li metal surface after the first charge in the (a) control cell and (b) experimental cell and after the 100th charge in the (c) control cell and (d) experimental cell. Cross-sectional EDS line scans (for sulfur) of the Li metal after the 100th charge in the (e) control cell and (f) experimental cell. The scale bars in (a–d) represent 100 μm. The scales in (e,f) represent the position in Li metal.

Source: Reprinted with permission from Ref. 121. Copyright (2014) American Chemical Society.

electrolytes can be found in Chapter 4. It is very clear that the tremendous improvements realized on the sulfur positive electrode now allow the focus to be shifted to other cell components. Regardless of the solvent, significant progress has been achieved on the Li side through systematic studies on additives in the presence of polysulfides in the electrolyte solution.

5.5. Morphology Control Upon Stripping and Plating

It is well known that the use of Li metal as the negative electrode may cause dendrite formation, although some argue that this problem is suppressed with the presence of polysulfides through the consumption of dendritic Li via (Equation (15)).[3,18,35] Nonetheless, observing Li shunts in Li–S cells is not uncommon, suggesting that Li may still pierce through the separator, hence causing short-circuit.[122] In a recent study, Han *et al.* investigated the effect of the Li:S ratio on the corrosion rate of the Li negative electrode upon cycling. Even with a ten fold excess in Li, they observed that the thickness of the corrosion film reached 65% of that of the pristine Li electrode after 50 cycles. This corrosion film corresponds to the SEI as well as the pulverized/mossy Li, and unfortunately it is not clear how much of the pristine Li thickness remained intact.[55] Quite similarly, Lv *et al.* had observed that electrolyte and polysulfides penetrated deeply and corroded almost half of the Li anode when using thick sulfur positive electrodes.[76] These two studies are, in fact, among the very few that address the issue of Li aging and/or failure in Li–S cells. In the following, different approaches to control the morphology of the Li electrode are presented. By limiting its electrochemically active surface area, the extent of interfacial reactions is reduced, hence prolonging the cycle life of the battery.

5.5.1. Approaches to prevent dendritic/mossy Li formation

5.5.1.1. *Reduction of Li roughness*

Sion Power demonstrated improvements in the cycle life of Li–S cells when reducing Li roughness, notably using a current collector while cycling Li at close to 100% of its depth of discharge (DOD).[3] Leading to this idea, was first the ability to distinguish morphology-limited cell failure from that caused by electrolyte depletion. Using Monte Carlo simulations combined with experimental observations, Mikhaylik *et al.* evidenced that cycle life (N) was a function of the inverse square of Li DOD, or thickness of stripped-plated Li (Th) as shown in Figure 5.7(a). By cycling only a thin layer of metallic Li at low DOD, they managed to approach 10^4 cycles, which they attributed to the minimal surface roughness under these conditions. However in this case, the substantial excess of Li would have a direct impact on the energy density. According to the simulations (Figure 5.7(b)), Li morphology can be controlled without Li excess, by going to the other extreme and cycling instead the Li electrode at 100% DOD. In this approach, a current collector is required to direct Li plating (see Section 4.2).

The second finding was that applying pressure onto the Li electrode could increase the cycle life, especially under high charge rates. The pressed electrode remained compact even after stripping and plating Li at 25% of its DOD for 50 cycles at C/3. Even when increasing the charge rate to 1 C the Li morphology did not deteriorate and little capacity fade was observed contrary to the unpressed Li electrode. By suppressing cycling-induced Li surface roughness, Mikhaylik et al. showed that the thermal runaway of Li–S cells was increased from about 140°C (cells cycled without pressure) up to the melting point of Li, around 181°C.[3]

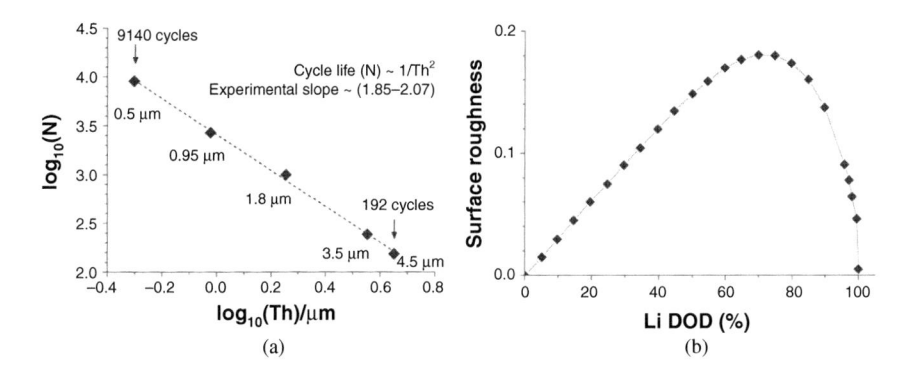

Figure 5.7 (a) Cycle life (N) of experimental Li–S cells at various thickness (Th) of stripped/plated Li. (b) Monte Carlo simulated surface roughness after the first discharge. *Source*: Reproduced from Ref. 3.

5.5.1.2. Li alloys to change Li deposition behavior

Li–B alloys, and Li$_7$B$_6$ in particular, were studied in Li–S batteries, not as alternative negative electrode materials but more so as Li host. In fact, Li-rich alloys were shown to maintain the fibrillary network framework of the Li$_7$B$_6$ constituent after removal of free Li.[123,124] While a Li metal electrode showed serious dendrite growth on the surface after 20 cycles in a Li–S battery, the Li–B alloy retained a smooth surface, and exhibited lower impedance under the same test conditions.[125] Zhang et al. varied the amount of Li–B alloy and obtained their best results for the alloy containing 40 wt.% B (corresponding to 7Li.Li$_7$B$_6$).[123] It would be really interesting to see future works focus on the mechanical stability of Li–B–S cells with high areal capacity and/or under high current densities.[12] The demonstration by Cheng et al. of low capacity fading (0.032% over 2000 cycles), and high CE (91–92%) in the absence of LiNO$_3$ additive for relatively high loading of sulfur (4.6 mg$_{(S)}$ cm^{-2}) is indeed extremely encouraging.[124]

5.5.1.3. Cs$^+$ additive for electrostatic shielding

Ding et al. recently demonstrated a self-healing electrostatic shield (SHES) mechanism that could influence the Li deposition process to obtain a dendrite-free Li electrode.[126] Contrary to SEI-forming approaches, this mechanism does not rely on the mechanical strength of a protection layer, but instead on an electrostatic shield formed by electrolyte additives. According to the Nernst equation, cesium ions (Cs$^+$) at low concentration have an effective reduction potential lower than that of Li$^+$ at 1.0 M concentration (e.g. −3.144 V vs. SHE for Cs$^+$ at 0.01 M compared to −3.040 V vs. SHE for Li$^+$ at 1.0 M) and a much lower chemical activity.[126] The idea is that during the Li deposition process, the Cs$^+$ electrostatic shield can repel the incoming Li$^+$ ions and force them to be deposited on the valley area instead of the peak area of the Li film to smooth it along the plating reaction. The additive cations with a reduction potential lower than Li$^+$ will not electroplate, but rather adsorb to regions with the greatest potential field. Exploiting this concept further, Kim et al. employed CsNO$_3$ as an additive in Li–S cells to both control Li dendrite growth on the negative electrode surface and reduce the shuttle effect of polysulfides species.[127] When comparing the crosssection of Li electrodes after the same three cycles, without additive or in the presence of either $LiNO_3$ or $CsNO_3$, they found that the length of Li dendrites was decreased from 115 to 25 μm and 6 μm, respectively. This confirms the positive effect of Cs^+ cations on the Li/electrolyte interface, over the impact of NO_3^- anions alone. Owing to the low cut-off voltage (1 V), nitrates are expected to be consumed over extended cycling (which was not reported in this publication), yet the effect of Cs^+ shielding should be maintained.

5.5.2. Lithium electrode design

Another route to mediate the growth behavior of deposited Li during the charge–discharge process is to tune the architectural design of metallic negative electrode.

5.5.2.1. *3D current collectors*

The inhomogeneous distribution of current density on the Li metal electrode, partly responsible for the formation of dendrites, can be alleviated by moving from a two-dimensional (2D) foil current collector to a 3D conductive substrate; the greater electroactive surface area not only lowers the real specific surface area current density, but it also makes it easier to cycle the Li electrode close to 100% DOD (see in Section 3.1.1) by minimizing volume fluctuation. The use of three-dimensional (3D) current collector to support Li was demonstrated using a graphene foam (GF),[97] or a porous copper foil with submicron skeleton.[128] Based on the textural parameters for planar Cu, Cu foam and 3D Cu skeleton, Yang *et al.* proposed that the ratio of electroactive surface area to geometric area of electrode (electroactive area ratio) dictates the fraction of Li deposited inside the 3D structure rather than plated on surface of the electrode.[128] In the case of a low electroactive area ratio, Li can be detached easily from the current collector upon stripping (especially under high current rate density) thereby resulting in electric disconnection, "dead lithium" and poor CE. In contrast, when Li-metal negative electrode was accommodated in the 3D current collector, they were able to suppress dendrite formation and remarkably improve the cycle life of Li-metal negative electrodes. The nanostructured graphene framework reported by Cheng et al. exhibited similar behavior,[97] and in both cases, LiNO$_3$ and Li$_2$S$_n$ additives were used in order to improve the Li/electrolyte interfacial properties.[35,57] Unfortunately, none of these works applied the concept of 3D nanostructured Li electrode to the Li/S system and instead assembled cells against a LiFePO$_4$ positive electrode. Although an impressive 97% efficiency was achieved using the 3D submicron Cu architecture, the voltage hysteresis in symmetrical cells showed a steady increase during the galvanostatic experiment.[128] This can be assigned to increasing charge-transfer resistance, the downside of the high electroactive surface which exacerbates the parasitic reduction of electrolyte salt and solvents[128] (see Section 1), so Li consumption will be a problem for long-term cycling.

5.5.2.2. Hybrid negative electrode designs

In a highly publicized study, Zheng et al. reported the deposition of a monolayer of interconnected amorphous hollow carbon nanospheres onto a copper substrate to help direct the Li metal plating inside the cavities of the carbon spheres and concomitantly facilitate the formation of a stable SEI.[129] The overall concept is illustrated in Figures 5.8(a) and 5.8(b), although it should be clarified that the protected electrode is not lithiated and therefore requires a pretreatment, or to be paired with a Li containing positive electrode. The authors experimentally determined that Li dendrites did not form up to a practical current density of 1 mA cm^{-2}, which would roughly correspond to 0.1 C in a Li/S cell with 5 mg$_{(S)}$ cm^{-2}.

A different approach to protect the Li negative electrode in Li–S cells was developed by Huang et al.[130] They assembled a hybrid electrode structure consisting of graphite externally connected to the Li metal (Figure 5.8(c)). The lithiated graphite functioned as an artificial SEI layer on Li metal that supplies the Li$^+$ ion on demand, while minimizing direct contact between soluble polysulfides and Li surface. Li–S cells incorporating the hybrid negative electrode delivered a specific capacity

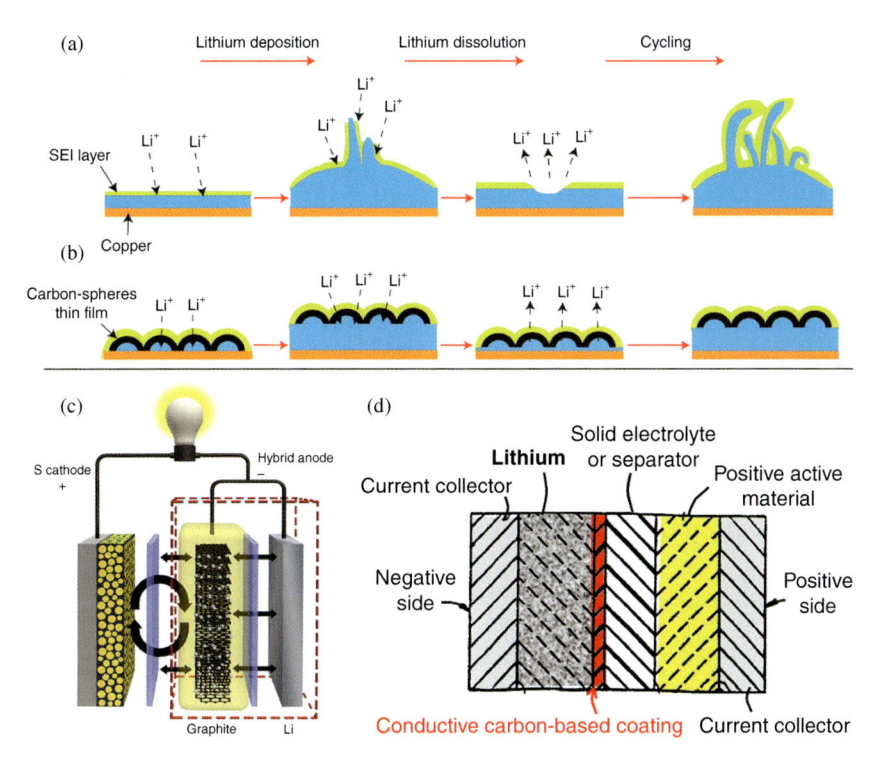

Figure 5.8 (a–b) Schematic diagrams of the different Li negative electrode structures. (a) A thin film of SEI layer forms quickly on the surface of deposited Li (blue). (b) Modifying the Cu substrate with a hollow carbon nanosphere layer creates a scaffold for stabilizing the SEI layer. Reprinted with permission from Macmillan Publishers Ltd: *Nature Communications* (Ref. 129), copyright 2014. (c) Schematic of the hybrid negative electrode design to manipulate the surface reactions in Li–S batteries. Reprinted with permission from Macmillan Publishers Ltd: *Nature Communications* (Ref. 130), copyright 2014. (d) Illustration of a cross section of a Li battery having a carbon-based layer between a negative electrode and an electrolyte separator.

Source: Adapted from Ref. 131, not subject to copyright.

of 800 mAh $g_{(S)}^{-1}$ after 400 cycles at 1 C with a CE exceeding 99%. A possible flaw, the feasibility of such complex cell design should be examined further.

In a potentially more practical system, layers of carbon between the Li negative electrode and the electrolyte separator have been used as an artificial SEI (Figure 5.8(d)). It is well known that the SEI formed on carbon has been long known to be better than that on Li. And in fact, this is the sole reason why commercial LIBs use graphite rather than metallic Li as negative electrodes. In the work of Koksbang, an interlayer between an active metal negative electrode (e.g. Li) and an electrolyte separator is comprised of a carbon-based material which is a conductor of both electrons and metal ions of the negative electrode.[131] The layer restricts penetration of electrolytic organic and anionic salt constituents and prevents the degradation of the negative electrode. This concept calls for a compromise, between the decrease in energy density associated to this extra inactive component and the amount of Li consumed to form an SEI on one side and the improvement in cycling efficiency and cycle life on the other. It should also be noted that the SEI formed on carbon in ether solvents and in the presence of polysulfides may not be as stable as the one optimized for traditional Li-ion systems.

Interestingly, the different approaches presented here are not all exclusive, so further improvements could potentially be achieved by considering a holistic approach. For instance, the reduction of Li surface roughness as an effect of applying non-isotropic pressure could be prolonged with the help of Cs^+ additive. Also, most of these methods are either applicable or actually designed for an "anode-less" cell configuration, so replacing sulfur by Li_2S at the positive electrode would not be an issue. Morphological control alone will unlikely enable the targeted Li cycling efficiency of 99.8%, however it could nicely complement any of the chemical and/or physical protection methods presented hereafter.

5.6. Chemical Protection

Contrary to the methods of forming passive SEI layers or controlled Li deposits *in situ* through the use of strategically selected electrolytes, additives, or cell architecture designs, chemical protection refers to coatings formed *ex situ* by direct chemical reaction of metallic Li with compounds before cells are assembled. Chemically formed layers protect mainly by changing the kinetics of SEI formation and ageing and, to some extent, avoiding direct contact of Li with the electrolyte.

5.6.1. Li-ion conductive coatings

5.6.1.1. *Li nitride*

Lithium nitride (Li$_3$N) has an exceptionally high Li$^+$-conductivity (up to ~10^{-3} S cm^{-1}) and has been proposed as a solid electrolyte candidate for LIBs.[132–134] When directly contacted with the Li metal electrode, a Li$_3$N layer can aid in preventing the side reactions between Li and the electrolyte, acting itself as a stable SEI.[133] One study by Ma *et al.* employs Li$_3$N as a protective layer to address the obstacles of the Li negative electrode in Li–S batteries, namely to prevent the contact between the polysulfides and the Li electrode and therefore suppress the undesired corrosive reaction (see Figure 5.9).[135] The Li$_3$N protective layer is fabricated on the surface of a Li electrode by the direct reaction between the Li and

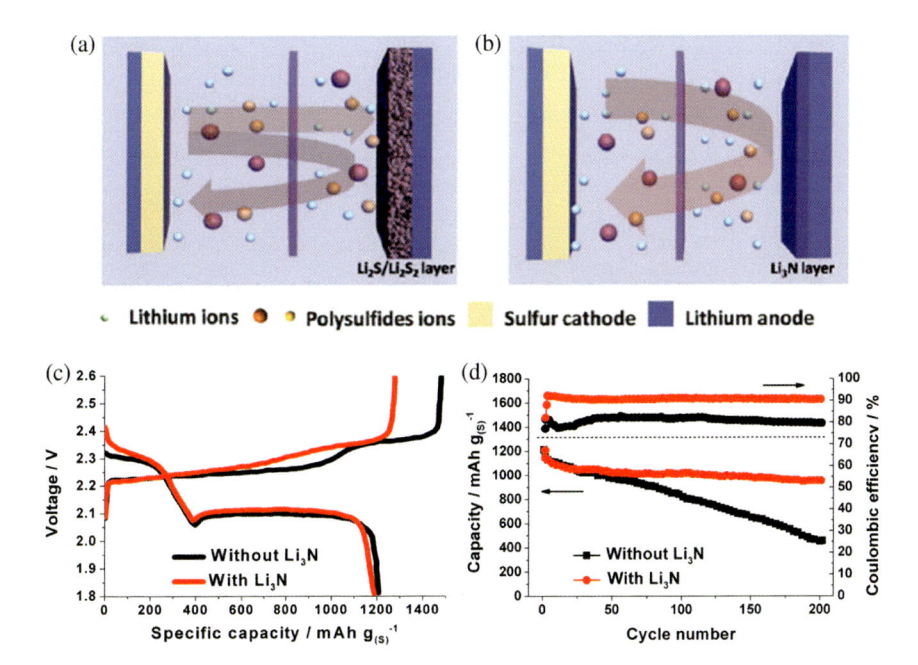

Figure 5.9 Schematic of the design of a Li metal electrode in Li–S batteries: (a) without a Li$_3$N layer and (b) with a Li$_3$N layer. (c) The initial charge–discharge profiles of Li–S batteries with and without Li$_3$N layers at 0.2 C. (d) The cycling performance and coulombic efficiencies of Li–S batteries with and without Li$_3$N layers at 0.2 C.

Source: Reproduced from Ref. 135 with permission of the Royal Society of Chemistry.

N_2 gas at room temperature. As a result, the migration of Li polysulfides back to the positive electrode and their reutilization in the subsequent cycles is possible. This inhibits capacity fading, and furthermore, the Li dendrites originating from a non-uniform deposition of Li can be suppressed by the Li_3N layer, improving the safety of the battery.

When AC impedance measurements were performed on Li|electrolyte| Li symmetrical cells, the initial impedance of the protected Li was higher than that of the primitive Li electrode, indicating the presence of the protective layer on the surface.[135] However, after ageing the battery for 240 hours, the impedance of the Li_3N-PLE stabilized, whereas the impedance of the unprotected Li electrode continued to rise. This proves that the undesired side-reactions between Li metal and the ether-based electrolyte are suppressed in the presence of the Li_3N protective barrier. Furthermore, cycling results for Li–S batteries with and without Li_3N layers show that the polysulfide shuttle is diminished by Li_3N. As displayed in Figure 5.9, for the cell fabricated with Li/Li_3N, the discharge capacity retention was 79.7% after 200 cycles with an average CE of 91.7%. This is in great contrast to the cell fabricated with unprotected Li, where the capacity retention and CE were only 37.2 and 80.7%, respectively.

5.6.1.2. Li phosphorous oxynitride

Another Li-ion conductor, Li phosphorous oxynitride (also known as LiPON) is common in the literature as a protective layer on Li metal or as a solid electrolyte.[136–138] The invention of Bates includes the use of LiPON to separate direct contact between battery electrolytes and Li electrodes.[139] Here, the LiPON layer is created by reactive magnetron sputtering of Li_3PO_4 onto Li in the presence of pure nitrogen gas or by rapid electron beam evaporation of Li_3PO_4 in the presence of N_2. It is claimed that a film of LiPON having a thickness between 0.1 and 0.5 μm provides satisfactory protection of the Li anode. By preventing direct contact of Li and the electrolyte, the likelihood of a film barrier-forming chemical reaction is significantly reduced, and the formation of dendritic growth of Li or the passivating "dead" Li upon the anode due to the cycling of the battery is also reduced. Additionally, by reducing the likelihood of the formation of a resistive film on Li (from side-reactions with electrolyte), the longer the battery will be capable of supplying the desired current at the rated voltage. Still further, by preventing the dendritic growth of Li and the formation of "dead" lithium on the anode surface, the efficiency of the battery will not be adversely affected by these factors. Although this approach has not been used specifically for Li–S batteries, the benefits a LiPON-protected Li metal electrode are all desirable when coupling lithium to sulfur-based electrodes.

5.6.2. Silane-based coatings

Silane coatings have been applied to Li metal to extend the storage shelf-life after exposure to electrolytes, decrease gas-phase reactions with atmospheric contaminants, and enhance the cycling performance of Li metal.[140–146] Silane layers are typically generated by the self-terminating reaction between a chloride substituent on the silane and a proton from the hydroxy-terminated layer on a clean lithium surface as displayed in Figure 5.10.

A detailed mechanism was proposed in the first publication on silane-based coatings.[140] With the use of multiple spectroscopic tools, including XPS, energy dispersive X-ray spectroscopy (EDX), and polarization modulated infrared reflection absorption spectroscopy (PM-IRRAS), the authors were able to show the disappearance of the native surface hydroxide species and the appearance of silicon and chloride on the Li surface from reaction with chlorotrimethylsilane (CTMS):

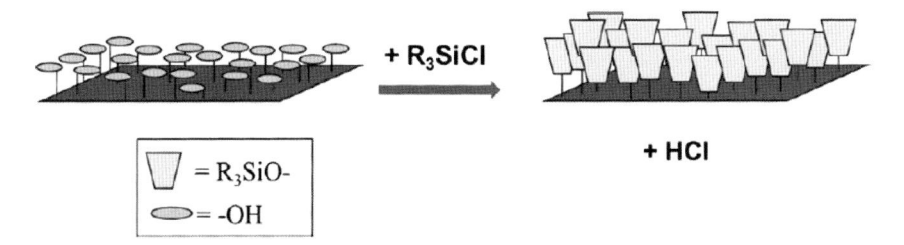

Figure 5.10 Schematic representation of the silane coating mechanism.
Source: Reproduced from Ref. 141.

$$\text{LiOH} + \text{ClSi(Me)}_3 \rightarrow \text{LiOSi(Me)}_3 + \text{HCl}, \qquad (5)$$

$$\text{LiOH} + 2\text{ClSi(Me)}_3 \rightarrow (\text{Me})_3\text{SiOSi(Me)}_3 + \text{HCl} + \text{LiCl}, \qquad (6)$$

$$\text{LiOSi(Me)}_3 + \text{ClSi(Me)}_3 \rightarrow \text{LiCl} + (\text{Me})_3\text{SiOSi(Me)}_3, \qquad (7)$$

$$2\text{HCl} + 2\text{Li} \rightarrow 2\text{LiCl} + \text{H}_2. \qquad (8)$$

It was proposed that similar reactions would occur in the presence of a native Li oxide surface. This concept was patented for "a chemically bonded protective layer formed by reacting a D or P block precursor with the oxygen containing layer".[146] Examples of D or P block precursors included silanes (CTMS), phosphines (chloro diisopropyl phosphine and chloro diethyl phosphine), and boranes (bromo dimethyl borane). All of these protective layers decrease the resistance of the Li compared to the untreated reference and were found to be ionically conductive.[140,146]

Since these preliminary studies, mainly focusing on CTMS, other silane-based compounds were used as precursors for direct reaction with Li surfaces. The effect of R-group size on the cycle life of Li was investigated.[141,144] It was found that very small R-groups (TMS) and R-groups bulkier than triphenyl show enhanced cycle life compared to control samples while R-groups between these in size show reduced cycle life. Also, Li/Li cell tests showed that coated electrodes, by reacting Li with tetraethoxysilane (TEOS), exhibit stable plating and stripping of Li at 1 mA cm^{-2} for 100 cycles without any change in cell impedance.[143] Most recently, Li coated with cyclopentadienyldicarbonyl iron (II) silanes (Fp-silanes) had up to a 500% enhancement in cycle life compared to uncoated lithium in coin cells.[142]

To the best of our knowledge, silane-based coatings have not yet been applied to Li metal electrodes for use in Li sulfur batteries. This seems like a logical research avenue to explore for protecting Li in polysulfide-containing electrolytes.

5.6.3. Pretreatments

Even when handled in a dry box, before submerging Li into an electrolyte, its surface typically develops a thin layer of oxide, hydroxide, carbonate, and nitride.[141,147] In addition to these salts, various spectroscopic measurements have indicated a tightly bound monolayer consisting of hydroxy or carboxylic acid groups on the surface.[140] With the exception of these tightly bound groups, the other surface contaminants are sitting on the surface rather than being bound to it.[140,144] Therefore, the buildup of these contaminants will affect the quality and performance of any coating layer.

Various methods to remove the surface contaminants were investigated by Neuhold *et al.*[141] The simplest method found was stirring in a liquid that is unreactive with Li metal. Alkali metals are frequently shipped submerged in alkanes because they do not react with them, or do so only very slowly. Also if a reaction does occur to form Li-alkyls, these are known to be soluble in hydrocarbons.[148] After investigating various solvents, n-pentane was chosen to clean the Li surface over other alkanes because its vapor pressure is low enough that it does not evaporate significantly while the Li is being cleaned but allows the Li surface to dry quickly when removed from the liquid.[141]

An invention by De Jonghe *et al.* relates to a composition comprising of a Li or other alkali or alkaline earth metal layer having a surface coated with a chemical protective layer.[147] Their protective layer is, at least transiently, physically and chemically stable in an ambient air environment and protects the Li metal from further chemical reaction. It is also covalently bonded to the metal surface, and conducts ions of the metal. Chemical compositions which fit this description include phosphates or carbonates. The protective layers may be formed by a liquid, vapor or gas phase surface treatment with a chemical precursor and in one example, this process simply consists of dipping Li metal into a solution of anhydrous phosphoric acid in dry DME to form Li phosphate. Other possible protective layers mentioned by De Jonghe *et al.* include Li metaphosphate ($LiPO_3$), Li dithionate ($Li_2S_2O_4$), Li fluoride (LiF), Li metasilicate (Li_2SiO_3), and Li orthosilicate (Li_2SiO_4), which could be reacted with the Li (or other) metal surface in the acid form, (e.g. HF + Li metal → Li–metal–LiF + H_2 evolution).[147]

As intuited by the short length of this section, chemical protection for the Li electrode in Li–S cells has been largely underexplored. The reason for that may be that chemically deposited layers are too thin and/or not dense enough to impede polysulfides in solution from reacting with the metal underneath. Their mechanical properties may also be insufficient to prevent the development of Li microstructures, so their use is often limited to temporary protection layers in combination with physical protection methods (see Section 6.3 and Figure 5.15 hereafter). In this case, it might be interesting to reinvestigate chemical protection layers as a way to functionalize a surface (see Figure 5.13) without sacrificing the energy density.

5.7. Physical Protection

Physical protection refers to rigid or semirigid layers (e.g. ceramic layers, *ex situ* formed polymer coatings, separators, etc.) that are in direct contact with Li without being chemically-bonded to it.[149–157] In many cases physical barriers are in fact chemically-bonded to Li at the surface, but maintain their chemical nature up to several hundred nanometers from the surface.[152,154] They protect the electrode mainly by avoiding detrimental side reactions through lack of direct contact with the electrolyte, and by spatially constricting the growth of dendrites. In Li–S batteries, the concept of physical protection can therefore be assimilated to a selective barrier preventing polysulfide from diffusing to and reacting with the surface of Li. Many different solutions have been proposed for the protection of Li anodes including coating the anode with interfacial or protective layers formed from polymers, ceramics, or glasses. The important characteristic of such interfacial or protective layers is the ability to conduct Li-ions.

5.7.1. Artificial SEI: Polymer or inorganic coatings

5.7.1.1. *Polymer coatings*

In one very early study, a protection layer was introduced to the surface of the Li electrode to enhance the charge/discharge performance of Li–S batteries.[151] The protection layer was formed by a cross-linking reaction of a curable monomer (poly(ethylene glycol) dimethacrylate) in the presence of liquid electrolyte and a photoinitiator (methyl benzoylformate) by the UV curing method. The thickness of the protection layer was about 10 μm. When the Li electrode was coated with the protection layer, the cells showed an enhanced charge/discharge performance as compared to cells without the protection layer, resulting in an average discharge capacity of 270 mAh $g^{-1}_{cathode}$ during 100 cycles.

The surface of a Li electrode was also modified by spin-coating a poly(VC-co-acrylonitrile) copolymer solution in DMSO onto the Li metal to form an ionically conductive P(VC–co–AN) layer as a SEI.[158] The thin protective polymer layer with strong adhesion to the Li electrode suppressed the corrosion of Li metal and stabilized the interface in prolonged contact with organic electrolyte. Not only storage, but the cycling performances of Li/LiCoO$_2$ cells were also improved by using the surface modified Li electrode. SEM analysis of Li electrodes after repeated cycling revealed that the surface modification effectively suppressed dendrite growth during cycling, which resulted in stable cycling characteristics compared to that of the cell with an unmodified Li electrode. Left in atmospheric conditions for 20 minutes, the unmodified Li metal corroded quickly to a dull silvery gray and then black tarnish, due to the formation of Li hydroxide (LiOH), Li nitride (Li$_3$N) and Li carbonate (Li$_2$CO$_3$) on the surface. On the contrary, the P(VC–co–AN) modified Li metal changed little with time, suggesting the effective protection from O$_2$, N$_2$ and H$_2$O in the atmosphere.

Although single ion conducting ionomers can induce more homogeneous Li electrodeposition by preventing Li$^+$ depletion at Li surface, currently available materials do not allow room temperature operation due to their low room temperature conductivities. In the work of Song et al. a highly conductive ionomer/liquid electrolyte hybrid layer was fabricated by laminating a few micron-thick Nafion layer on Li metal electrode followed by soaking in 1 M LiPF$_6$ in EC:DEC (1:1) electrolyte.[149] It is well known that Nafion is a cation conductor, yet it only exhibits sufficient ionic conductivities when membranes are hydrated or swollen with an organic solvent to enable such cationic movement. The hybrid layer could realize stable Li electrodeposition at high current densities up to 10 mA cm^{-2} and permitted room temperature operation of Li/Li symmetric cells with low polarizations for more than 2000 hours, which corresponds to more than five fold enhancement compared with bare Li metal electrodes. These results demonstrate that the hybrid strategy successfully combines the advantages of bionic liquid electrolyte (fast Li$^+$ transport) and single ionic ionomer (prevention of Li$^+$ depletion). While this approach is surely anticipated to assist in even deposition of Li and limit the formation of dendritic structures, one issue envisioned is the ability of the swollen Nafion to protect Li from reaction between polysulfides and liquid electrolyte solvent molecules.

Such ionically-conductive organic films mentioned above have represented a step in the right direction, yet the utilization of a surface layer which is both ionically and electronically conductive provides additional benefits. This concept was first proposed by Fauteux *et al.*[150] where the *electronic* conductivity facilitates a uniform attraction of the particular alkali metal ions (such as Li-ions) through the surface layer and onto the anode during electrodeposition to suppress dendrite growth and, in turn, substantially increase the cycle life of the cell. In this invention, the surface layer applied to the negative electrode includes a polymer which will form radical anions upon contact with the negative electrode — such as vinylnaphthalene — and a second compound which will facilitate ionic conductivity such as polyethylene oxide.

The inventions by Skotheim *et al.* also relate to novel stabilized negative electrodes consisting of a Li metal coated with a thin film of an electroactive polymer capable of transmitting alkali metal ions interposed between the Li negative electrode and a polymer or liquid electrolyte.[159,160] Certain conjugated polymers are suggested, such as polyacetylene and poly(p-phenylene), which can be doped with Li-ions to be electrically conductive by direct contact of thin films with Li metal. The Li metal dissolves Li-ions which diffuse into the polymer structure up to a certain maximum concentration. The Li-doped polymer film provide a constant potential across the Li metal surface due to its high electrical conductivity, thereby providing thermodynamically more favorable stripping conditions without pitting of the electrode surface. Similarly, the plating takes place solely at the Li conducting polymer interface where a uniform potential is maintained. Therefore, no preferential deposition occurs at protruding areas of the Li surface. The thickness of the electroconducting film is 0.01–10 μm, preferably 0.1–5 μm. Evaporation in a vacuum is the preferred method of deposition of the electroconductive film because it provides dense films and complete surface coverage. Useful starting of polymers for the formation of the electroconductive polymer film may be any conjugated structure which is capable of being doped electrically conductive by Li-ions, such as poly(p-phenylene), polyacetylene, poly(phenylene vinylene), polyazulene, poly(perinaphthalene), polyacenes and poly(naphthalene-2,6-diyl).

In Li–S batteries, coating layers of conductive polymers, such as poly(3,4-ethylenedioxythiophene) (PEDOT),[161] polypyrrole (PPy),[162,163] and polyaniline (PANI)[164,165] have been applied to sulfur *cathodes* to improve the performance by confining the soluble polysulfide species at the cathode. Recently, Li *et al.* found that the capability of PEDOT in improving long-term cycling stability and high-rate performance of the sulfur cathode is the best among the three polymers.[166]

In one study however, PPy was coated via a planetary milling process onto Li powder to inhibit the chemical reaction with soluble polysulfides.[152] Li powder was used as the negative electrode material over a typical foil to further improve the cycling efficiency and diminish dendrite growth, as previously reported.[167–169] As displayed in Figures 5.11(a) and 5.11(b), the PPy-coated Li powder seems to possess a core–shell structure having about a 2 μm thick layer. The electrochemical performance of the PPy-coated Li powder anode cell was compared with that of an uncoated Li powder anode cell and found to increase the capacity retention of Li–S batteries (Figure 5.11(c)). The major role of the PPy coating is to suppress a reduction reaction between dissolved Li polysulfides in the liquid electrolyte and Li powder, which minimizes the overcharge and reduces capacity fading and the loss of active materials.

Poly(3,4-ethylenedioxythiophene) (PEDOT) has a high electrical conductivity and good electrochemical stability,[170–172] however, PEDOT is not ionically conductive. Thus, as a protective coating layer on Li metal, ion-conductive copolymers such as poly(3,4-ethylenedioxythiophene)–*co*–poly(ethylene glycol) (PEDOT–*co*–PEG)have been investigated.[153,154] In the copolymer, PEG is a highly ion-conductive polymer that transports Li-ions. However, the use of PEG homopolymer as a surface coating layer is not proper, since PEG is easily dissolved in the electrolyte solution. As a result, The PEDOT-*co*-PEG copolymer hardly dissolves in organic electrolyte and exhibits higher ionic conductivity.[153] Additionally, the PEDOT-*co*-PEG copolymer possesses strong adhesive properties to a Li surface, which makes it effective for mechanically suppressing Li-dendrite growth.

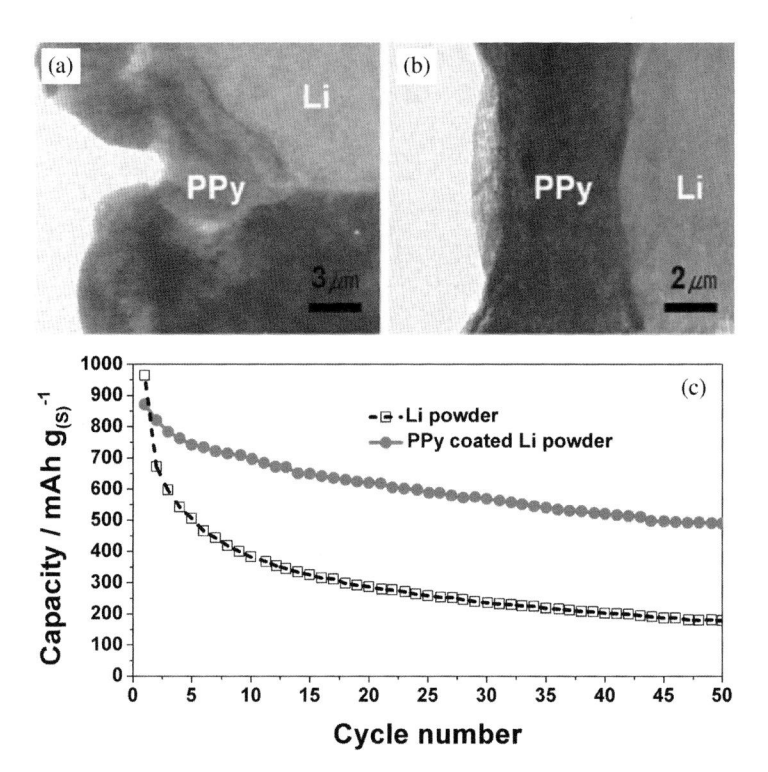

244 *Li–S Batteries: The Challenges, Chemistry, Materials and Future Perspectives*

Figure 5.11 TEM images of the PPy-coated Li powder (a, b) and the cycle performance of Li/S cells incorporating the uncoated Li powder and the PPy-coated Li powder (c).
Source: Reproduced from Ref. 152.

PEDOT–*co*–PEG was coated onto Li metal from a 1 wt.% dispersion in nitromethane by directly spin-coating in a dry box filled with argon gas and examined as a protective layer in Li/LiCoO$_2$ cells.[153] The thin conductive polymer with strong adhesion to the Li electrode caused the capacity retention of the Li/LiCoO$_2$ cell to increase from 9.3 to 87.3% after 200 cycles compared to the cell with the pristine Li electrode. The improvement in cycling stability was attributed to the conductive polymer coating suppressing Li dendrite growth and the deleterious reaction between the Li electrode and the electrolyte solution during cycling.

In a Li–S battery, a stable and less resistive SEI is formed between the ether-based electrolyte and the Li anode with a 10 μm thick

PEDOT–*co*–PEG protective layer,[154] which can effectively inhibit the corrosion reaction between the Li negative electrode and Li polysulfides. Particularly, with approximately 2.5–3 mg cm^{-2} sulfur loading on the electrode and commercial electrolyte, the discharge capacity remains at 815 mAh g^{-1} after 300 cycles at 0.5 C with an average CE of 91.3%. This type of polymer appears to be one of the most promising for protecting Li metal in Li–S batteries and warrants further examination. Particularly, spin coating was used to prepare films on Li in both cases,[153,154] yet it would be interesting to see the effect of different preparation techniques and thickness/homogeneity of the films.

5.7.1.2. Al_2O_3 coatings

Alumina (Al_2O_3) is an electrochemically inactive ceramic material with a large natural abundance and low cost.[155] For Li–S batteries, Al_2O_3 coatings have been applied to sulfur cathodes[173–175] and separators[176] to adsorb soluble polysulfides and limit their diffusion to the negative electrode. Recently, Al_2O_3 has been coated onto the Li negative electrode to restrict the side reactions between soluble polysulfides and Li by acting as a physical separation barrier.[155]

Jing *et al.* used a spin-coating technique to coat Li electrodes with porous Al_2O_3 layers.[155] Although this is a great example of how such protective layers can aid in blocking contact between electrolytes and polysulfides with Li metal, it appears that optimizations of layer thickness and perhaps different fabrication techniques are required to ensure a more homogeneous coverage. In this case, it seems obvious that solvent molecules and/or polysulfides have the potential to diffuse through the "porous" Al_2O_3 layer. Additionally, since Al_2O_3 is inert to electron and ion conduction, the layers must be thin enough not to completely shut down electrochemical pathways for Li stripping and deposition. It should be noted that if the layer is thin enough, and lithiated, the $Li_xAl_2O_3$ alloy phase is ionically conductive.[177]

Atomic layer deposition (ALD) is a modified chemical vapor deposition (CVD) process in which precursors react with the substrate in self-limiting half reactions, pulsed cyclically to build up coatings one atomic layer at a time.[156] ALD is ideally suited for Li metal protection due to its unique properties of angstrom-scale thickness control, pinhole-free conformal films, and low temperature deposition below the melting point of Li (180°C).[157] ALD coatings have proven to be effective passivation layers for metals such as Cu,[178] Mg,[179] and steel[180] from corrosion in electrolytes and even on reactive metals such as Ca.[181] In batteries, thin ALD coatings have been applied to non-metallic anodes[182–185] and cathodes[186–188] to improve battery cycling performance; however, to maintain high ionic conductivities without increasing cell impedance, these ALD coatings are typically less than 2 nm thick.

In the work of Kozen et al. much care was taken in ensuring a uniform coverage of Al_2O_3 directly on Li metal foil using a unique ultrahigh vacuum (UHV) system and thickness of the layers were optimized.[157] The authors claim that for adequate protection in a Li–S environment, 14 nm thick layers are required. With this 14 nm thick Al_2O_3 coating on Li, the capacity retention for Li–S cells was ~90% after 100 cycles, vs. only 50% without the coating (Figure 5.12).

A follow up study was done on much thinner Al_2O_3 layers (~2 nm) to specifically gather an understanding of how ALD coatings of Al_2O_3 onto

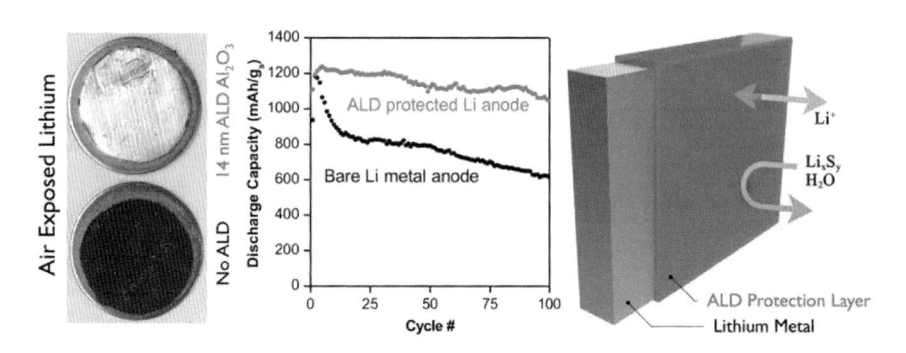

Figure 5.12 14 nm Al_2O_3 coatings on Li metal electrodes by ALD. The left panel displays the protected Li in comparison to unprotected Li after exposure to atmosphere at 25°C and 40% relative humidity for 20 hours. The middle panel shows the cycle performance of Li–S cells with protected and unprotected Li anodes. A schematic of the function of the Al_2O_3 protection layer in inhibiting contact between Li and contaminant species is displayed in the right panel.

Source: Reprinted with permission from Ref. 157. Copyright (2015) American Chemical Society.

Li aid in the suppression of dendrite formation and their effect on cycling Li/electrolyte/Li symmetrical cells.[156] Both groups show improvements with Al_2O_3 coatings in their own systems of study. The 2 nm thick ALD coatings on Li were able to extend the cycle life of Li vs. Li cells cycled at 1 mA cm^{-2} from approximately 700 cycles for non-coated Li to 1,259 cycles.[156] In this study, even 3–4 nm coatings were deemed to be too thick and did not result in significant improvement over the non-coated electrodes. This discrepancy in the optimum thickness indicates that testing and optimizations need to be performed in the specific battery type of interest. For example, Li protection methods which work well in LIBs might not perform well in Li–S batteries where the presence of polysulfide species needs to be taken into consideration.

5.7.2. Solid electrolytes

A priori, a straightforward approach to the crossover of sulfur redox species to the negative electrode in Li–S batteries would be to rely on a non-liquid electrolyte to ensure Li^+ conduction. In addition to eliminating the polysulfide shuttle, a solid electrolyte is also expected to enable stable cycling of metallic Li by preventing dendrite formation. As demonstrated with conventional Li^+ insertion chemistry, the removal of any liquid component from Li metal or even LIBs is a game changer in terms of reliability and safety for practical applications.[189,190]

5.7.2.1. *Solid polymer electrolytes with inorganic fillers*

Polymer electrolytes can be divided between gel polymer electrolytes (GPEs) and SPEs. Typically, a GPE consists of a polymer matrix swollen by a non-aqueous electrolyte so that the transport properties are fairly similar to that of the neat liquid electrolyte. On the other hand, SPEs are solvent-free electrolytes in which Li salts are dissolved without any liquid component in high molecular weight polymers.[191] In the pioneer work of Jeon *et al.* a dry poly(ethylene oxide) (PEO) based-SPE was incorporated in sulfur or Li_2S composite positive electrodes, yet the pure electrolyte layer still contained TEGDME.[192] In fact, the poor cycling performance was assigned to morphology changes in the sulfur electrode, and in particular, to sulfur aggregation upon repeated dissolution/precipitation, confirming that GPEs do not offer a suitable physical barrier to sulfur intermediates. Unfortunately, the ionic conductivity of SPEs is generally in the order of 10^{-4} S cm^{-1} at 70–90°C (or 10^{-7} S cm^{-1} at room temperature vs. 10^{-2} S cm^{-1} for liquid non-aqueous electrolytes) so that only thin films can be used to limit the cell impedance, which in turns can reduce the physical barrier between electrodes (and resistance to internal shorting).[193,194]

To improve the conductivity and reduce the interfacial resistance between the Li electrode and the electrolyte, nanometric inorganic fillers have been demonstrated in Li-ion and Li–S applications.[32,193] In one of the first reports of GPE applied to Li–S batteries, Hassoun and Scrosati were using a hot-pressed SPE membrane of poly(ethylene oxide)-Li triflate (PEO-LiCF$_3$SO$_3$) containing finely dispersed zirconia nanoparticles and Li sulfide as an additive.[195] While the effect of the Li$_2$S additive was not clear, the nano ZrO$_2$ filler had been previously demonstrated by the same group to stabilize the Li/electrolyte interface by lowering the kinetics of SEI formation and preventing dendrite growth.[196] At 70°C, the charge–discharge CE of the all-solid-state Li/S cells approached 100%; yet lower current densities showed lower CE, so that parasitic redox shuttle of polysulfide intermediates cannot be ruled out even in such solvent-free system.[195] Among other representative examples of GPE-containing Li/S cells, Liang et al. later studied a nano SiO$_2$-filled PEO-LiTFSI polymer electrolyte associated to a S/mesoporous-carbon (MPC) positive electrode.[197] Their all-solid-state polymer battery showed promising cycling performance with a specific capacity of about 800 mAh g$_{(S)}^{-1}$ at 70°C after 25 cycles, yet no CE was reported. Instead, EDX analyses of the sulfur content on the surface of Li electrodes were carried out after 10 cycles against different sulfur positive electrodes. The amount of sulfur detected on the Li surface cycled against the S/MPC electrode was much lower than that of the Li cycled against unconfined sulfur, which validates the positive electrode design yet suggests that the SPE was offering an insufficient barrier to sulfur intermediates diffusion. In fact, polyethers — PEO and its derivatives in particular — constitute the most studied polymer hosts in Li metal polymer cell configurations. The dissolution and dissociation of Li salts are similar to glymes such as TEGDME, which share the sequential oxyethylene groups of PEO.[30] Therefore, it is not really surprising that such polymer electrolytes can also solvate Li polysulfides.[30,198]

5.7.2.2. Triblock single Li$^+$-conductor polymer electrolytes

Recently, Wang et al. proposed to use mixture of nanostructured block copolymer electrolytes of polystyrene-b-poly(ethylene oxide) (SEO) and LiTFSI to mitigate polysulfide diffusion within SPEs.[198] Although they showed that in Li–S cells, SEO minimized the migration of polysulfides relative to PEO, this suppression was still inadequate for practical application. Other works by Bouchet et al. had shown that BAB triblock polymer electrolyte consisting of a polyelectrolyte based on poly(styrene trifluoromethanesulphonylimide of Li) P(STFSILi) (block "B") associated with a central PEO block ("A") demonstrate single Li$^+$ ion conduction ($t^+ > 0.85$), with a markedly improved mechanical strength and ionic conductivity,[194] compared to the neutral triblock copolymer such as SEO.[199] When used in Li metal batteries, such anionic block copolymer electrolytes effectively slow down the dendrite growth by an increased control of the interfacial reactions at the negative electrode, while also extending the electrochemical stability window (ESW) up to 5 V vs. Li$^+$/Li.[194] It is therefore likely that anionic SPEs would inhibit Li$_2$S$_n$ dissociation and impede polysulfide migration towards the Li electrode.

5.7.2.3. Ceramic ionic conductors and their stability at low potential

The second prominent family of solid electrolytes developed for metal and metal-ion batteries corresponds to inorganic glassy or ceramic materials, sharing the specificity of being single-ion (Li$^+$) conductors. This implies, in contrast with polymer electrolytes, that the Li$^+$ conduction mechanism is radically different from liquid electrolytes. First of all, there are no counter anions or solvent molecules to diffuse along, the Li⁺ transference number is therefore near unity, and the ionic conductivity at room temperature can be much higher. Of particular interest for the Li–S system, there is no possibility to solvate S_n^{2-} and conduction pathways inside the solid cannot accommodate large anions, even less allow their transport. The main criteria for electrolytes remains to be ionic conductivity, and in this respect, materials such as Li_2S–P_2S_5 glass-ceramics,[200,201] thio-LISICONs (Li_2S–GeS_2–P_2S_5),[202,203] thio-phosphates (LAGP: $Li_{1+x}Ge_{2-y}Al_y(PO_4)_3$ and LATP: $Li_{1.3}Al_{0.3}Ti_{1.7}(PO_4)_3$),[204,205] or garnets (LLZO: $Li_7La_3Zr_2O_{12}$) have been reported to exceed 10^{-4} S cm⁻¹ at room temperature. The record is held by $Li_{10}GeP_2S_{12}$ (1.2 10⁻² S cm⁻¹) owing to its 3D structure,[206] however its stability towards metallic Li is limited, similarly to LAGP, due to the reactivity of germanium centers.[207] Another issue faced at potentials close to 0 V — even for electrolyte materials stable against metallic Li such as the Li_2S–P_2S_5 glass-ceramics — is the dendritic growth (see Section 4). At the solid/solid interface, this phenomenon still occurs through grain boundaries and interconnected pores across the solid electrolyte even if the density of the layer is as high as 96%, or as a consequence of cracks generated along Li plating.[69,208] For these two reasons, ceramic electrolytes are generally coupled instead with Li alloys such as Li-Al or even Li-In, which exhibit higher potentials (0.38 V[209] and 0.62 V[206] vs. Li⁺/Li, respectively), and therefore, compromise the energy density of the resulting Li/S cells with an average discharge voltage of 1.3 V.[210] Nonetheless, as anticipated from the immobilization of sulfur species within a Li⁺-single conductor network, the CE reported for these all-solid-state Li–S cells was always close to 100%. Also, the discharge curves all show only one plateau at *ca.* 2.1 V (reported to Li⁺/Li) suggesting that the ideal electrochemical reaction of $2Li + S \leftrightarrows Li_2S$ proceeds in the solid state while spectroscopic studies did not show any evidence for polysulfide formation.[211] As mentioned above, achieving a proper interface between metallic Li and one of these superionic conductors would considerably increase the energy density of these systems by stepping-up the voltage (compared to a cell using a Li–In negative electrode); considering stable discharge capacities of 1100–1300 mAh $g_{(S)}^{-1}$ were obtained in an all-solid-state configuration,[203,210] there is no doubt such cells would find their way to the market.

In fact, chemical and/or electrochemical instability towards metallic Li can be alleviated by the incorporation of a thin buffer layer between the bulk ceramic electrolyte and the Li negative electrode. For instance, Sahu et al. successfully achieved chemical passivation of an arsenic-substituted Li_4SnS_4 solid electrolyte surface[212] using a Li-stable $3LiBH_4$–LiI composite.[213] Note that due to the low redox potential of halides in solids, these materials tend to develop a highly resistive layer at the positive electrode interface and therefore demonstrated poor cyclability when used as bulk electrolytes in Li–S cells.[214,215] Nevertheless, their high deformability is an asset when it comes to engineering a solid-solid interface and in that work, the buffer layer was applied by simple dip-coating in a 5 wt.% $3LiBH_4$–LiI solution in THF.

Shown in Figure 5.13(a) is a comparison of galvanostatic cycling for symmetrical Li|$Li_{3.833}Sn_{0.833}As_{0.166}S_4$|Li cells before and after the protective halide coating. The pristine electrolyte is not compatible with the metallic Li electrodes, and interfacial reaction between the solid electrolyte and the newly deposited Li causes spikes in the cell voltage. In contrast, a smooth cell voltage was achieved after $3LiBH_4$–LiI was coated. The compatibility of the coated electrolyte with metallic Li was further evidenced by CV on a Li|$Li_{3.833}Sn_{0.833}As_{0.166}S_4$|Pt cell (Figure 5.13(b)). In particular, the

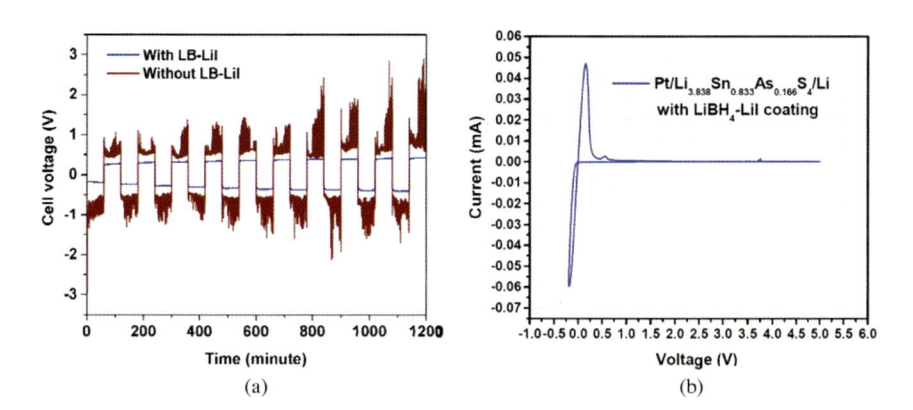

Figure 5.13 (a) Symmetrical cell test before and after $3LiBH_4$–LiI coating. (b) Cyclic voltammogram of a Pt|$Li_{3.838}Sn_{0.838}As_{0.166}S_4$|Li cell with $LiBH_4$–LiI passivation layer. The small peak observed at 0.53 V is attributed to the dealloying of Li–Pt.

Source: Reproduced from Ref. 212 with permission of The Royal Society of Chemistry.

cathodic current was observed right at 0 V, indicating that no side reaction happened during Li deposition. A sharp anodic peak was observed between 0 and 0.3 V referring to Li dissolution. Although this example does not concern the Li–S system, this proves the concept of a buffer layer for solving the compatibility issue of Li-ion conductors with metallic Li.

5.7.2.4. Ceramic glass membranes used as a separator

A similar idea consists of coating a GPE onto the rigid ceramic electrolyte. In this case, the buffer layer is also expected to confer its flexibility to the solid electrolyte assembly, to better accommodate volume changes at the negative electrode.[216,217] One important example of such a flexible composite material was developed by Ohara Corporation, composed of particles of a Li-ion conducting glass-ceramic material, such as described above, and a SPE based on PEO–Li salt complexes. Based on the same idea, Lee et al. assembled a flexible LATP–PTFE–LiPF$_6$–EC/DMC composite electrolyte which almost maintained the room temperature ionic conductivity of the LATP ceramic (2.94 10^{-4} and 8.36 10^{-4} S cm^{-1}, respectively), while the thickness of the film could be decreased from 0.26 to 0.1 μm.[218] Decreasing the thickness of ceramic-based films while maintaining good mechanical properties is a major achievement, however, such composite membranes cannot be used in Li–S batteries, since the liquid component would either dissolve polysulfides and allow their migration towards the Li negative electrode or even react with the nucleophilic intermediates in the case of carbonate solvents. Meanwhile, Aetukuri et al. most recently demonstrated that one-ceramic-particle-thick–polymer composite membranes could prevent the growth of dendrites.[219] In brief, 75–90 μm large Li$_{1.6}$Al$_{0.5}$Ti$_{0.95}$Ta$_{0.5}$(PO$_4$)$_3$ particles were embedded in a cyclo-olefin polymer matrix with their top and bottom surfaces exposed to allow for ionic transport. Despite its low shear modulus, there is no driving force for dendrite formation through such polymer matrix because it is a Li-ion insulator. Besides, it exhibits immeasurably low solubility and swelling in most battery electrolyte solvents, so that the positive and negative electrode sides can potentially be fully isolated. The drawback of this approach is that it does not address the chemical incompatibility of the ceramic electrolyte with metallic Li, nor does it provide good solid-solid interfaces with the electrodes. This may be circumvented, for example, by placing a porous separator infiltrated with a liquid or gel electrolyte between each electrode and the solid Li-ion-conductive membrane,[220] in which case the electrolyte in contact with metallic Li must be tailored to prevent dendrite growth and minimize interfacial reactions (see Section 3).

5.7.3. The protected lithium electrode: Myth or reality?

The concept of a PLE, although very generically termed, designates the coupling of a dense, pinhole-free Li$^+$ solid electrolyte with a barrier layer comprised of an artificial SEI (see above Li$_3$N) or polymer electrolyte to isolate the Li electrode both chemically and physically. In contrast with all the approaches mentioned previously, the PLE architecture provides a hermetic enclosure for the active metal at the negative electrode, therefore enabling the use of electrolytes stable to nucleophiles but not Li metal,[41,62] or even aqueous electrolytes[221] in the positive electrode compartment.

5.7.3.1. Principle: Buffer layer + dense ceramic membrane

As described above, the ceramic electrolyte most commonly used in PLE assemblies is the LATP produced by Ohara Inc. in Japan and Corning in the USA. Visco *et al.* were the first to solve the problem of chemical incompatibility with Li by introducing a barrier layer (a solid layer such as Cu$_3$N, LiPON, or non-aqueous electrolyte) between the Li metal and the Ohara glass.[222] Polyplus has since patented numerous negative electrode architectures consisting of layered assemblies,[223,224] some of which include polymer adhesive barrier to hermetically seal the joint between the metal electrode (including its protective membrane) and the container.[221,225] They demonstrated these protected negative electrode architectures in Li–S cells,[225] as well as in both non-aqueous and aqueous Li/air cells.[221]

The barrier layer can be made of a glass (Li silicate, phosphate, phosphorus oxynitride, etc.), as described in a previous section or made from an organic polymeric material such as a nitrogen or phosphorus containing polymer. In a preferred battery of Koksbang, the protective layer contains polyvinylpyridine (PVP) or derivatives thereof — preferably poly–2–vinylpyridine or poly–2–vinylquinoline — and iodine complexed with the PVP for reducing passivation of the lithium-containing negative electrode (see Figure 5.14(b)).[226] From these numerous patents, it emerges that the role of this layer is two fold: (i) To ensure chemical compatibility between the Li metal and the solid electrolyte, as seen in Figure 5.14(a) for Cu_3N, and (ii) ensure proper contact between the Li metal and the solid electrolyte. This is realized either with the help of a foil bonding layer (capable of forming a bond with the active metal) such as aluminum (Al), or by hot pressing the active metal to the barrier layer.[227] Note that bonding the active metal to the barrier layer typically involves evaporating or sputtering the active metal onto the barrier layer laminate as shown

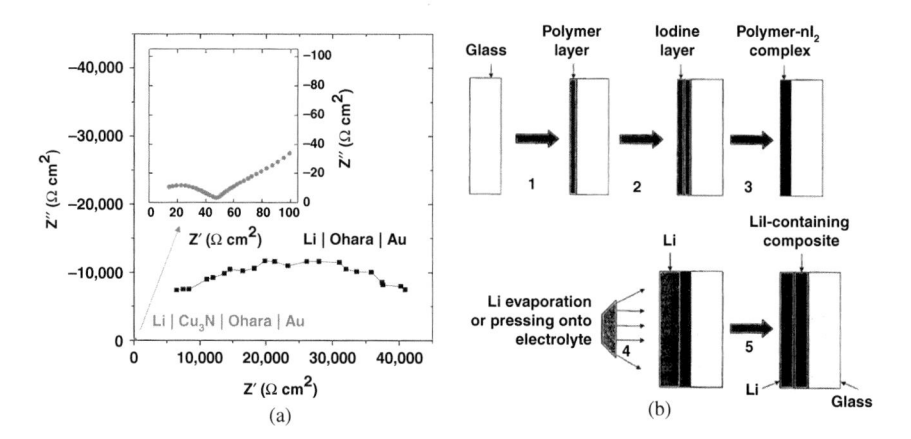

Figure 5.14 (a) Impedance plots for cells having unprotected and Cu_3N-protected Ohara material, showing the performance benefits of ionically protective layer. (b) One of the possible method of making an electrochemical device structure incorporating the protective layer.

Source: Adapted and reproduced from Ref. 223, not subject to copyright.

in Figure 5.14(b). In fact, the methods proposed for fabricating a PLE are of primary importance. Most commercially available Li foils display a surface roughness of the same order of magnitude as the thickness of the barrier layer (0.1–5 μm), so that most deposition processes would not form an adherent gap-free barrier layer on the Li surface.

5.7.3.2. Multilayered assemblies

For the cells to display long cycle life, high Li cycling efficiency and high-energy density, a multilayered structure is favorable to the PLE (Figure 5.15). By increasing the number of layers, a tortuous pathway is created to prevent the Li electrode to be accessed by unwanted species through pinholes as easily as when using a single layer. Some of the layers, and especially the outer layer might be tailored to increase the barrier properties by decreasing the transport of water, electrolyte or redox species towards the Li electrode.

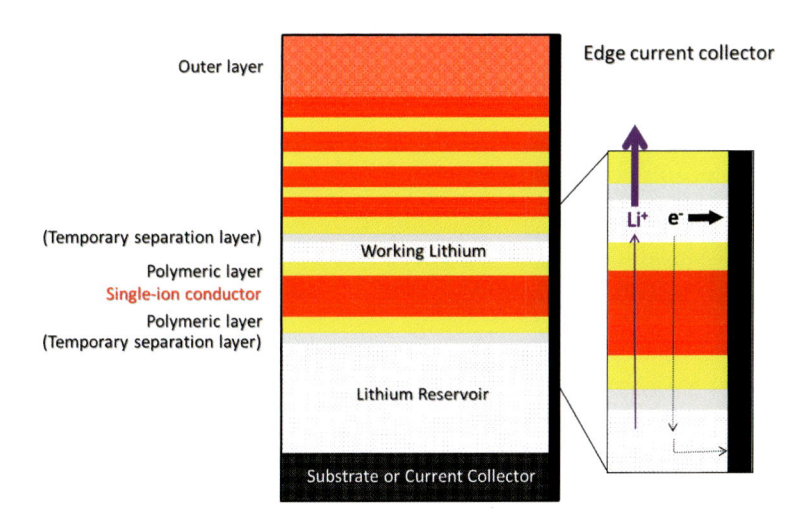

Figure 5.15 Structure for use in an electrochemical cell, including several multilayered structures, including an embedded layer in between the working Li layer and the Li reservoir. The inset shows the lowest resistance path for electron current (in black) through the edge current collector. Li$^+$ flow across the embedded layer only to replace the small percentage of Li lost in the working Li layer.

Source: Adapted from Ref. 228, not subject to copyright.

In addition to isolating the negative electrode from environments that are typically unsuitable for Li, the performance of a PLE in an electrochemical cell depends on the ionic conductivity and thickness of both the polymeric barrier and the single-ion conductor layer to control the flux of Li$^+$ from the electrolyte to the Li electrode. Where the conductivity is about 10^{-7} S cm^{-1}, the thickness should be limited from 0.25 to 1 micron, whereas materials exhibiting conductivity of 10^{-4} to 10^{-3} S cm^{-1} can afford to be laminated at 10–1000 microns — in particular for the dense glass or ceramic single-ion conductor layer — and more preferably between 10 and 100 microns.[222] Besides, Affinito *et al.* recently proposed to introduce a Li stabilization layer positioned within the PLE assembly to control depletion and replating of Li upon charge and discharge of a battery.[228] As shown in Figure 5.15, the thickness of the working Li layer is selected such as this layer is lost upon full discharge, while the Li-ion conductor layers embedded underneath shield the bottom Li reservoir from the damages of high Li$^+$ flux. In this example, the embedded layer does not display any substantial electronic conductivity and therefore most of the current passes through the working Li layer to the edge current collector (Figure 5.15 inset). Only when the working Li layer loses active Li during cycling is this Li replaced by Li$^+$ flowing from the reservoir so that this one does not degenerate into moss or islands. In the concept of "hybrid negative electrode" (see Section 5.2.2), only the interfacial reactions are displaced from the Li electrode while the present configuration fulfills an additional role which is to displace the plating and stripping away from the bulk Li electrode.

5.7.3.3. *Reality check: The attainable gravimetric and volumetric energy density*

Very recently, McCloskey published an interesting study in which he calculated the techno-economic constraints applied to the Li metal solid protective separator in terms of thickness (hence the final specific energy and energy density) but also in terms of cost, in order to make Li–S cells competitive against state-of-the-art Li-ion systems.[229] Indeed, assuming the issue of Li instability solved, sulfur utilization and aging phenomena in a PLE/sulfur cell should be mostly dictated by sulfur electrochemistry in the cathode compartment. Because the electronic and ionic conductivities of elemental sulfur and Li$_2$S are so low, all-solid-state Li–S systems require large amounts of carbon and solid electrolyte, while catholyte-type cells typically use a very high electrolyte-to-sulfur ratio to support appropriate polysulfide dissolution. In either case, practical specific energy (Wh kg^{-1}) and energy density (Wh L^{-1}) are severely diminished. And yet, any cathode/catholyte optimization would be rendered unavailable if the PLE itself is too thick and/or too expensive.

Using the calculation tables provided by McCloskey,[229] we show in Figure 5.16 that a "realistic" sulfur positive electrode (75 wt.% S, 1000 mAh g$_{(S)}^{-1}$) cannot deliver the energy density of an optimized Graphite/LiCoO$_2$ cell,[230] even for extremely high ariel loading (15 mg$_{(S)}$ cm^{-2}) and without considering any PLE (i.e. for thickness = 0, Figure 5.16(a)). Note that we consider here a striving electrolyte to sulfur ratio (3 µL mg$_{(S)}^{-1}$),[33] which in the absence of the PLE would rapidly dry-out owing to Li morphology development and side-reactions (see Section 1).[3] However, the same sulfur electrode can offer a step-up in specific (gravimetric) energy, providing the solid electrolyte separator is thin enough (Figure 5.16(b)). In fact, the 150 µm thick LICGC™ membranes commercialized by Ohara Inc.[231] would bring the Li–S cell to its "break-even point" with the Li-ion reference. The break-even thickness for the solid electrolyte separator depends on its density and therefore sulfides and polymeric protection layers offer the best options (Figure 5.16(c)). Last but not the least, cost considerations play an important role in the development of Li–S systems owing to the inexpensiveness of sulfur (0.22$ kg^{-1}). Yet, it appears from McCloskey's study that the solid separator/electrolyte would need to be less than 10$ m^{-2} for Li–S to be cost-competitive against LIBs (Figure 5.16(d)).[229] At this stage, for research purposes, 4 cm^2 LICGC™ membranes sell at around $250 each.

Unsurprisingly, the thinner the solid separator in a PLE configuration, the higher the attainable specific energy for the whole PLE/S cell (and energy density to a lesser extent). What these simple calculations do not show is the importance of the ionic conductivity and the shear modulus of the protection layer to allow the homogeneous current distribution to the Li metal and prevent dendrite growth. Deposition of ultra-thin ceramic layers onto Li (or vice versa, see Figure 5.14(b)) is still a challenge and

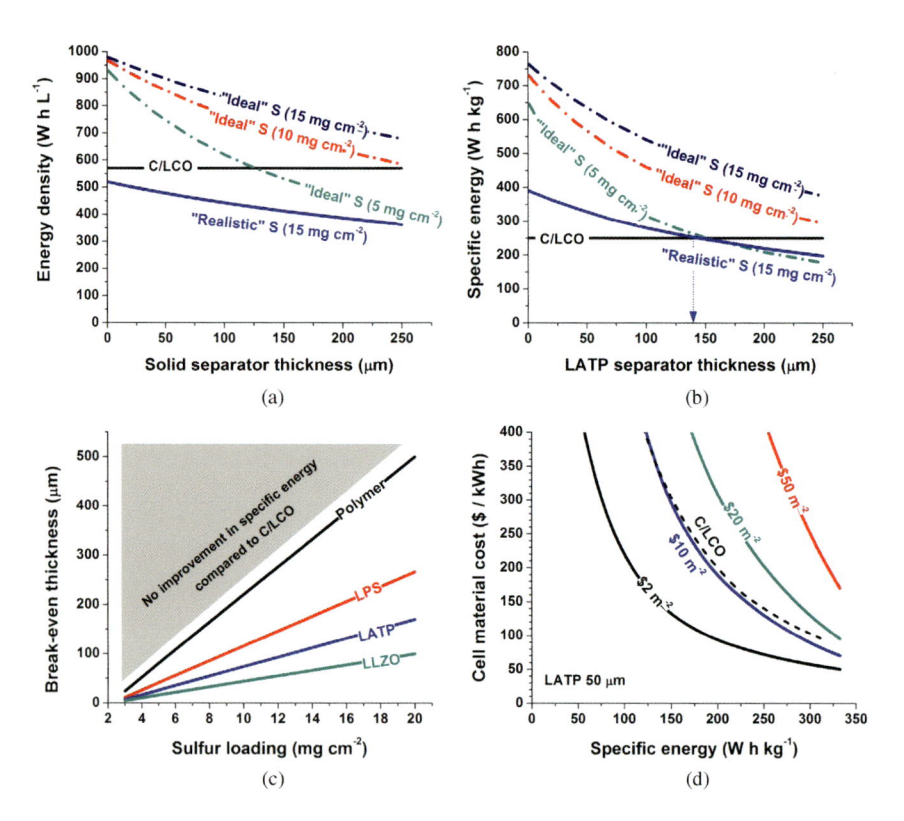

Figure 5.16 Attainable volumetric (a) and specific (b) energy densities for "ideal" sulfur electrodes (100 wt.% sulfur, 100% utilization) and a thick "realistic" positive electrode (75 wt.% sulfur, 60% utilization) as functions of the solid separator thickness. For the specific energy, a density of 3.0 corresponding to LATP is used. Dotted horizontal lines correspond to a state-of-the-art Graphite/LiCoO$_2$ cell, so that intersection with a calculated curve gives the "break-even" thickness for each cell configuration. (c) Break-even thickness for different solid separators calculated vs. sulfur loading. (d) Cell material costs for Li–S cells depending on the solid separator cost for a 50 μm thick LATP. A 20% excess Li at the negative electrode and an electrolyte to sulfur ratio of 3 μL mg$_{(S)}^{-1}$ are considered in all cases. All calculations are original and were performed using spreadsheets available in.[229]

even the one-ceramic-particle-thick–polymer composite membranes described above were about 100 μm thick.[219] Regarding the attainable specific energy, it is very clear that light polymeric protection layers should be preferred; (Figure 5.16(c)), which are also more likely to be cheaper and easier to scale-up. An interesting research avenue would be to combine classes of polymers which showed interesting properties such as electronically conductive PPy[152] and BAB anionic block copolymer electrolyte[194] within a multilayer assembly (Figure 5.15). The outer layer should be tailored to be stable against polysulfide intermediates as well as not swellable by the liquid electrolyte.

5.8. Conclusions

If the corrosion and consumption of Li by polysulfides (and electrolyte solvents/salts) can be avoided, Li deposition and stripping can approach 100% CE. This is ideal for engineering high-energy density Li–S batteries which require little to no excess of Li metal in the negative electrode. Overviewed in this chapter were the various methods used for protecting metallic Li electrodes from corrosion by chemical reaction with electrolytes and polysulfides produced during the discharge reaction of Li–S batteries. Additionally, certain battery failure mechanisms, including short-circuits and thermal run-away, by dendrite formation during charge of the Li–S battery can be avoided with proper protection strategies.

The "holy-grail" of Li sulfur batteries is to achieve near-theoretical specific energies and energy densities (2500 Wh kg^{-1} and 2800 Wh L^{-1}, respectively) with long-term cycle life. In theory, if problems at the negative electrode can be overcome, no excess Li metal would be required. This would be a major step forward in achieving such high-energy densities and prolonged cycling. In fact, by starting with Li$_2$S in the positive electrode, no active negative electrode material (i.e. Li metal) would be required during assembly of Li–S batteries. However, this proposed design of Li–S batteries necessitates all issues related to every component of the cell (negative electrode, positive electrode, electrolyte/separator, and current collectors) be tackled first. This chapter was designed to summarize work that has already been performed and provide guide for future work on lithium metal electrode in Li–S batteries; one of the many challenges of this promising high-energy storage system.

Considering all the challenges of the Li metal electrode discussed in this chapter, other cell configurations that either employ a prelithiated non-Li-metal anode or a Li-containing cathode (typically, Li$_2$S) coupled with a metal-free anode might appear like a tempting option to solve the Li-related problems.[77,94] Unfortunately, alternative negative electrode materials either possess low specific capacity such as graphite (360 mAh g^{-1} for graphite, vs. 3860 mAh g^{-1} for metallic Li),[4,5] or have problems of their own that limit their capacity practically: discharge of silicon is typically cut-off to 1000 mAh g^{-1} out of the 3580 mAh g^{-1} theoretically achievable to ensure reversible cycling.[232,233] In terms of cell voltage, a 200 mV shift by replacing Li by silicon for instance, results in an additional 10% decrease in energy. Moreover, sulfur is usually used as a positive electrode material in its elemental form (i.e. charged state), so that prelithiation of the negative electrode is required, hence bringing additional manufacturing challenges. Overall, the coupling of a "practical" silicon negative electrode with a realistic sulfur positive electrode,[229] (at infinite thickness) would bring the specific energy down to 300 Wh kg^{-1}, i.e. close to that of using a Li metal electrode with 300% excess Li (350 Wh kg^{-1}).

It is not the scope of this chapter to discuss the concept of Si–Li–S cells,[234] yet we would like to draw attention to the similarities and differences between the Li and the silicon electrodes, in order to highlight that what we learned here on Li can apply to silicon or other metal-free electrodes. For instance, it was found that in order to achieve prolonged cycle life for Si negative electrodes and minimize the cumulative capacity loss, it was necessary to prelithiate the silicon and also to limit the capacity to 1000 mAh g^{-1} in order to keep it working at low potential (e.g. 20–600 mV vs. Li$^+$/Li). At these potentials, self-discharge and polysulfide redox shuttle would occur just as readily as on metallic Li, so that surface passivation must be achieved as well. Some of the chemical and physical methods described here, such as the silane pretreatment or the ALD deposition of Al$_2$O$_3$ are, in fact, easier to realize on SiO$_2$/SiOH rich surface than on Li metal. In summary, with the exception of morphological control of Li plating/stripping processes, almost all of the techniques outlined in the chapter to increase the CE and form a stable SEI can be extended from Li metal to alternative negative electrode materials.

Bibliography

1. J. Akridge, Y. Mikhaylik and N. White, *Solid State Ionics*, **175** (2004) 243–245.
2. P. G. Bruce, S. A. Freunberger, L. J. Hardwick and J.-M. Tarascon, *Nature Materials*, **11** (2011) 19–29.
3. Y. V. Mikhaylik, I. Kovalev, R. Schock, K. Kumaresan, J. Xu, and J. Affinito, *ECS Transactions*, **25** (2010) 23–34.
4. B. Diouf and R. Pode, *Renewable Energy*, **76** (2015) 375–380.
5. J.-M. Tarascon and M. Armand, *Nature*, **414** (2001) 359–367.
6. E. Peled, *Journal of the Electrochemical Society*, **126** (1979) 2047.
7. D. Aurbach, *Journal of the Electrochemical Society*, **143** (1996) 3525.
8. D. Aurbach, I. Weissman, A. Zaban and O. Chusid, *Electrochimica Acta*, **39** (1994) 51–71.
9. A. Kominato, E. Yasukawa, N. Sato, T. Ijuuin, H. Asahina and S. Mori, *Journal of Power Sources*, **68** (1997) 471–475.
10. D. Aurbach, *Solid State Ionics*, **148** (2002) 405–416.
11. D. Aurbach, *Journal of the Electrochemical Society*, **142** (1995) 2873.
12. D. Lv, Y. Shao, T. Lozano, W. D. Bennett, G. L. Graff, B. Polzin, J. Zhang, M. H. Engelhard, N. T. Saenz, W. A. Henderson, P. Bhattacharya, J. Liu and J. Xiao, *Advanced Energy Materials*, **5** (2015) 1400993.
13. M.-C. Lin, M. Gong, B. Lu, Y. Wu, D.-Y. Wang, M. Guan, M. Angell, C. Chen, J. Yang, B.-J. Hwang and H. Dai, *Nature*, **520** (2015) 324–328.
14. Z. Rong, R. Malik, P. Canepa, G. Sai Gautam, M. Liu, A. Jain, K. Persson and G. Ceder, *Chemistry of Materials*, **27** (2015) 6016–6021.
15. S.-W. Kim, D.-H. Seo, X. Ma, G. Ceder and K. Kang, *Advanced Energy Materials*, **2** (2012) 710–721.
16. W. Xu, J. Wang, F. Ding, X. Chen, E. Nasybulin, Y. Zhang and J.-G. Zhang, *Energy & Environmental Science*, **7** (2014) 513.
17. R. Cao, W. Xu, D. Lv, J. Xiao and J.-G. Zhang, *Advanced Energy Materials*, **5** (2015) 1402273.
18. Y. V. Mikhaylik and J. R. Akridge, *Journal of the Electrochemical Society*, **151** (2004) A1969–A1976.
19. M. R. Busche, P. Adelhelm, H. Sommer, H. Schneider, K. Leitner and J. Janek, *Journal of Power Sources*, **259** (2014) 289–299.
20. K. Xu, *Chemical Reviews*, **104** (2004) 4303–4418.
21. E. Peled, *Journal of the Electrochemical Society*, **144** (1997) L208.
22. T. Fujieda, N. Yamamoto, K. Saito, T. Ishibashi, M. Honjo, S. Koike, N. Wakabayashi and S. Higuchi, *Journal of Power Sources*, **52** (1994) 197–200.

23. K. Kanamura, *Journal of the Electrochemical Society*, **144** (1997) 1900.
24. D. Aurbach, *Journal of the Electrochemical Society*, **136** (1989) 1611.
25. M. Odziemkowski, *Journal of the Electrochemical Society*, **139** (1992) 3063.
26. D. Aurbach, *Journal of Power Sources*, **89** (2000) 206–218.
27. M. Ishikawa and M. Morita, Current issues of metallic lithium anode, In: G.-A. Nazri and G. Pistoia (Eds.), *Lithium Batteries*, Springer US, (2003) pp. 297–312.
28. D. Aurbach, M. Daroux, P. Faguy and E. Yeager, *Journal of Electroanalytical Chemistry and Interfacial Electrochemistry*, **297** (1991) 225–244.
29. J. Gao, M. A. Lowe, Y. Kiya and H. D. Abruña, *The Journal of Physical Chemistry C*, **115** (2011) 25132–25137.
30. T. Yim, M.-S. Park, J.-S. Yu, K. J. Kim, K. Y. Im, J.-H. Kim, G. Jeong, Y. N. Jo, S.-G. Woo, K. S. Kang, I. Lee and Y.-J. Kim, *Electrochimica Acta*, **89** (2013) 737–743.
31. T. Yim, M.-S. Park, J.-S. Yu, K. J. Kim, K. Y. Im, J.-H. Kim, G. Jeong, Y. N. Jo, S.-G. Woo, K. S. Kang, I. Lee and Y.-J. Kim, *Electrochimica Acta*, **107** (2013) 454–460.
32. S. Zhang, K. Ueno, K. Dokko and M. Watanabe, *Advanced Energy Materials*, **5** (2015) 1500117.
33. M. Hagen, D. Hanselmann, K. Ahlbrecht, R. Maça, D. Gerber and J. Tübke, *Advanced Energy Materials*, **5** (2015) 1401986.
34. Y. V. Mikhaylik and J. R. Akridge, *Journal of the Electrochemical Society*, **150** (2003) A306.
35. D. Aurbach, E. Pollak, R. Elazari, G. Salitra, C. S. Kelley and J. Affinito, *Journal of the Electrochemical Society*, **156** (2009) A694.
36. D. Aurbach, O. Youngman, Y. Gofer and A. Meitav, *Electrochimica Acta*, **35** (1990) 625–638.
37. D. Aurbach, *Journal of the Electrochemical Society*, **145** (1998) 2629.
38. D. Aurbach, *Journal of the Electrochemical Society*, **145** (1998) 1421.
39. P. Dan, E. Mengeritski, Y. Geronov, D. Aurbach and I. Weisman, *Journal of Power Sources*, **54** (1995) 143–145.
40. E. Mengeritsky, *Journal of the Electrochemical Society*, **143** (1996) 2110.
41. M. Cuisinier, C. Hart, M. Balasubramanian, A. Garsuch and L. F. Nazar, *Advanced Energy Materials*, **5** (2015) 1401801.
42. E. Peled, *Journal of the Electrochemical Society*, **136** (1989) 1621.
43. D.-R. Chang, S.-H. Lee, S.-W. Kim and H.-T. Kim, *Journal of Power Sources*, **112** (2002) 452–460.

44. C. Barchasz, J.-C. Lepretre, S. Patoux and F. Alloin, *Journal of the Electrochemical Society*, **160** (2013) A430–A436.
45. D. Aurbach and E. Granot, *Electrochimica Acta*, **42** (1997) 697–718.
46. J. Qian, W. A. Henderson, W. Xu, P. Bhattacharya, M. Engelhard, O. Borodin and J.-G. Zhang, *Nature Communications*, **6** (2015) 6362.
47. D. Aurbach, O. Chusid, I. Weissman and P. Dan, *Electrochimica Acta*, **41** (1996) 747–760.
48. W. A. Henderson, *The Journal of Physical Chemistry B*, **110** (2006) 13177–13183.
49. S. S. Zhang, *Journal of Power Sources*, **231** (2013) 153–162.
50. R. Younesi, G. M. Veith, P. Johansson, K. Edström and T. Vegge, *Energy & Environmental Science*, **8** (2015) 1905–1922.
51. S. Kim, Y. Jung and S.-J. Park, *Electrochimica Acta*, **52** (2007) 2116–2122.
52. J.-W. Park, K. Yamauchi, E. Takashima, N. Tachikawa, K. Ueno, K. Dokko and M. Watanabe, *The Journal of Physical Chemistry C*, **117** (2013) 4431–4440.
53. R. Demir-Cakan, M. Morcrette, Gangulibabu, A. Guéguen, R. Dedryvère and J.-M. Tarascon, *Energy & Environmental Science*, **6** (2013) 176.
54. Y. Diao, K. Xie, S. Xiong and X. Hong, *Journal of Power Sources*, **235** (2013) 181–186.
55. Y. Han, X. Duan, Y. Li, L. Huang, D. Zhu and Y. Chen, *Materials Research Bulletin*, **68** (2015) 160–165.
56. H. S. Ryu, H. J. Ahn, K. W. Kim, J. H. Ahn, K. K. Cho and T. H. Nam, *Electrochimica Acta*, **52** (2006) 1563–1566.
57. S. Xiong, K. Xie, Y. Diao and X. Hong, *Journal of Power Sources*, **246** (2014) 840–845.
58. D. Aurbach, *Journal of the Electrochemical Society*, **136** (1989) 3198.
59. J. Howe, L. Boatner, J. Kolopus, L. Walker, C. Liang, N. Dudney and C. R. Schaich, *Journal of Materials Science*, **47** (2012) 1572–1577.
60. M. Lécuyer, J. Gaubicher, M. Deschamps, B. Lestriez, T. Brousse and D. Guyomard, *Journal of Power Sources*, **241** (2013) 249–254.
61. S. Wenzel, D. A. Weber, T. Leichtweiss, M. R. Busche, J. Sann and J. Janek, *Solid State Ionics*, **286** (2016) 24–33.
62. Y. Gorlin, A. Siebel, M. Piana, T. Huthwelker, H. Jha, G. Monsch, F. Kraus, H. A. Gasteiger and M. Tromp, *Journal of the Electrochemical Society*, **162** (2015) A1146–A1155.
63. M. Zier, F. Scheiba, S. Oswald, J. Thomas, D. Goers, T. Scherer, M. Klose, H. Ehrenberg and J. Eckert, *Journal of Power Sources*, **266** (2014) 198–207.

64. Y. Cui, A. Abouimrane, J. Lu, T. Bolin, Y. Ren, W. Weng, C. Sun, V. A. Maroni, S. M. Heald and K. Amine, *Journal of the American Chemical Society*, **135** (2013) 8047–8056.
65. P. Abellan, B. L. Mehdi, L. R. Parent, M. Gu, C. Park and W. Xu, Y. Zhang, I. Arslan, J.-G. Zhang, C.-M. Wang, J. E. Evans and N. D. Browning, *Nano Letters*, **14** (2014) 1293–1299.
66. R. L. Sacci, N. J. Dudney, K. L. More, L. R. Parent, I. Arslan, N. D. Browning and R. R. Unocic, *Chemical Communications*, **50** (2014) 2104.
67. R. L. Sacci, J. M. Black, N. Balke, N. J. Dudney, K. L. More and R. R. Unocic, *Nano Letters*, **15** (2015) 2011–2018.
68. B. L. Mehdi, J. Qian, E. Nasybulin, C. Park, D. A. Welch, R. Faller, H. Mehta, W. A. Henderson, W. Xu, C. M. Wang, J. E. Evans, J. Liu, J.-G. Zhang, K. T. Mueller and N. D. Browning, *Nano Letters*, **15** (2015) 2168–2173.
69. M. Nagao, A. Hayashi, M. Tatsumisago, T. Kanetsuku, T. Tsuda and S. Kuwabata, *The Physical Chemistry Chemical Physics*, **15** (2013) 18600.
70. H. Cheng, C. B. Zhu, M. Lu and Y. Yang, *Journal of Power Sources*, **174** (2007) 1027–1031.
71. J. Maibach, C. Xu, S. K. Eriksson, J. Åhlund, T. Gustafsson, H. Siegbahn H. Rensmo, K. Edström and M. Hahlin, *Review of Scientific Instruments*, **86** (2015) 044101.
72. N. M. Trease, T. K. Köster and C. P. Grey, *Electrochemical Society Interface*, **20** (2011) 69.
73. K. A. See, M. Leskes, J. M. Griffin, S. Britto, P. D. Matthews, A. Emly, A. Van der Ven, D. S. Wright, A. J. Morris, C. P. Grey and R. Seshadri, *Journal of the American Chemical Society*, **136** (2014) 16368–16377.
74. R. Bhattacharyya, B. Key, H. Chen, A. S. Best, A. F. Hollenkamp and C. P. Grey, *Nature Materials*, **9** (2010) 504–510.
75. J. Xiao, J. Z. Hu, Honghao chen, M. Vijayakumar, J. Zheng and H. Pan, J. Xiao, J. Z. Hu, H. Chen, M. Vijayakumar, J. Zheng, H. Pan, E. D. Walter, M. Hu, X. Deng, J. Feng, B. Y. Liaw, M. Gu, Z. D. Deng, D. Lu, S. Xu, C. Wang and J. Liu, *Nano Letters*, **15** (2015) 3309–3316.
76. D. Lv, J. Zheng, Q. Li, X. Xie, S. Ferrara, Z. Nie , L. B. Mehdi, N. D. Browning, J.-G. Zhang, G. L. Graff, J. Liu and J. Xiao, *Advanced Energy Materials*, **5** (2015) 1402290.
77. S. Chandrashekar, N. M. Trease, H. J. Chang, L.-S. Du, C. P. Grey and A. Jerschow, *Nature Materials*, **11** (2012) 311–315.
78. M. Rosso, C. Brissot, A. Teyssot, M. Dollé, L. Sannier, J.-M. Tarascon, R. Bouchet and S. Lascaud, *Electrochimica Acta*, **51** (2006) 5334–5340.

79. K. J. Harry, X. Liao, D. Y. Parkinson, A. M. Minor and N. P. Balsara, *Journal of the Electrochemical Society*, **162** (2015) A2699–A2706.
80. W. Li, H. Yao, K. Yan, G. Zheng, Z. Liang, Y.-M. Chiang and Y. Cui, *Nature Communications*, **6** (2015) 7436.
81. A. J. Smith, J. C. Burns and J. R. Dahn, *Electrochemical and Solid-State Letters*, **13** (2010) A177.
82. J. C. Burns, D. A. Stevens and J. R. Dahn, *Journal of the Electrochemical Society*, **162** (2015) A959–A964.
83. B. Wang, X.-L. Wu, C.-Y. Shu, Y.-G. Guo and C.-R. Wang, *Journal of Materials Chemistry*, **20** (2010) 10661.
84. N. A. Cañas, K. Hirose, B. Pascucci, N. Wagner, K. A. Friedrich and R. Hiesgen, *Electrochimica Acta*, **97** (2013) 42–51.
85. J.-J. Woo, V. A. Maroni, G. Liu, J. T. Vaughey, D. J. Gosztola and K. Amine, *Journal of the Electrochemical Society*, **161** (2014) A827–A830.
86. M. Cuisinier, P.-E. Cabelguen, S. Evers, G. He, M. Kolbeck, A. Garsuch, T. Bolin, M. Balasubramanian and L. F. Nazar, *Journal of Physical Chemistry Letters*, **4** (2013) 3227–3232.
87. M. A. Lowe, J. Gao and H. D. Abruña, *RSC Advances*, **4** (2014) 18347.
88. J. Nelson, S. Misra, Y. Yang, A. Jackson, Y. Liu, H. Wang, H. Dai, J. C. Andrews, Y. Cui and M. F. Toney, *Journal of the American Chemical Society*, **134** (2012) 6337–6343.
89. C.-N. Lin, W.-C. Chen, Y.-F. Song, C.-C. Wang, L.-D. Tsai and N.-L. Wu, *Journal of Power Sources*, **263** (2014) 98–103.
90. C. Barchasz, F. Molton, C. Duboc, J.-C. Leprêtre, S. Patoux and F. Alloin, *Analytical Chemistry*, **84** (2012) 3973–3980.
91. S. Xiong, K. Xie, Y. Diao and X. Hong, *Electrochimica Acta*, **83** (2012) 78–86.
92. Y. Han, X. Duan, Y. Li, L. Huang, D. Zhu and Y. Chen, *Ionics*, **22** (2016) 151.
93. R. Xu, J. C. M. Li, J. Lu, K. Amine and I. Belharouak, *Journal of Materials Chemistry A*, **3** (2015) 4170–4179.
94. S. S. Zhang, *Electrochimica Acta*, **70** (2012) 344–348.
95. M.-K. Song, Y. Zhang and E. J. Cairns, *Nano Letters*, **13** (2013) 5891–5899.
96. A. Rosenman, R. Elazari, G. Salitra, E. Markevich, D. Aurbach and A. Garsuch, *Journal of the Electrochemical Society*, **162** (2015) A470–A473.
97. X.-B. Cheng, H.-J. Peng, J.-Q. Huang, R. Zhang, C.-Z. Zhao and Q. Zhang, *ACS Nano*, **9** (2015) 6373–6382.

98. K. Dokko, N. Tachikawa, K. Yamauchi, M. Tsuchiya, A. Yamazaki, E. Takashima, *Journal of the Electrochemical Society*, **160** (2013) A1304–A1310.
99. W. Weng, V. G. Pol and K. Amine, *Advanced Materials*, **25** (2013) 1608–1615.
100. N. Azimi, W. Weng, C. Takoudis and Z. Zhang, *Electrochemistry Communications*, **37** (2013) 96–99.
101. M. L. Gordin, F. Dai, S. Chen, T. Xu, J. Song, D. Tang, N. Azimi, Z. Zhang and D. Wang, *ACS Applied Materials & Interfaces*, **6** (2014) 8006–8010.
102. E. S. Shin, K. Kim, S. H. Oh and W. I. Cho, *Chemical Communications*, **49** (2013) 2004–2006.
103. L. Suo, Y.-S. Hu, H. Li, M. Armand and L. Chen, *Nature Communications*, **4** (2013) 1481.
104. Y. Yamada, M. Yaegashi, T. Abe and A. Yamada, *Chemical Communications*, **49** (2013) 11194.
105. Y. Yamada, K. Furukawa, K. Sodeyama, K. Kikuchi, M. Yaegashi, Y. Tateyama and A. Yamada, *Journal of the American Chemical Society*, **136** (2014) 5039–5046.
106. R. Miao, J. Yang, X. Feng, H. Jia, J. Wang and Y. Nuli, *Journal of Power Sources*, **271** (2014) 291–297.
107. K. Sodeyama, Y. Yamada, K. Aikawa, A. Yamada and Y. Tateyama, *The Journal of Physical Chemistry C*, **118** (2014) 14091–14097.
108. L. E. Camacho-Forero, T. W. Smith, S. Bertolini and P. B. Balbuena, *The Journal of Physical Chemistry C*, **119** (2015) 26828–26839.
109. M. Cuisinier, P.-E. Cabelguen, B. D. Adams, A. Garsuch, M. Balasubramanian and L. F. Nazar, *Energy & Environmental Science*, **7** (2014) 2697.
110. Z. Lin, Z. Liu, W. Fu, N. J. Dudney and C. Liang, *Advanced Functional Materials*, **23** (2013) 1064–1069.
111. Z. Lin, Z. Liu, N. J. Dudney and C. Liang, *ACS Nano*, **7** (2013) 2829–2833.
112. T. Hakari, M. Nagao, A. Hayashi and M. Tatsumisago, *Journal of Power Sources*, **293** (2015) 721–725.
113. K. Xu, S. Zhang and R. Jow, *Journal of Power Sources*, **143** (2005) 197–202.
114. H. Kaneko, K. Sekine and T. Takamura, *Journal of Power Sources*, **146** (2005) 142–145.
115. S. Xiong, X. Kai, X. Hong and Y. Diao, *Ionics*, **18** (2012) 249–254.

116. J.-H. Song, J.-T. Yeon, J.-Y. Jang, J.-G. Han, S.-M. Lee and N.-S. Choi, *Journal of the Electrochemical Society*, **160** (2013) A873–A881.
117. V. Etacheri, O. Haik, Y. Goffer, G. A. Roberts, I. C. Stefan, R. Fasching and D. Aurbach, *Langmuir*, **28** (2012) 965–976.
118. M. Ishikawa, *Journal of the Electrochemical Society*, **141** (1994) L159.
119. Y. Matsuda, M. Ishikawa, S. Yoshitake and M. Morita, *Journal of Power Sources*, **54** (1995) 301–305.
120. M. Wu, Z. Wen, J. Jin and Y. Cui, *Electrochimica Acta*, **103** (2013) 199–205.
121. C. Zu, A. Manthiram, *Journal of Physical Chemistry Letters*, **5** (2014) 2522–2527.
122. M. Hagen, E. Quiroga-González, S. Dörfler, G. Fahrer, J. Tübke, M. J. Hoffmann, H. Althues, R. Speck, M. Krampfert, S. Kaskel and H. Föll, *Journal of Power Sources*, **248** (2014) 1058–1066.
123. X. Zhang, W. Wang, A. Wang, Y. Huang, K. Yuan, Z. Yu, J. Qiu and Y. Yang, *Journal of Materials Chemistry A*, **2** (2014) 11660.
124. X.-B. Cheng, H.-J. Peng, J.-Q. Huang, F. Wei and Q. Zhang, *Small*, **10** (2014) 4257–4263.
125. B. Duan, W. Wang, H. Zhao, A. Wang, M. Wang, K. Yuan, Z. Yu and Y. Yang, *ECS Electrochemistry Letters*, **2** (2013) A47–A51.
126. F. Ding, W. Xu, G. L. Graff, J. Zhang, M. L. Sushko, X. Chen, Y. Shao, M. H. Engelhard, Z. Nie, J. Xiao, X. Liu, P. V. Sushko, J. Liu and J.-G. Zhang, *Journal of the American Chemical Society*, **135** (2013) 4450–4456.
127. J.-S. Kim, T. H. Hwang, B. G. Kim, J. Min and J. W. Choi, *Advanced Functional Materials*, **24** (2014) 5359–5367.
128. C.-P. Yang, Y.-X. Yin, S.-F. Zhang, N.-W. Li and Y.-G. Guo, *Nature Communications*, **6** (2015) 8058.
129. G. Zheng, S. W. Lee, Z. Liang, H.-W. Lee, K. Yan, H. Yao, H. Wang, W. Li, S. Chu and Y. Cui, *Nature Nanotechnology*, **9** (2014) 618–623.
130. C. Huang, J. Xiao, Y. Shao, J. Zheng, W. D. Bennett, D. Lu, L. V. Saraf, M. Engelhard, L. Ji, J. Zhang, X. Li, G. L. Graff and J. Liu, *Nature Communications*, **5** (2014) 3015.
131. R. Koksbang, US Patent (1995) **US5387479**.
132. U. V. Alpen, A. Rabenau and G. H. Talat, *Applied Physics Letters*, **30** (1977) 621.
133. M. Wu, Z. Wen, Y. Liu, X. Wang and L. Huang, *Journal of Power Sources*, **196** (2011) 8091–8097.

134. Y. Yan, J. Y. Zhang, T. Cui, Y. Li, Y. M. Ma, J. Gong and X. Wu, *The European Physical Journal B*, **61** (2008) 397–403.
135. G. Ma, Z. Wen, M. Wu, C. Shen, Q. Wang, J. Jin and X. Wu, *Chemical Communications*, **50** (2014) 14209–14212.
136. S. H. Jee, M.-J. Lee, H. S. Ahn, D.-J. Kim, J. W. Choi, S. J. Yoon, S. C Nam, S. H. Kim and Y. S. Yoon, *Solid State Ionics*, **181** (2010) 902–906.
137. S. Jacke, J. Song, L. Dimesso, J. Brötz, D. Becker and W. Jaegermann, *Journal of Power Sources*, **196** (2011) 6911–6914.
138. J. F. Ribeiro, R. Sousa, J. P. Carmo, L. M. Gonçalves, M. F. Silva, M. M. Silva and J. H. Correia, *Thin Solid Films*, **522** (2012) 85–89.
139. J. B. Bates, US Patent (1994) **US5314765**.
140. F. Marchioni, K. Star, E. Menke, T. Buffeteau, L. Servant, B. Dunn and F. Wudl, *Langmuir*, **23** (2007) 11597–11602.
141. S. Neuhold, D. J. Schroeder and J. T. Vaughey, *Journal of Power Sources*, **206** (2012) 295–300.
142. S. Neuhold, J. T. Vaughey, C. Grogger and C. M. López, *Journal of Power Sources*, **254** (2014) 241–248.
143. G. A. Umeda, E. Menke, M. Richard, K. L. Stamm, F. Wudl and B. Dunn, *Journal of Material Chemistry*, **21** (2011) 1593–1599.
144. R. S. Thompson, D. J. Schroeder, C. M. López, S. Neuhold and J. T. Vaughey, *Electrochemistry Communications*, **13** (2011) 1369–1372.
145. B. Chaloner-Gill, O. K. Chang, N. Golovin and E. Saidi, *Enhanced Lithium Surface*, US5354631, 1994.
146. K. Star, J. Muldoon, F. Marchioni, F. Wudl, B. Dunn and M. N. Richard *et al.*, US Patent (2014) **US8840688**.
147. L. De Jonghe, S. J. Visco, Y. S. Nimon and A. M. Sukeshini, US Patent (2005) **US6911280**.
148. V. D. Braun, W. Betz and W. Kern, *Die Makromolekulare Chemie*, **42** (1960) 89–95.
149. J. Song, H. Lee, M.-J. Choo, J.-K. Park and H.-T. Kim, *Scientific Reports*, **5** (2015) 14458.
150. D. G. Fauteux, M. Van Buren, J. Shi and M. Rona, US Patent (1995) **US5434021**.
151. Y. M. Lee, N.-S. Choi, J. H. Park and J.-K. Park, *Journal of Power Sources*, **119–121** (2003) 964–972.
152. S. Jung Oh, W. Young Yoon, *International Journal of Precision Engineering and Manufacturing*, **15** (2014) 1453–1457.
153. I. S. Kang, Y.-S. Lee and D.-W. Kim, *Journal of the Electrochemical Society*, **161** (2013) A53–A57.

154. G. Ma, Z. Wen, Q. Wang, C. Shen, J. Jin and X. Wu, *Journal of Material Chemistry A*, **2** (2014) 19355–19359.
155. H.-K. Jing, L.-L. Kong, S. Liu, G.-R. Li and X.-P. Gao, *Journal of Material Chemistry A*, **3** (2015) 12213–12219.
156. E. Kazyak, K. N. Wood and N. P. Dasgupta, *Chemistry of Materials*, **27** (2015) 6457–6462.
157. A. C. Kozen, C.-F. Lin, A. J. Pearse, M. A. Schroeder, X. Han, L. Hu, S.-B. Lee, G. W. Rubloff and M. Noked, *ACS Nano*, **9** (2015) 5884–5892.
158. S. M. Choi, I. S. Kang, Y.-K. Sun, J.-H. Song, S.-M. Chung and D.-W. Kim, *Journal of Power Sources*, **244** (2013) 363–368.
159. T. A. Skotheim, US Patent (1997) **US5648187**.
160. T. A. Skotheim, G. L. Soloveichik and A. B. Gavrilov, US Patent (1999) **US5961672**.
161. H. Chen, W. Dong, J. Ge, C. Wang, X. Wu, W. Lu and L. Chen, *Scientific Reports*, **3** (2013) 1910.
162. C. Wang, W. Wan, J.-T. Chen, H.-H. Zhou, X.-X. Zhang, L.-X. Yuan and Y.-H. Huang, *Journal of Material Chemistry A*, **1** (2013) 1716–1723.
163. J. Shao, X. Li, L. Zhang, Q. Qu and H. Zheng, *Nanoscale*, **5** (2013) 1460.
164. W. Zhou, Y. Yu, H. Chen, F. J. DiSalvo and H. D. Abruña, *Journal of the American Chemical Society*, **135** (2013) 16736–16743.
165. L. Xiao, Y. Cao, J. Xiao, B. Schwenzer, M. H. Engelhard, L. V. Saraf, Z. Nie, G. J. Exarhos and J. Liu, *Advanced Materials*, **24** (2012) 1176–1181.
166. W. Li, Q. Zhang, G. Zheng, Z. W. Seh, H. Yao and Y. Cui, *Nano Letters*, **13** (2013) 5534–5540.
167. S.-T. Hong, J.-S. Kim, S.-J. Lim and W. Y. Yoon, *Electrochimica Acta*, **50** (2004) 535–539.
168. W.-S. Kim and W.-Y. Yoon, *Electrochimica Acta*, **50** (2004) 541–545.
169. J. H. Chung, W. S. Kim, W. Y. Yoon, S. W. Min and B. W. Cho, *Journal of Power Sources*, **163** (2006) 191–195.
170. D. Lepage, C. Michot, G. Liang, M. Gauthier and S. B. Schougaard, *Angewandte Chemie International Edition*, **50** (2011) 6884–6887.
171. L. Zhan, Z. Song, J. Zhang, J. Tang, H. Zhan, Y. Zhou and C. Zhan, *Electrochimica Acta*, **53** (2008) 8319–8323.
172. Y.-S. Lee, K.-S. Lee, Y.-K. Sun, Y. M. Lee and D.-W. Kim, *Journal of Power Sources*, **196** (2011) 6997–7001.
173. X. Han, Y. Xu, X. Chen, Y.-C. Chen, N. Weadock, J. Wan, H. Zhu, Y. Liu, H. Li, G. Rubloff, C. Wang and L. Hu, *Nano Energy*, **2** (2013) 1197–1206.
174. M. Yu, W. Yuan, C. Li, J.-D. Hong and G. Shi, *Journal of Materials Chemistry A*, **2** (2014) 7360.

175. X. Li, A. Lushington, J. Liu, R. Li and X. Sun, *Chemical Communications*, **50** (2014) 9757.
176. Z. Zhang, Y. Lai, Z. Zhang, K. Zhang and J. Li, *Electrochimica Acta*, **129** (2014) 55–61.
177. S. C. Jung and Y.-K. Han, *Journal of Physical Chemistry Letters*, **4** (2013) 2681–2685.
178. A. I. Abdulagatov, Y. Yan, J. R. Cooper, Y. Zhang, Z. M. Gibbs and A. S. Cavanagh et al., *ACS Applied Materials & Interfaces*, **3** (2011) 4593–4601.
179. P. C. Wang, Y. T. Shih, M. C. Lin, H. C. Lin, M. J. Chen and K. M. Lin, *Thin Solid Films*, **518** (2010) 7501–7504.
180. E. Marin, L. Guzman, A. Lanzutti, W. Ensinger and L. Fedrizzi, *Thin Solid Films*, **522** (2012) 283–288.
181. J. A. Bertrand and S. M. George, *Journal of Vacuum Science & Technology A: Vacuum, Surfaces, and Films*, **31** (2013) 01A122.
182. W. Xu, S. S. S. Vegunta and J. C. Flake, *Journal of Power Sources*, **196** (2011) 8583–8589.
183. E. Memarzadeh Lotfabad, P. Kalisvaart, K. Cui, A. Kohandehghan, M. Kupsta, B. Olsen and D. Mitlin, *Physical Chemistry Chemical Physics*, **15** (2013) 13646.
184. L. A. Riley, A. S. Cavanagh, S. M. George, S.-H. Lee and A. C. Dillon, *Electrochemical and Solid-State Letters*, **14** (2011) A29.
185. L. A. Riley, A. S. Cavanagh, S. M. George, Y. S. Jung, Y. Yan and S.-H. Lee et al., *ChemPhysChem*, **11** (2010) 2124–2130.
186. J.-T. Lee, F.-M. Wang, C.-S. Cheng, C.-C. Li and C.-H. Lin, *Electrochimica Acta*, **55** (2010) 4002–4006.
187. L. A. Riley, S. Van Atta, A. S. Cavanagh, Y. Yan, S. M. George, P. Liu, A. C. Dillon and S.-H. Lee, *Journal of Power Sources*, **196** (2011) 3317–3324.
188. H. Kim, J. T. Lee, D.-C. Lee, A. Magasinski, W. Cho and G. Yushin, *Advanced Energy Materials*, **3** (2013) 1308–1315.
189. K. Takada, *Acta Materialia*, **61** (2013) 759–770.
190. A. Aboulaich, R. Bouchet, G. Delaizir, V. Seznec, L. Tortet, M. Morcrette, P. Rozier, J.-M. Tarascon, V. Viallet and M. Dollé, *Advanced Energy Materials*, **1** (2011) 179–183.
191. J. Y. Song, Y. Y. Wang and C. C. Wan, *Journal of Power Sources*, **77** (1999) 183–197.
192. B. H. Jeon, J. H. Yeon, K. M. Kim and I. J. Chung, *Journal of Power Sources*, **109** (2002) 89–97.

193. D. T. Hallinan and N. P. Balsara, *Annual Review of Materials Research*, **43** (2013) 503–525.
194. R. Bouchet, S. Maria, R. Meziane, A. Aboulaich, L. Lienafa, J.-P. Bonnet, T. N. T. Phan, D. Bertin, D. Gigmes, D. Devaux, R. Denoyel and M. Armand, *Nature Materials*, **12** (2013) 452–457.
195. J. Hassoun and B. Scrosati, *Advanced Materials*, **22** (2010) 5198–5201.
196. F. Croce, S. Sacchetti and B. Scrosati, *Journal of Power Sources*, **161** (2006) 560–564.
197. X. Liang, Z. Wen, Y. Liu, H. Zhang, L. Huang and J. Jin, *Journal of Power Sources*, **196** (2011) 3655–3658.
198. D. R. Wang, K. H. Wujcik, A. A. Teran and N. P. Balsara, *Macromolecules*, **48** (2015) 4863–4873.
199. G. M. Stone, S. A. Mullin, A. A. Teran, D. T. Hallinan, A. M. Minor, A. Hexemer and N. P. Balsara, *Journal of the Electrochemical Society*, **159** (2012) A222–A227.
200. A. Hayashi, R. Ohtsubo, T. Ohtomo, F. Mizuno and M. Tatsumisago, *Journal of Power Sources*, **183** (2008) 422–426.
201. M. Nagao, A. Hayashi and M. Tatsumisago, *Electrochimica Acta*, **56** (2011) 6055–6059.
202. T. Kobayashi, Y. Imade, D. Shishihara, K. Homma, M. Nagao, R. Watanabe, T. Yokoi, A. Yamada, R. Kanno and T. Tatsumi, *Journal of Power Sources*, **182** (2008) 621–625.
203. M. Nagao, Y. Imade, H. Narisawa, T. Kobayashi, R. Watanabe, T. Yokoi, T. Tatsumi and R. Kanno, *Journal of Power Sources*, **222** (2013) 237–242.
204. H. Aono, *Journal of the Electrochemical Society*, **137** (1990) 1023.
205. J. Fu, *Solid State Ionics*, **104** (1997) 191–194.
206. N. Kamaya, K. Homma, Y. Yamakawa, M. Hirayama, R. Kanno, M. Yonemura, T. Kamiyama, Y. Kato, S. Hama, K. Kawamoto and A. Mitsui, *Nature Materials*, **10** (2011) 682–686.
207. J. K. Feng, L. Lu and M. O. Lai, *Journal of Alloys and Compounds*, **501** (2010) 255–258.
208. Y. Ren, Y. Shen, Y. Lin, C.-W. Nan, *Electrochemistry Communications*, **57** (2015) 27–30.
209. R. Kanno, M. Murayama, T. Inada, T. Kobayashi, K. Sakamoto, N. Sonoyama, A. Yamada and S. Kondo, *Electrochemical and Solid-State Letters*, **7** (2004) A455.
210. H. Nagata and Y. Chikusa, *Journal of Power Sources*, **263** (2014) 141–144.

211. T. Takeuchi, H. Kageyama, K. Nakanishi, M. Tabuchi, H. Sakaebe, T. Ohta, H. Senoh, T. Sakai and K. Tatsumi, *Journal of the Electrochemical Society*, **157** (2010) A1196.
212. G. Sahu, Z. Lin, J. Li, Z. Liu, N. Dudney and C. Liang, *Energy & Environmental Science*, **7** (2014) 1053–1058.
213. H. Maekawa, M. Matsuo, H. Takamura, M. Ando, Y. Noda, T. Karahashi and S.-I. Orimo, *Journal of the American Chemical Society*, **131** (2009) 894–895.
214. A. Unemoto, S. Yasaku, G. Nogami, M. Tazawa, M. Taniguchi, M. Matsuo, T. Ikeshojo and S.-I. Orimo, *Applied Physics Letters*, **105** (2014) 083901.
215. A. Unemoto, C. Chen, Z. Wang, M. Matsuo, T. Ikeshoji and S. Orimo, *Nanotechnology*, **26** (2015) 254001.
216. J. Christensen, P. Albertus, R. S. Sanchez-Carrera, T. Lohmann, B. Kozinsky, R. Liedtke, J. Ahmed and A. Kojic, *Journal of the Electrochemical Society*, **159** (2012) R1.
217. W.-B. Luo, S.-L. Chou, J.-Z. Wang, Y.-M. Kang, Y.-C. Zhai and H.-K. Liu, *Chemical Communication*, **51** (2015) 8269–8272.
218. S. S. Lee, Y. J. Lim, H. W. Kim, J.-K. Kim, Y.-G. Jung and Y. Kim, *Solid State Ionics*, **284** (2016) 20–24.
219. N. B. Aetukuri, S. Kitajima, E. Jung, L. E. Thompson, K. Virwani, M.-L. Reich, M. Kunze, M. Schneider, W. Schmidbauer, W. W. Wilcke, D. S. Bethune, J. C. Scott, R. D. Miller and H.-C. Kim, *Advanced Energy Materials*, **5** (2015) 1500265.
220. J. H. Gordon and J. J. Watkins, US Patent (2014) **US8771879**.
221. S. J. Visco, V. Y. Nimon, A. Petrov, K. Pridatko, N. Goncharenko and E. Nimon, L. De Jonghe, Y. M. Volfkovich and D. A. Bograchev, *Journal of Solid State Electrochemistry*, **18** (2014) 1443–1456.
222. S. J. Visco, Y. S. Nimon and B. D. US Patent (2007) **US7282302**.
223. S. J. Visco, Y. S. Nimon and B. D. Katz, US Patent (2008) **US7390591**.
224. S. J. Visco and Y. S. Nimon, US Patent (2012) **US8202649**.
225. S. J. Visco, Y. S. Nimon, L. C. De Jonghe, B. D. Katz and A. Petrov, US Patent (2013) **US8445136**.
226. R. Koksbang, US Patent (1996) **US5487959**.
227. M. Y. Chu, S. J. Visco and L. C. DeJonghe, US Patent (2002) **US6413285**.
228. J. D. Affinito, Y. V. Mikhaylik, Y. M. Geronov and C. J. Sheehan, US Patent (2015) **US9040201**.
229. B. D. McCloskey, *Journal of Physical Chemistry Letters*, **6** (2015) 4581–4588.
230. M. S. Whittingham, *Proceedings of the IEEE*, **100** (2012) 1518–1534.

231. K. Nakajima, T. Katoh, Y. Inda and B. Hoffman, Lithium ion conductive glass ceramics: Properties and application in lithium metal batteries, In: Oak Ridge National Laboratory, USA, 2010. Retrieved from http://oharacorp.com/pdf/ohara-presentation-ornl-symposium-10-08-2010.pdf.
232. J. B. Goodenough and K.-S. Park, *Journal of the American Chemical Society*, **135** (2013) 1167–1176.
233. U. Kasavajjula, C. Wang and A. J. Appleby, *Journal of Power Sources*, **163** (2007) 1003–1039.
234. R. Elazari, G. Salitra, G. Gershinsky, A. Garsuch, A. Panchenko and D. Aurbach, *Electrochemistry Communications*, **14** (2012) 21–24.

Chapter 6

Analytical Techniques for Lithium–Sulfur Batteries

Manu U.M. Patel[*,‡] and Robert Dominko[†,§]

[*]*Department of Chemistry, Technische Universität München, Lichtenbergstrasse 4, 85748 Garching, Germany*
[†]*National Institute of Chemistry, Hajdrihova 19, 1000 Ljubljana, Slovenia*
[‡]*manu_patel.ubrani@tum.de*
[§]*robert.dominko@ki.si*

6.1. Introduction

A lithium sulfur (Li–S) cell like any other battery system is composed of a negative electrode (Li metal) and a positive electrode, which is typically a carbon/sulfur composite; these electrodes are separated by a thin electron insulating layer wetted with an electrolyte. The carbon in the cathode gives the required electronic conductivity for sulfur (S); in contrast, the electrolyte acts as a medium for the conduction of Li ions (Li$^+$) between the anode and cathode electrode and enables the propagation of the electrochemical conversion of elemental S to Li$_2$S through the soluble polysulfide intermediates during battery operation.[1,2]

Even after decades of research, the possibility of commercialization of Li–S systems remains unrealized, due to certain critical issues to be addressed.[1-4] Sulfur exists in polyatomic molecular states with different structures in nature; however, the most stable allotrope form of S at room temperature is a cyclic ring S_8. When S in the cathode electrode is discharged, the cyclo-S_8 undergoes a series of structural and morphological changes. The electrochemical reduction of S during the discharge occurs in two stages; in the first stage, the ring structure of S is opened, and long-chain polysulfides are formed. In the second stage of discharge, these long-chain polysulfides are further reduced to medium and short-chain polysulfides.[3-6] This process is supported by the evidence of the shape of discharge and charge curves of the battery galvanostatic cycle, which shows two characteristic plateaus. The first plateau occurs in the potential range of 2.4–2.25 V and involves the reduction of the elemental S_8 to Li to long-chain polysulfides Li_2S_x with $4 \leq x \leq 8$. Long-chain polysulfides have high solubility in the ether-based electrolytes that are typically used in Li–S batteries. The second plateau corresponds to another phase change in which polysulfides precipitate into insoluble Li_2S. Even though the electrochemical steps in high and low voltage plateau seem to be well defined, there is a complicated equilibrium between polysulfides and two end members of the discharge/charge process which is a function of the chemical environment in the cell; it remains under discussion in many scientific and technical papers.

Not all analytical techniques used in the Li-ion battery system can be directly applied to the Li–S system due to the specificity of Li–S battery operation, in which polysulfides are dissolved in the electrolyte. The reduction of sulfur leads to different equilibrium states with polysulfide anions or radicals and corresponds to different sulfur reduction mechanisms that are reported in the literature,[7,8] and the authors have differences of opinion about the intermediate species that might be formed during the electrochemical process. Many research groups have conducted analytical studies aimed at better understanding of the S electrochemical redox reaction occurring in the Li–S rechargeable cells. Due to active material dissolution, the investigation of the discharge/charge mechanism consists of the electrolyte composition characterization as well as changes in the cathode composition. For this purpose, different analytical techniques have been used, enabling characterization of the dissolved, solid Li polysulfides produced in the electrolyte and cathode during cycling. Some of the most commonly used advanced analytical techniques for Li–S systems are UV–Visible (UV–Vis) spectroscopy[9–11], X-Ray Absorption Spectroscopy[12–16] (XAS), X-Ray diffraction[17] (XRD), X-ray Diffraction and Transmission X-ray Microscopy,[18] Raman spectroscopy,[19] X-Ray Photoelectron Spectroscopy[20] (XPS), Rotating Ring Disc Electrode[21] (RRDE), Modified 4-electrode Swagelok cell[22] and Solid State Nuclear Magnetic Resonance spectroscopy[12,23] (NMR) in different modes (*in situ* or *ex situ*). In this chapter, the discussion is focused on the UV–Vis spectroscopy, X-ray absorption spectroscopy and 4-electrode Swagelok cell by means of cyclic voltammetry (CV) measurements.

6.2. UV–Vis Spectroscopy as an Analytical Technique for Li–S Batteries

6.2.1. Working principle

The basic or general interaction between the UV–Vis electromagnetic radiation and molecules coming in contact can be explained by interactions between the radiation and matter or molecules, which induce a number of processes, including reflection, scattering, absorbance, fluorescence/phosphorescence (absorption and reemission), and photochemical reaction.[24] Using this as a foundation, we can use UV–Vis as an analytical tool to monitor polysulfides formed in the Li–S battery system.

The working principle of UV–Visible spectroscopy is based on Beer-Lambert law, which states that the absorbance of incident light is directly proportional to the path length (l) of the sample (the length or distance the light travels through the sample) and concentration (c) of the sample.

$$A = a(\lambda) \times l \times c, \tag{1}$$

where A is the measured absorbance, $a(\lambda)$ is a wavelength-dependent absorptivity coefficient, l is the path length, and c is the analyte concentration.

In general, while measuring UV–Vis spectra, we consider only the absorbance of energy from the electromagnetic radiation by the molecules. Since light is a form of energy, the absorption of light by molecules causes an increase in the energy content of the molecules (or atoms). It can be said that the molecules absorb the energy from the light and undergo excitation from their normal state (ground state) to a higher energy state (excited state).

In some molecules and atoms, photons of UV and visible light have enough energy to cause transitions between the different electronic energy levels. The wavelength of absorbed light has the required energy to move an electron from a lower energy level to a higher energy level. These transitions should result in very narrow absorbance bands at wavelengths highly characteristic to the difference in energy levels of the absorbing species.[27]

However, in the case of molecules, vibrational and rotational energy levels are superimposed on the electronic energy levels. Consequently, many transitions with different energies can lead to a broadening of bands. The broadening is even greater in solutions owing to solvent solute interaction.

When a beam of light passes through or is reflected back from a sample, the amount of light absorbed is the difference between the incident radiation (I_0) and the transmitted radiation (I). The amount of light absorbed is expressed as either transmittance or absorbance. Transmittance usually is given in terms of a fraction of 1 or as a percentage and is defined as follows:

$$T = \frac{I}{I_0} \text{ or } \%T = \left(\frac{I}{I_0}\right)100. \qquad (2)$$

Absorbance can be defined as follows:

$$A = -\log T. \qquad (3)$$

For most applications, absorbance values have been used, since the relationship between absorbance and both concentration and path length normally is linear.

A UV–Vis spectrum generally shows only a few broad absorbance bands. Compared with techniques such as infrared spectroscopy, which produces many narrow bands, UV–Vis spectroscopy provides a limited amount of qualitative information.[27] Most absorption by organic compounds results from the presence of π bonds (i.e. unsaturated). A chromophore is a molecular group usually containing a π bond that produces a compound with absorption between 185 and 1000 nm. Color is an important property of any substance, as the color of matter is related to its absorptivity or reflectivity.

Most of the inorganic polysulfides are colored in their natural state; thus, UV–Visible spectroscopy is becoming a highly effective way of detecting polysulfides, both quantitatively and qualitatively. Polysulfides of different metal ions, different stoichiometry and varying concentration have a different range of absorption. Similar to the organic chromophore molecules, the inorganic polysulfides also have an absorption range between 400 and 800 nm.[25–27] The shift in the absorption range may arrive due to the changing chain length or due to changing concentrations of polysulfides. This phenomenon will be explained in detail in the case of Li polysulfide intermediates formed during *in situ* battery operation as well as *ex situ* measurements on chemically synthesized lithium polysulfides with different stoichiometry.[10,28]

6.2.2. In operando UV–Vis spectroscopy

Electrochemical characteristics of the Li–S cells (capacity retention and degradation) are largely influenced by the type of the electrolyte. The observed differences are related to the solubility and mobility of polysulfides in electrolytes. Comparison of the same type of the cathode composite in different electrolytes typically results in a difference in the obtained specific capacity and in the electrochemical stability. Figure 6.1 shows an example in the cycling performance of PRINTEX XE2/S composite cycled with two different electrolytes. Batteries cycled with 1 M LiTFSI in sulfolane electrolyte typically had a higher initial capacity, which was followed by faster capacity fading. A much slower capacity drop was observed with a battery cycled with 1 M LiTFSI in TEGDME:DOL electrolyte. Because S cathodes were from the same batch and both batteries contained the same amount of electrolyte normalized per mg of S, we can ascribe the observed difference to the mechanism occurring during charge and discharge. In order to understand this mechanism, we used

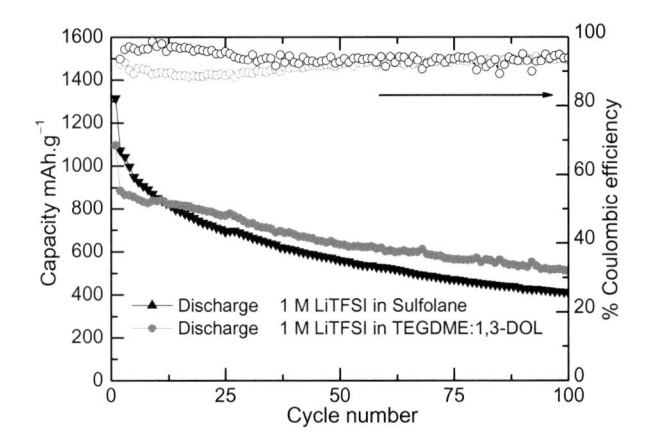

Figure 6.1 Cycling behavior of Li–S composite in two different electrolyte systems. *Source*: Reproduced with permission from Ref. 10.

UV–Vis spectroscopy *in operando* mode for the determination of the process leading to different capacity fading characteristics.

To enable qualitative and quantitative deconvolution of UV–Vis spectra measured in the operando mode, calibration curves should be measured. That requires synthesis of polysulfides, preferably with different chain lengths. One way to obtain different polysulfides is chemical synthesis by using a different stoichiometric ratio between lithium and sulfur, which can be dissolved in the electrolyte with different concentrations. Figure 6.2 shows photographs of wetted separators prepared with dissolved polysulfides in the 1 M LiTFSI in sulfolane and in the 1 M LiTFSI in TEGDME:DOL electrolytes (the concentration of catholyte was 50 mM in all photographs). The photographs show a visible difference between different stoichiometric mixtures (polysulfides) and between electrolytes comparing the same type of polysulfide.

Dissolved polysulfides have characteristic shifts in the reflection curve whereas short-chain polysulfides can be found at the shorter wavelengths while longer polysulfides are detected more often in the longer wavelengths (Figure 6.3(a)). There is also a strong dependency of the depths of the shift with concentration (Figure 6.3(b)).

In order to find a methodology for discrimination between different catholyte solutions, we focused on the behavior of the first derivative of

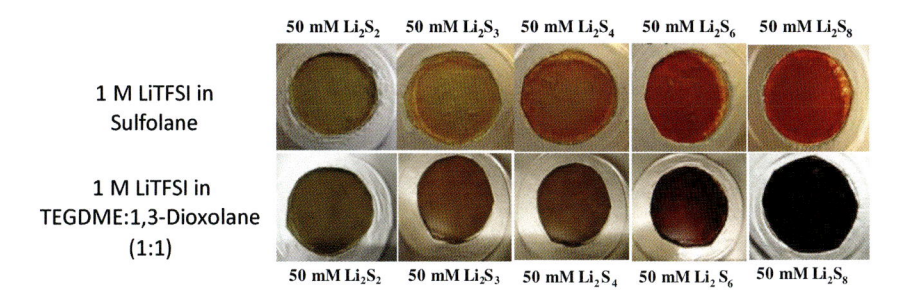

Figure 6.2 Photographs of six different catholyte solutions in two different electrolytes used (the stoichiometric ratio between Li and S is presented at the top and bottom of each picture).

Source: Reproduced with permission from Ref. 10.

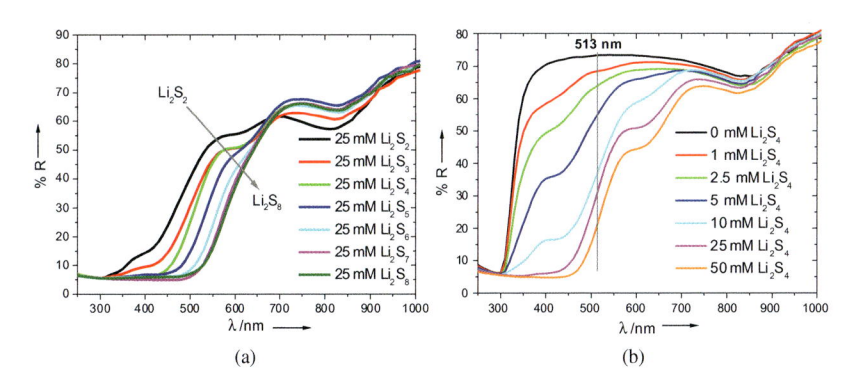

Figure 6.3 UV–Vis spectra of dissolved polysulfides in the 1 M LiTFSI TEGDME:DOL electrolyte: (a) UV–Vis spectra of full range stoichiometric mixtures Li_2S_x with $x = 2....8$ and (b) UV–Vis spectra of catholyte solutions containing dissolved Li_2S_4 stoichiometric mixture with different concentrations.

Source: Reproduced with permission from Ref. 10.

UV–Vis spectra. As presented in Figures 6.4 and 6.5, the derivatives of UV–Vis spectra for each composition showed a typical position of the first order maximums, regardless of the concentration used for the measurement. We have determined five characteristic wavelengths for both electrolyte systems. The catholyte solutions in the 1 M LiTFSI in sulfolane electrolyte showed maximums of derivatives at 570 nm (long-chain

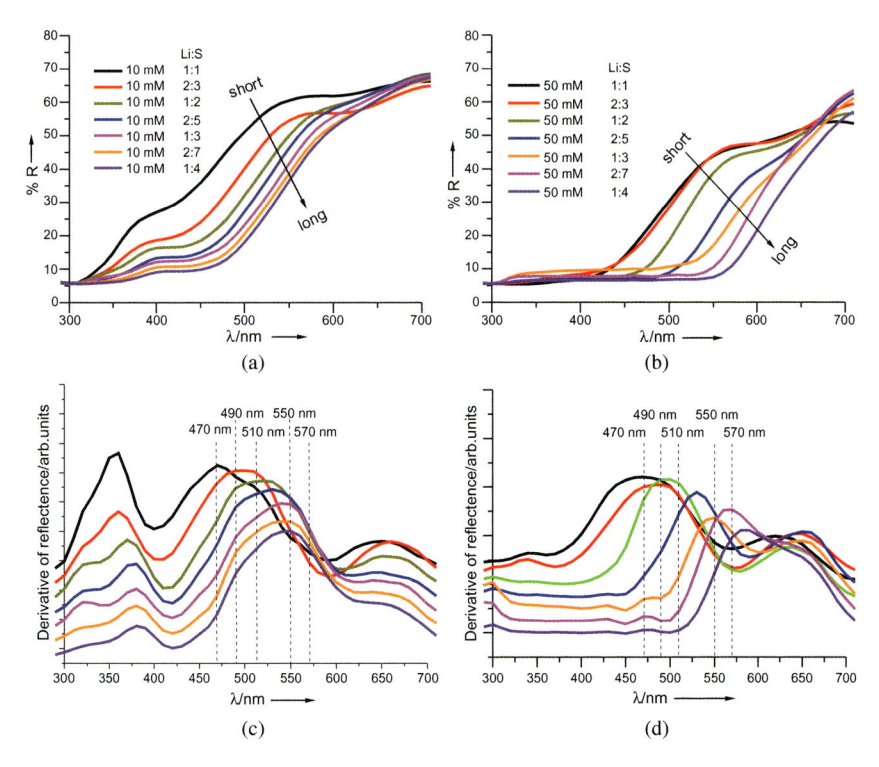

282 *Li–S Batteries: The Challenges, Chemistry, Materials and Future Perspectives*

Figure 6.4 UV–Vis spectra of different stoichiometric mixtures between Li and S for (a) 10 and (b) 50 mm concentrations in 1M LiTFSI in sulfolane electrolyte and (c, d) the corresponding first-order derivative curves.

Source: Reproduced with permission from Ref. 10.

polysulfides), at 550 and 510 nm (mid-chain polysulfides) and 490 and 470 nm (short-chain polysulfides) as shown in Figure 6.4. A similar trend was observed with catholyte solutions in 1 M LiTFSI in TEGDME:DOL electrolyte with a difference in the position of derivative peaks. We found the derivative for long-chain polysulfides at 620 nm, derivatives for the mid-chain polysulfides at 580 and 550 nm, and derivatives for short-chain polysulfides at 520 and 450 nm, as shown in Figure 6.5. In a similar way, six different concentrations (1, 2.5, 5 mM, 10, 25, and 50 mM) were measured.

The obtained values were correlated to the natural logarithm of the concentration and a linear correlation was observed for each selected wavelength. Figures 6.6 and 6.7 present the dependences of the measured

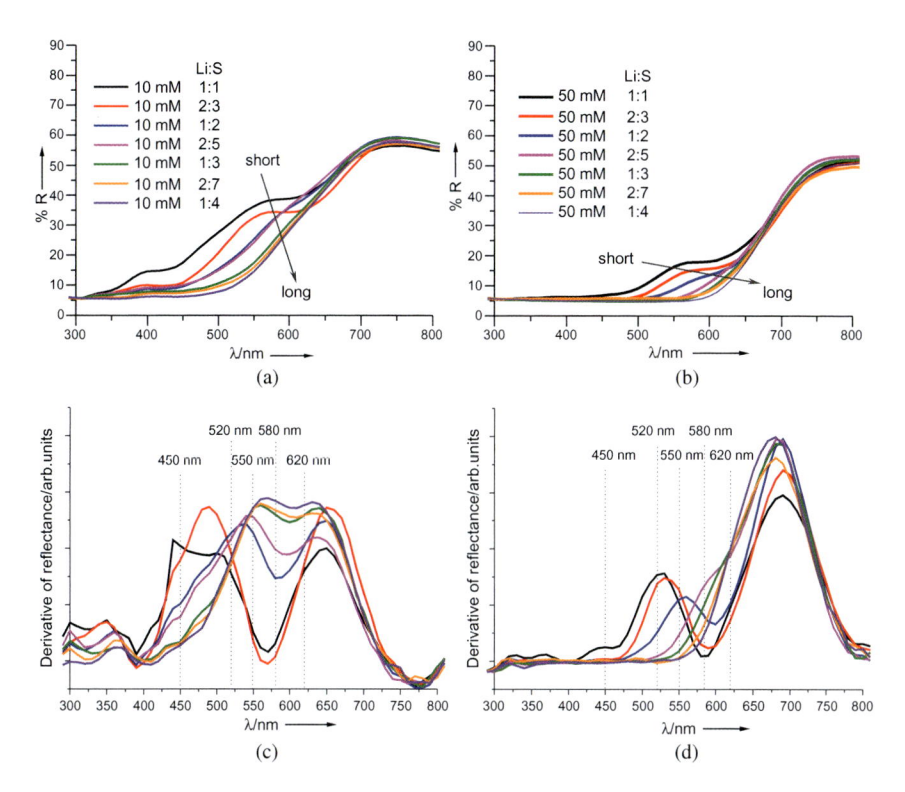

Analytical Techniques for Lithium–Sulfur Batteries 283

Figure 6.5 UV–Vis spectra of different stoichiometric mixtures between Li and S for (a) 10 and (b) 50 mm concentrations in 1M LiTFSI in TEGDME:DOL electrolyte and (c, d) the corresponding first-order derivative curves.

Source: Reproduced with permission from Ref. 10.

normalized reflectance with the concentrations of polysulfides in the sulfolane and TEGDME:DOL electrolyte, respectively.

The standards and battery in operando mode were measured in the polymer pouch bag with a sealed glass cover (Figure 6.8). It is a typical pouch cell with a Li electrode that has a hole that enables measurements of UV–Vis spectra of the catholyte composition in the separator.

This configuration enables the detection of polysulfides in the separator as well as from the backside of the cathode composite (if the cathode is pressed onto an Al mesh). Spectra were measured in a way that UV light is directly focused on the glass cover of the cell. Measurements were done in the reflection mode during the battery discharge/charge.

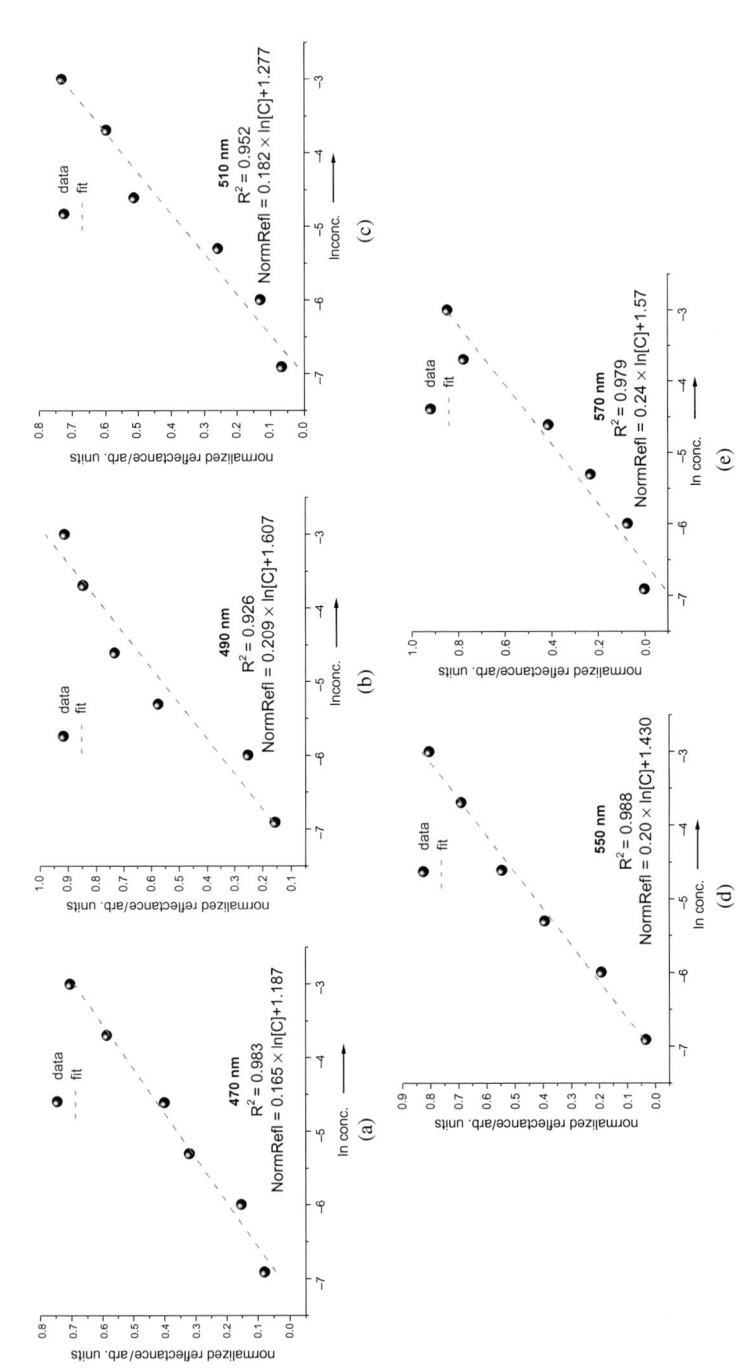

Figure 6.6 Linear fits of normalized intensities measured from catholyte solutions in the 1 M LiTFSI in sulfolane electrolyte with different concentrations of short-chain polysulfides (470 and 490 nm), of the mid-chain polysulfides (510 and 550 nm) and of the long-chain polysulfides (570 nm).

Source: Reproduced with permission from Ref. 10.

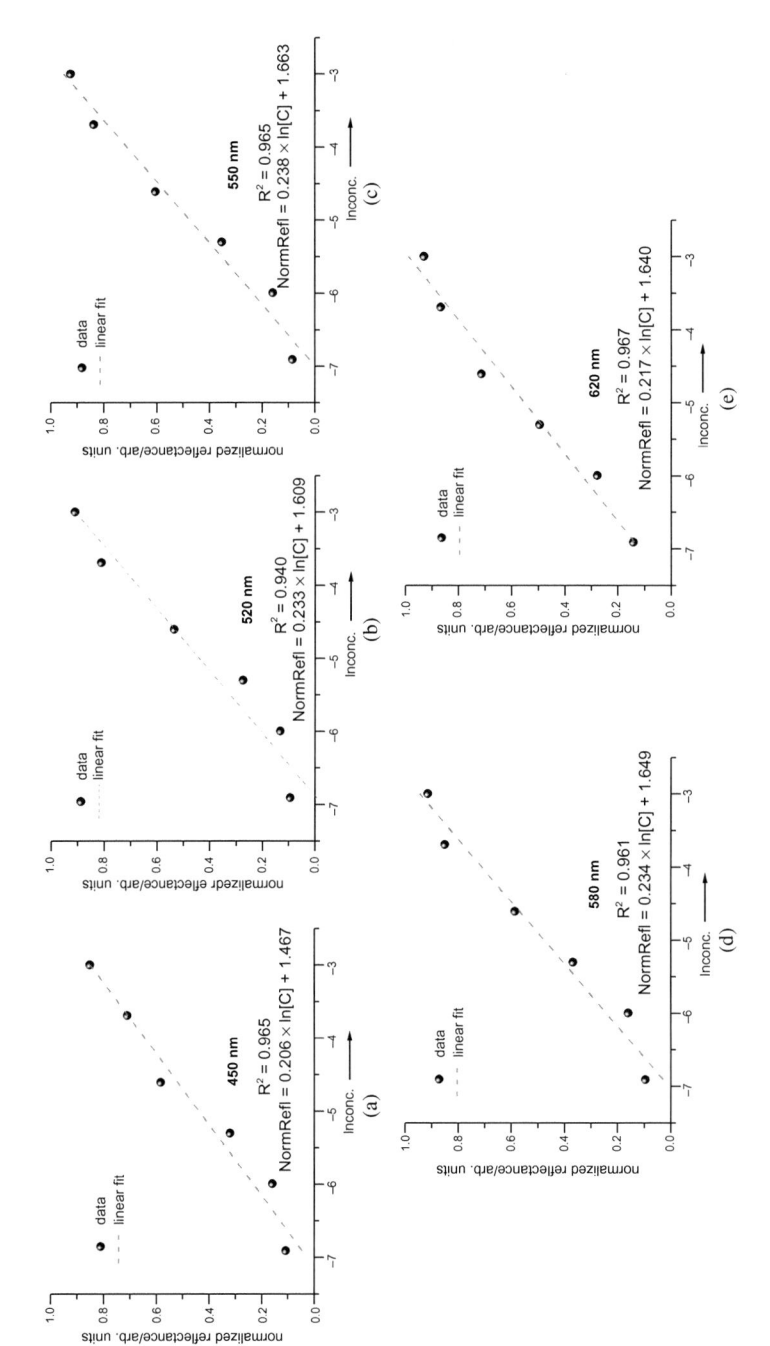

Figure 6.7 Linear fits of normalized intensities measured from catholyte solutions in the 1 M LiTFSI in sulfolane electrolyte with different concentrations of short-chain polysulfides (450 and 520 nm), of the mid-chain polysulfides (550 and 580 nm) and of the long-chain polysulfides (620 nm).

Source: Reproduced with permission from Ref. 10.

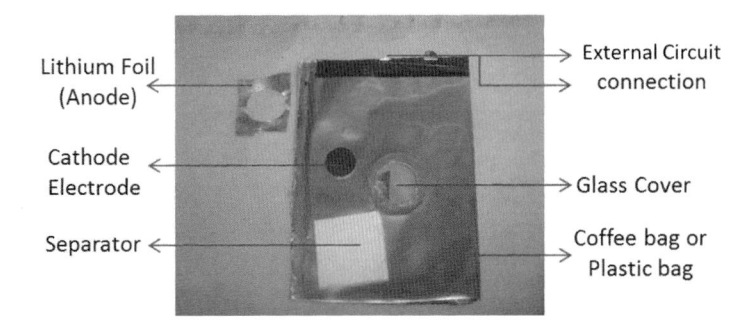

286 *Li–S Batteries: The Challenges, Chemistry, Materials and Future Perspectives*

Figure 6.8 A photograph of the pouch cell used in our work.
Source: Reproduced with permission from Ref. 9.

The battery configuration enables simultaneous measurements of UV–Vis spectra without stopping the battery and even without taking any part of the electrolyte out of the battery. During battery operation, spectra were measured in the range between 2000 and 250 nm. Figures 6.9 and 6.10 show the measured UV–Vis spectra from the Li–S batteries in the two different types of electrolytes. The first spectra taken during discharge showed almost no change in the reflection up to 350 nm, and this is in agreement with the measurement of the cell with the pure electrolyte. Due to the formation of soluble polysulfides during discharge, the absorption is shifted to higher wavelengths, which is in agreement with the predicted formation of long-chain polysulfides. During continued discharging, the absorption curve is shifted towards shorter wavelengths, which indicates the formation of short-chain polysulfides (shift of colors from red to green to blue in Figures 6.9(a) and 6.10(a)). The opposite shift of the absorption curve could be observed during charging, as can be seen in Figures 6.9(b) and 6.10(b). Again, a change of colors from red to green to blue indicates a shift of the spectra during the charging process. An even clearer picture of the dynamics within the separator can be obtained from the first-order derivatives (Figures 6.9(c) and 6.9(d) and Figures 6.10(c) and 6.10(d)) of the measured UV–Vis spectra. A continuous shift of the derivative peaks from long to short wavelengths can be observed in both electrolytes. We focused on the wavelength range between 650 and 400 nm since we observed changes in the derivative

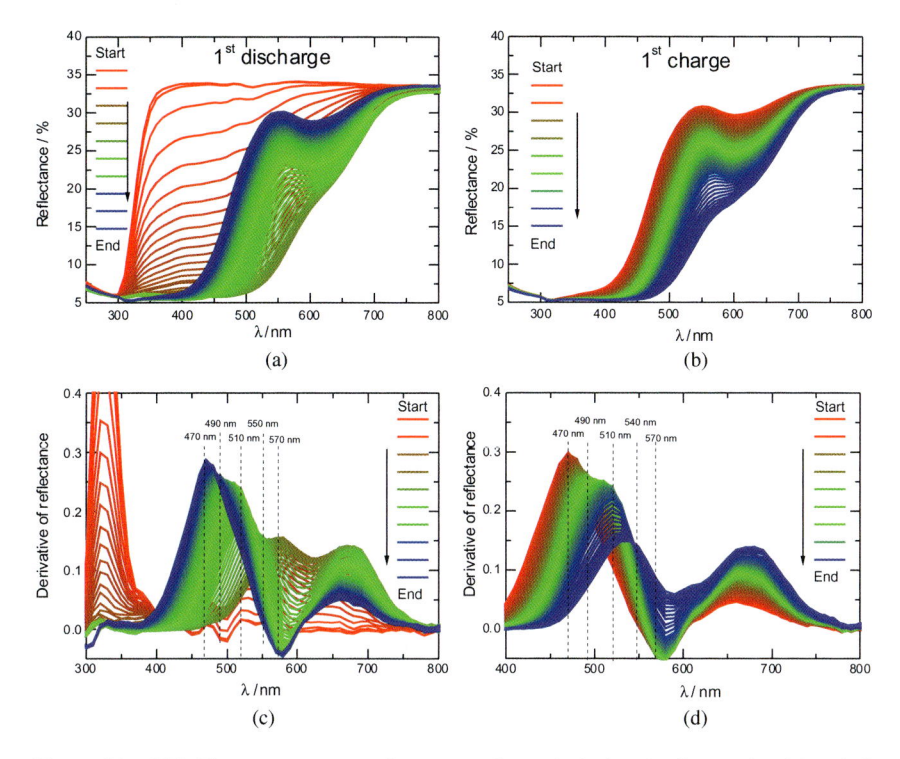

Analytical Techniques for Lithium–Sulfur Batteries 287

Figure 6.9 UV–Vis spectra measured *in operando* mode during the first cycle of the Li–S battery with 1 M LiTFSI in sulfolane electrolyte: (a) all spectra measured during discharge, (b) all spectra measured during charge, and (c, d) corresponding first-order derivatives of the UV–Vis spectra (color in all figures changes from red to green to blue).

Source: Reproduced with permission from Ref. 10.

peak position in that region. The major positions of derivative peaks were in agreement with peaks observed from the derivatives of catholyte solutions (Figures 6.4(c) and 6.4(d) and Figures 6.5(c) and 6.5(d)). We have to remember this pattern of long-chain polysulfide formation at the beginning of the discharge, which appears or shows observance in the long wavelength region of the UV–Vis spectrum. This is followed by the formation of mid- and short-chain polysulfides in the later stages of the discharge, which show an observance in the short wavelengths of the UV–Vis spectrum. This pattern of polysulfide observance is also applicable to the first derivative of the spectra.

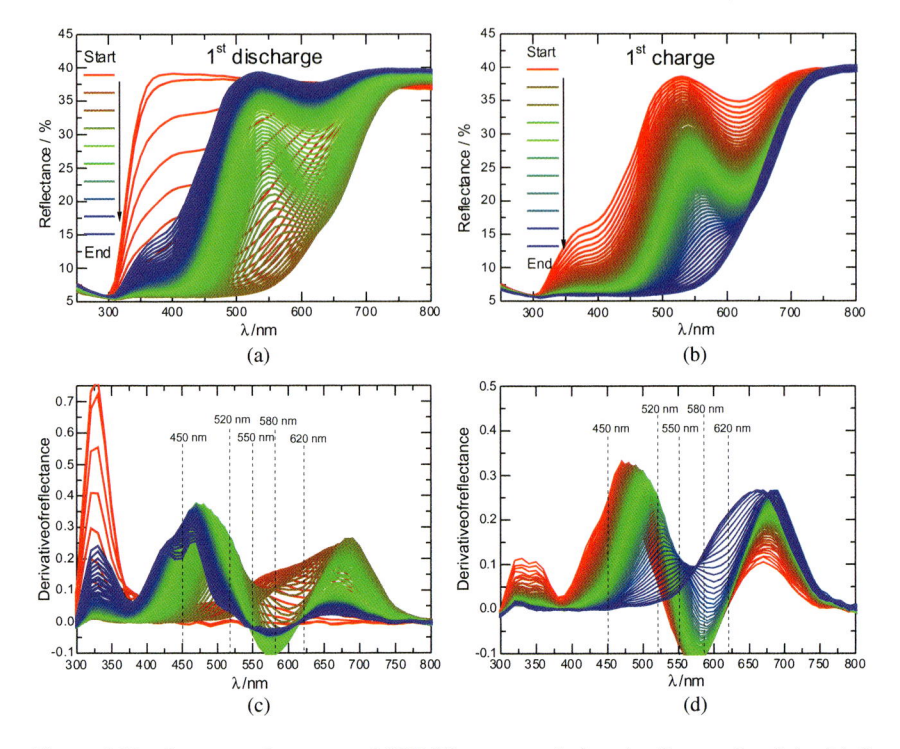

Figure 6.10 *In operando* measured UV–Vis spectra during the first cycle of the Li–S battery with 1 M LiTFSI in TEGDME:DOL electrolyte: (a) all spectra measured during discharge, (b) all spectra measured during charge, and (c, d) related 1st order derivatives of the UV–Vis spectra (color in all figures changes from red to green to blue).

Source: Reproduced with permission from Ref. 10.

Although the measured UV–Vis spectra showed a significant difference between the two electrolytes, our focus was on the qualitative and quantitative determination of polysulfides during *in operando* measurements.

Correlations between concentrations and normalized reflectances obtained from measurements with different stoichiometric equilibrium (Figures 6.6 and 6.7) were used for the development of a simple procedure that allows a direct comparison between different sets of measurements.

Figure 6.11 shows a comparison of the polysulfide evolution in the separator for both electrolyte systems. Batteries were measured at cycling rate C/20 during the first discharge and charge. Typical

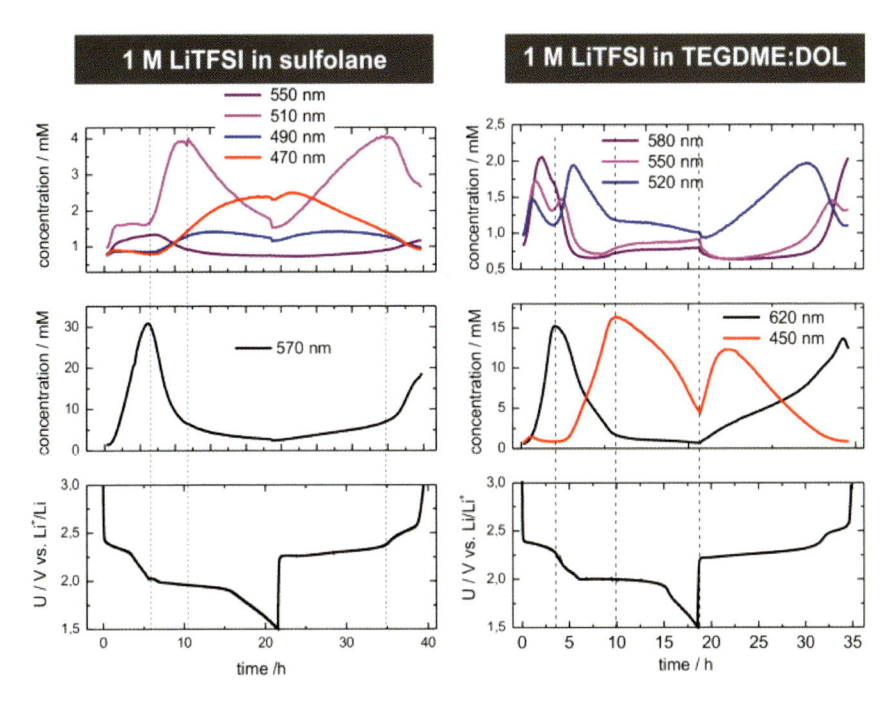

Analytical Techniques for Lithium–Sulfur Batteries 289

Figure 6.11 Recalculated concentrations of short-, mid-, and long-chain polysulfides detected in the separator for both types of electrolytes compared in this work. The left column corresponds to the measurement in 1 M LiTFSI in sulfolane electrolyte, and the right column corresponds to the measurement in 1 M LiTFSI in TEGDME:DOL electrolyte.

Source: Reproduced with permission from Ref. 10.

electrochemical curves with capacities around 800 mAhg^{-1} were obtained. Deconvolution of the UV–Vis spectra measured in the 1 M LiTFSI in sulfolane electrolyte showed saturation of the electrolyte with long-chain polysulfides (recalculated concentration form intensities at 570 nm). The maximum concentration of ~30 mM was achieved at the beginning of the low-voltage plateau. With a continuous discharge, we observed a decrease in the concentration of long-chain polysulfides and an increase in the concentration of mid-chain polysulfides (recalculated concentrations from intensities at 550 and 510 nm). The concentration of mid-chain polysulfides reached the maximum value (~5 mM) in the middle of the low voltage plateau, probably due to the conversion of long-chain polysulfides into shorter chains. With the maximum of mid-chain polysulfides, we observed the increase of short-chain polysulfides (recalculated concentrations from intensities at 490 and 470 nm) reaching the maximum value of ~2 mM at the end of the discharge. We observed the symmetrically opposite evolution of polysulfides in the separator during charging process with a difference in the concentration of long-chain polysulfides at the end of charge. The remaining long and mid-chain polysulfides in the separator indicated the incomplete reoxidation of polysulfides into S, and they can be correlated to the difference between the charge used for reduction and the charge used for oxidation.

Deconvolution of the UV–Vis spectra measured in the 1 M LiTFSI in TEGDME:DOL electrolyte showed the different evolution of polysulfides in comparison to the evolution in sulfolane-based electrolytes. We observed an increase in the concentration of long-chain polysulfides (recalculated from intensities at 620 nm) at the beginning of the discharge reaching the maximum at the end of the high voltage plateau. The recalculated concentration was 15 mM. With the decrease in the concentration of long-chain polysulfides we detected very small amounts of mid-chain polysulfides in the separator (intensities at 580 and 550 nm). A much higher concentration of short-chain polysulfides was observed (intensities at 520 and 450 nm). The increase of short-chain polysulfide concentration recalculated from intensities at 450 nm started with the decrease of concentration of long-chain polysulfides. The maximum concentration of short-chain polysulfides during discharge was equivalent to the concentration of long-chain polysulfides. With a continuous discharge at low voltage plateau, polysulfides were consumed and at the end of the discharge the detected concentration of short-chain polysulfides in the separator was approximately 5 mM.

During charging, the concentration of detected polysulfides in the separator was not exactly the same as during discharging; however, we were able to detect the higher formation of short-chain polysulfides, which was accompanied by the formation of long-chain polysulfides. At the end of charging, short-chain polysulfides were consumed completely, and the concentration of long-chain polysulfides was similar as in the experiment in which the battery was cycled in the sulfolane-based electrolyte.

Comparison of the polysulfide formation in the separator for two different types of electrolytes shows the difference in the mechanism of polysulfide formation/dissolution in the different electrolytes. If we want to design components for Li–S cells, it is necessary to understand the solubility and diffusivity of polysulfides in different electrolytes.

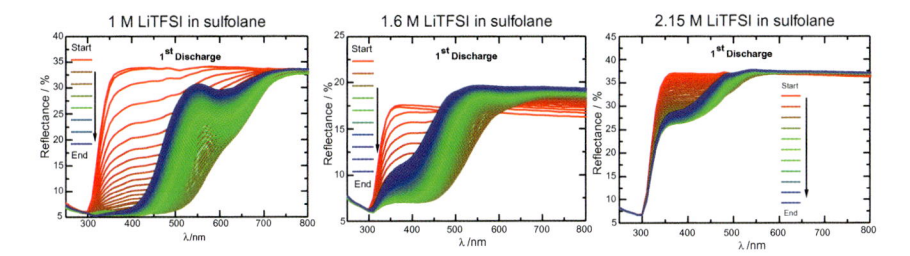

Figure 6.12 The first discharge UV–Vis spectra Li–S batteries in electrolytes with different concentrations of LiTFSI in sulfolane (color in all figures changes from red to green to blue). Not subject to copyright.

Other examples presenting the power of the proposed technique are measurements of electrolytes with increased concentrations of salt. Figure 6.12 shows the measured UV–Vis spectra of three different concentrations of salt in the TEGDME:DOL mixture of solvent. Researchers at CAS showed that more concentrated electrolytes lead to better cycling properties of Li–S batteries.[29] Improved cyclability was connected with a lower solubility of polysulfides due to increased concentrations of salts in the solvent. Our measurements with UV–Vis showed that a considerably lower amount of polysulfides in the electrolyte could be detected with increased concentration of LiTFSI in sulfolane. Complete quantitative determination of the amount of polysulfides was not done due to the large amount of work that necessitated the preparation of different standards required for the calibration curves. A massive amount of measurements for every change in the electrolyte required a great deal of systematic work, and this makes in operando UV–Vis spectroscopy less attractive. Nevertheless, it serves as a basic understanding of polysulfide evolution in different stages of the Li–S battery cycling in different environments.

6.3. Configuration of a 4-Electrode Swagelok Cell and Its Use as an Analytical Tool in Li–S Batteries

A 4-electrode Swagelok cell can also be used as an *in situ* analytical tool for the detection of soluble polysulfides in the electrolyte during Li–S battery operation. In addition to the standard Swagelok configuration, this

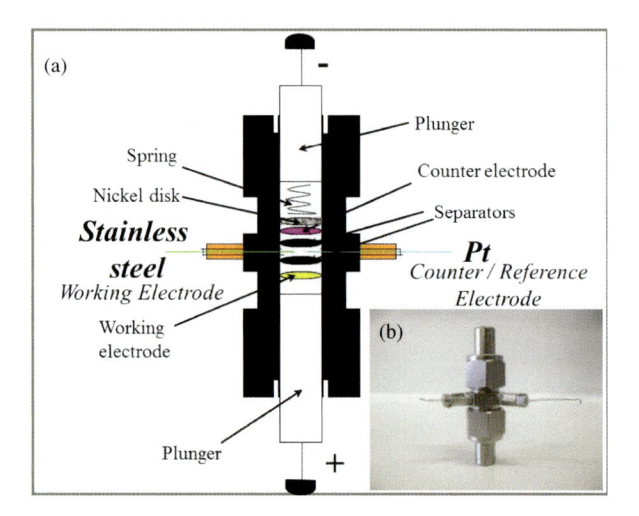

Figure 6.13 (a) A schematic representation of the 4-electrode Swagelok cell. (b) Photograph of the 4-electrode cell.

Source: Reproduced with permission from Ref. 22.

cell has two additional perpendicular electrodes (wires) placed between two separators, as shown in Figure 6.13(a). Soluble polysulfides in the electrolyte are reduced on the stainless steel (SS) wire, and the cumulative charge in the potential range between 2.25 and 1.5 V vs. platinum is used for the quantitative analysis. The cell can be used as an effective and reliable *in situ* analytical tool, which can significantly help in studying the impact of different solvents, salts, additives, and electrode architectures on the stability of the Li–S battery.[25]

Electrochemistry in the 4-electrode cell was tested simultaneously on two channels of galvanostat/potentiostat. Prior to starting the operation of the battery, a so-called background cyclovoltammogram between SS (working electrode) and platinum (counter electrode) wires needs to be measured (typical scan rate 2 mVs^{-1} in the potential range between 2.5 and 0.5 V). During this measurement, the Li–S battery was on OCV mode, as soon as the CV scan had been completed, the galvanostatic measurement of the Li–S battery was initiated for two hours with a current density corresponding to C/20 cycling rate. Measurements were repeated in this sequence until the battery reached the cut-off voltage of 1 V vs. metallic Li.

Here, we show complementary measurements to the UV–Vis measurements shown in Figure 6.13. The same types of electrolytes were also measured with the 4-electrode Swagelok cell, and a similar finding has been obtained. All the batteries that were measured in this set of experiments had the same amount of S in the composite, and the volume of the electrolyte was quantified to 75 μL per mg of S present in the electrode. The results obtained from 1, 1.6, and 2.15 M LiTFSI in sulfolane electrolytes are shown in Figure 6.14. The 1 M LiTFSI in sulfolane Figure 6.14(a) (solid line) shows the discharge curve in the first cycle, and the corresponding CVs measured during the battery relaxation are shown in Figure 6.14(b). A reduction peak at Up ≈1.8 V vs. platinum is observed in every CV measurement. An integration of this peak in the 1.5–2.25 V range gave the cumulative charge that is plotted in Figure 6.14(a) (red spheres). These results confirm predictions that soluble polysulfides are formed and diffuse into the separator at the beginning of the discharge. The partial cumulative charge associated with soluble polysulfides is increased at the beginning; the maximum is reached at a nominal composition of $Li_{0.3}S$, and then is gradually decreased. The maximum of the determined polysulfides is in the agreement with the peak position of polysulfides determined by in operando UV–Vis spectroscopy. This proves that both techniques are complementary with a difference; by using a 4-electrode modified Swagelok cell we can obtain only the quantity of polysulfides detected in the separator, while with UV–Vis spectroscopy we can distinguish between the chain length of polysulfides formed during battery cycling. As also observed with UV–Vis spectroscopy in the first discharge profile of Sulfur reduction in the battery with 1.6 M LiTFSI in sulfolane, the detected cumulative charges in this experiment (Figure 6.14(c) and blue spheres and corresponding CVs in Figure 6.14(d)) were far lower than the one in the 1 M LiTFSI electrolyte (0.75 μAh vs. 0.17 μAh). Measurements obtained with the battery with 2.15 M LiTFSI in the electrolyte are presented in Figures 6.14(e) and 6.14(f), respectively. Interestingly, we did not observe any change in the reduction layer of the CVs measured during the battery relaxation (Figure 6.14(f)). The CVs obtained during battery cycling remained in the same position, and there was no increase in the intensity of current in the region between 2.25 and 1.5 V, as a result of which the sum of all cumulative charges detected for this battery did not change. Comparing this

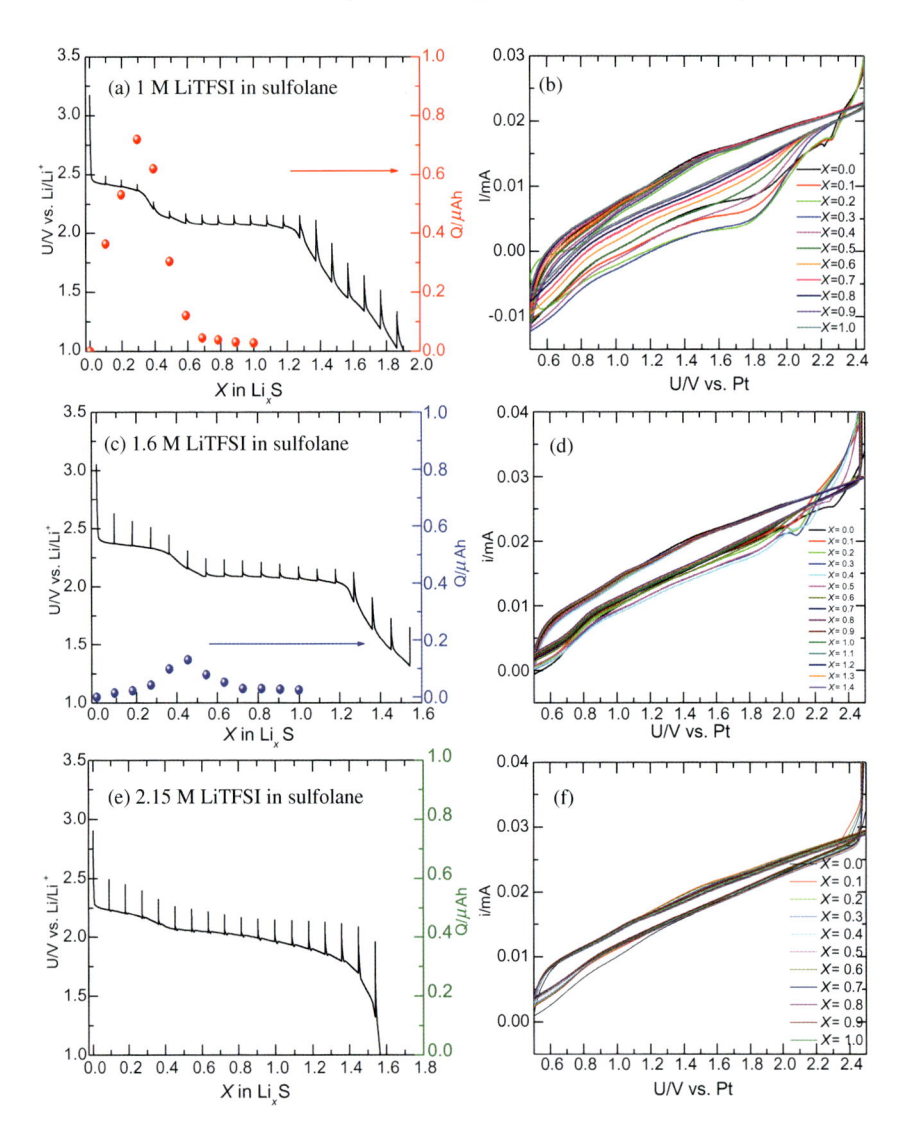

294 *Li–S Batteries: The Challenges, Chemistry, Materials and Future Perspectives*

Figure 6.14 (a, c, e) Electrochemical behavior during the first reduction of carbon/S composite with 1, 1.6, 2.15 M LiTFSI in sulfolane electrolyte and obtained cumulative charge by the integration of individual CV scans. (b, d, f) Corresponding CV scans measured for each $\Delta x = 0.1$ change of composition in Li_xS. Not subject to copyright.

result with measurements by using UV–Vis (Figure 6.12(c)), we can conclude that the sensitivity of UV–Vis spectroscopy is much higher compared to a 4-electrode Swagelok cell, although the use of the latter is much simpler. Nevertheless, both techniques confirmed that by increasing the concentration of LiTFSI salt in the electrolyte, the diffusion of soluble polysulfides could be suppressed.

Another example where a combined study with UV–Vis spectroscopy and a 4-electrode modified Swagelok cell helped us to understand the mechanism was during the introduction of an ion-selective separator based on fluorinated reduced graphene.

Both analytical tools have shown that soluble polysulfides (particularly long-chain) could be detected within both measurements. While the results obtained by using a glass fiber separator are in agreement with results discussed above (Figures 6.9 and 6.14), the introduction of an interlayer based on fluorinated graphene oxide (GO) shows a remarkable difference. The appearance of polysulfides during the high voltage plateau is very similar to the measurement with a bare glass fiber separator but the observed concentration of polysulfides also remained constant during low voltage plateau, which in our opinion indicates that polysulfides were restrained on the side of the positive electrode, since they cannot diffuse through the separator and react with metallic Li at the negative electrode (see Figure 6.15). The observed differences between separators can be explained by the hydrophobicity of the rGO–F interlayer.

6.4. X-ray Absorption Spectroscopy (XAS)

The interest in XAS as an analytical tool for Li–S batteries is based on its relative simplicity to determine the chemical state and local atomic structure for different oxidation states of sulfur. The XAS spectrum is typically divided into two regimes: X-ray absorption near edge spectroscopy (XANES) and extended X-ray absorption fine structure spectroscopy (EXAFS). Though the two have the same physical origin, this distinction is convenient for the interpretation. XANES is strongly sensitive to the formal oxidation state and the coordination chemistry (e.g. octahedral, tetrahedral coordination) of the absorbing atom, while EXAFS is used to determine the distances, coordination number, and the neighboring species of the absorbing atom in XANES. The valance state of the selected type of atom in the sample and the local symmetry of its unoccupied orbitals can be deduced from the information contained in the shape and the energy shift of the X-ray absorption edge.[31–34]

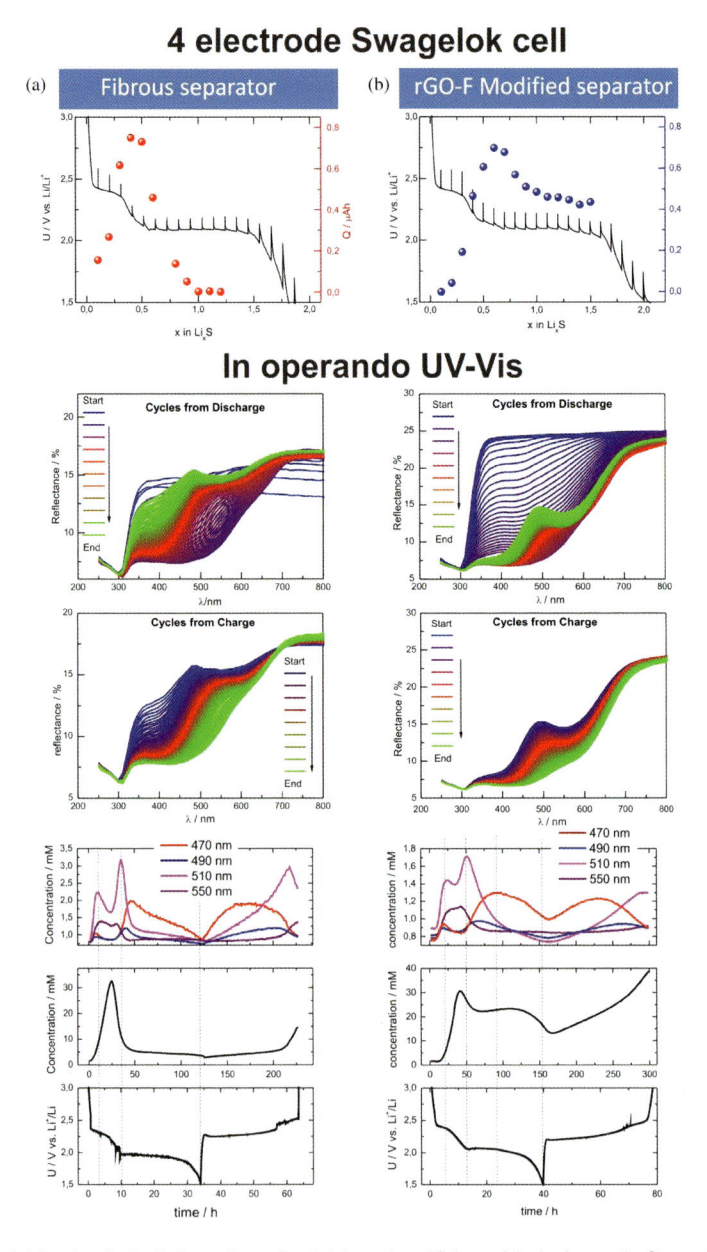

296 *Li–S Batteries: The Challenges, Chemistry, Materials and Future Perspectives*

Figure 6.15 Analytical detection of soluble polysulfides with 4-electrode Swagelok cell and in operando UV–Vis spectroscopy in the two different sets of Li–S batteries assembled with a glass fiber separator, with or without a fluorinated graphene layer.[30]

6.4.1. S K-edge XANES spectroscopy

S K-edge XANES spectroscopy is very sensitive to the oxidation state with a range of ~15 eV between reduced S compounds in metal complexes (S^{2-}) and S in the oxidation state S^{6+} as sulfate, sulfonate. Furthermore, several types of Li polysulfides have a unique pattern of transitions on the absorption edge. This makes S K-edge XANES very suitable for qualitatively determining the speciation of S compounds in samples with the complex composition in the battery.

At the energies of the S K-edge (2469–2485 eV), the total absorption is very high; thus, S K-edge XANES spectroscopy studies are preferably performed in diluted samples in fluorescence mode, either in a vacuum or in pure helium, in order to minimise absorption and scattering from the atmosphere.

The experimental set-up available at Elettra synchrotron (Basovizza, Trieste)[36] is shown in Figure 6.16. It enables measurements of the X-ray absorption spectra at sulfur K-edge in fluorescence-detection mode.

Using the set shown in Figure 6.16, it is possible to perform separate XAS measurements on the individual components of a Li–S battery and to study the local environment of the Li–S battery measured *in operando* during battery operation.

6.4.2. XAS of the reference compounds

S K-edge XANES spectra of the battery components (electrolyte, S, polysulfide, and Li sulfide) are shown in Figure 6.17(a). The energy positions of the S edge and preedge resonances are characteristic of each compound and can be ascribed to transitions of S 1s core electrons to unoccupied p-type molecular orbitals in molecules or empty bands in crystalline compounds.[41] The amount of S in the samples was adjusted to avoid excessive self adsorption and to enable corrections if necessary. A typical component in the Li–S batteries is LiTFSI (Li bis(tri-fluoromethanesulfonyl)imide), in which S is in the form of sulfates (S^{6+}), and the main edge peak appears at 2478 eV. Another form of S in the Li–S battery is pure elemental S. The energy position of the dominant preedge resonance appears at 2472 eV. A reduced form of S with an oxidation state of S^{2-} exists in the equilibrium

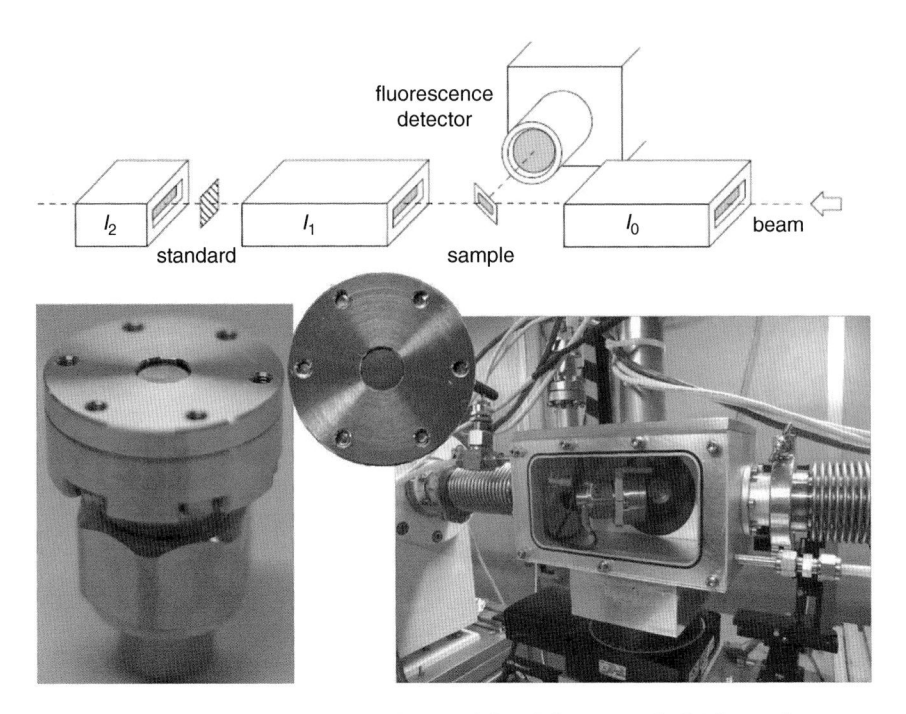

Figure 6.16 Experimental setup and Swagelok® cell for *operando* XAS experiments. Source: Reproduced with permission from Ref. 16.

of different polysulfides, in which the numbers of terminal and bonding atoms of S determine the length of the polysulfide chain. In the XANES spectra, polysulfides can be identified by a characteristic prepeak at 2470.2 eV together with an S peak at 2472 eV (Figure 6.17(a)). The relative intensity of the preedge at 2470.2 eV is proportional to the relative amount of Li in Li_2S_x (Figure 6.17(b)). A similar pre-edge peak at 2470.2 eV is observed in the XANES spectra of Na_2S_x compounds, with the same intensity variation correlated to the relative amount of Na in the Na_2S_x compound (Figure 6.17(c)). This characteristic feature in XANES spectra of the polysulfides can be used in the analysis of intermediate products during battery operation. The XANES spectrum of the crystalline form of Li_2S was found to be significantly different from those of long- or short-chain polysulfides (Figure 6.17(a)). Contrary to Li polysulfides, which have a molecular structure given by a chain of S atoms ending with two Li

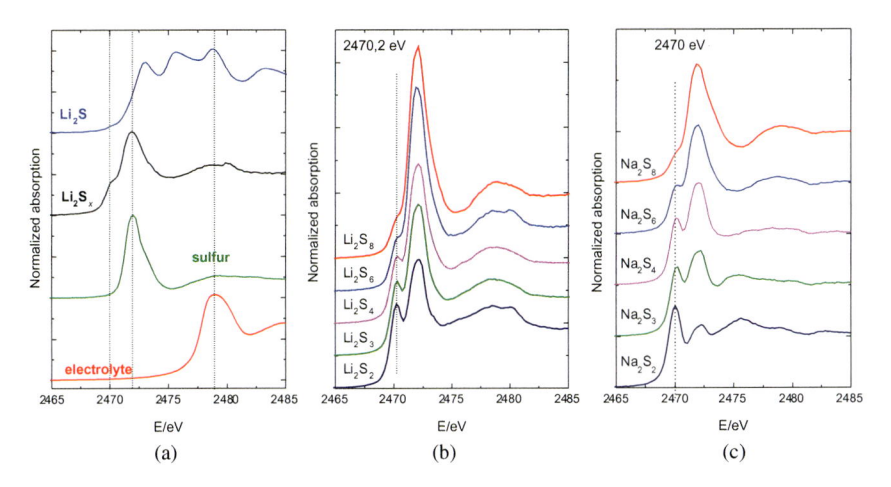

Figure 6.17 (a) S K-edge XANES spectra of S compounds present in the Li–S battery (electrolyte, S, polysulfides, and Li sulfide) mixed with BN (Li$_2$S$_x$ represents a typical spectrum for all polysulfide spectra). Spectra are shifted vertically for clarity. The vertical dotted lines at 2470.2, 2472.0, and 2478.0 eV mark characteristic prepeaks in measured sulfur reference compounds. (b, c) S K-edge XANES spectra of different (a) Li and (b) sodium polysulfides. Spectra are shifted vertically for clarity.

Source: Reproduced with permission from Ref. 12.

atoms, Li$_2$S has a crystalline anifluorite type structure, the local geometry of which gives rise to an entirely different XANES pattern.[12]

6.4.3. In operando mode XAS

In Operando Mode XAS can be considered to be a powerful tool that can be used to understand the mechanism of Li–S battery operation in the different chemical environments (solvents, additives, host matrices, etc.). An example of successful and complete XAS analysis in which both the XANES and EXAFS parts of XAS spectra were analyzed was performed with electrolyte by using LiTDI salt and ether-based solvents. With that, only signal of active sulfur components (sulfur, polysulfides, and Li$_2$S) is measured. Figure 6.18(a) shows a change of shape of the sulfur K-edge XANES spectra measured in operando mode during the discharge of the battery. A significant change in the composition can be derived from

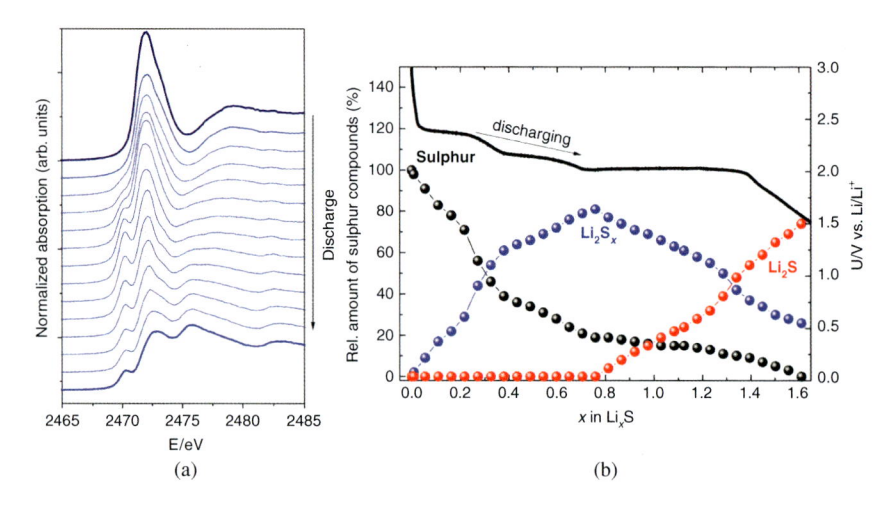

Figure 6.18 (a) *Operando* S K-edge XANES spectra of a Li–S battery measured during discharge (only selected spectra are presented due to clarity), (b) Electrochemical curve obtained during discharge and the relative amount of the three sulfur components (sulfur, polysulfides-Li$_2$S$_x$, and Li$_2$S).

Source: Reproduced with permission from Ref. 16.

the measurement. All components shown in Figure 6.17(a) can be identified by the characteristic energy position of the sulfur edge and preedge resonances. A dominant edge resonance at 2472 eV is from the elemental state of the sulfur. Polysulfides (Li$_2$S$_x$) can be recognized with two edge resonances, beside resonance at 2472 eV an additional preedge resonance also appears at 2470.2 eV. In the case of radicals, additional components are expected at lower energies, around 2468 eV,[35] and they were not observed in our measurements. For polysulfides, the relative intensity of the two resonances at 2470.2 and 2472 eV is proportional to the relative amount of terminal vs. internal sulfur atoms, i.e. proportional to the length of the S$_x^{2-}$ chain. In theory, the position of the prepeak also depends on the stoichiometry,[36] and it was not observed in any experimental work, most probably due to the complicated equilibriums of different chain length polysulfides. Nevertheless, in shorter chains, the pre-peak is more expressed than in longer chains, which enables at least the determination of long- and short-chain polysulfides. In general, the length of polysulfides can be determined from XANES spectra based on the ratio between the intensities (peak areas) of the peaks at 2470.2 and 2472 eV. Again, due to the complicated equilibriums in the polysulfide mixtures (chemically or electrochemically prepared), there is no correlation between intensities and predicted stoichiometry. As mentioned above, the XANES spectrum of crystalline Li_2S with antifluorite structure, in contrast, was found to be significantly different compared to long- or short-chain polysulfides or elemental sulfur.

All sulfur K-edge XANES spectra measured on the battery in operando mode (Figure 6.18) can be deconvoluted by using three components (sulfur, polysulfides, and Li_2S). While the application of Li_2S and elemental sulfur is straightforward, more complications are expected in fitting the correct polysulfide composition. There can be several reference spectra required for the successful fitting or in some cases only one composition can be applied. Another strategy is to extract reference spectra from the measurement. For instance, at the end of a high voltage plateau, XANES spectra are combined only from the remaining sulfur, which is a minor component, and polysulfides. At this point, we can consider the relative homogenous composition of polysulfides, which is observed as a peak composition in the deconvolution of UV–Vis spectra.

The third reference in the spectra is crystalline Li_2S which can be chemically synthesized, and its XANES spectra correspond to the spectra observed at the end of discharge; it is in agreement with the theoretically predicted spectra. Reference compounds can be further used for the description of the intermediate state of the battery. Figure 6.18(b) shows the electrochemical curve obtained during the discharge of the Li–S cell and the relative amounts of the three sulfur components. The quality of the linear combination fits (LCF) is demonstrated in Figure 6.19. The result shows the evolution of three components during the battery discharge. In the first part of the discharge that corresponds to the high voltage plateau, only sulfur is consumed and transformed into polysulfides. Sulfur content at the beginning of the low voltage plateau is approximately 20%, and the rest are polysulfides. During the high voltage plateau, two different rates of sulfur reduction can be observed. This can be explained that at the beginning only sulfur is reduced to polysulfides while in the latter stage of the reduction process polysulfides are also reduced to their shorter chain counterparts. Again, this is within expectation that at the end of the high voltage plateau, mid-chain polysulfides are mainly present in

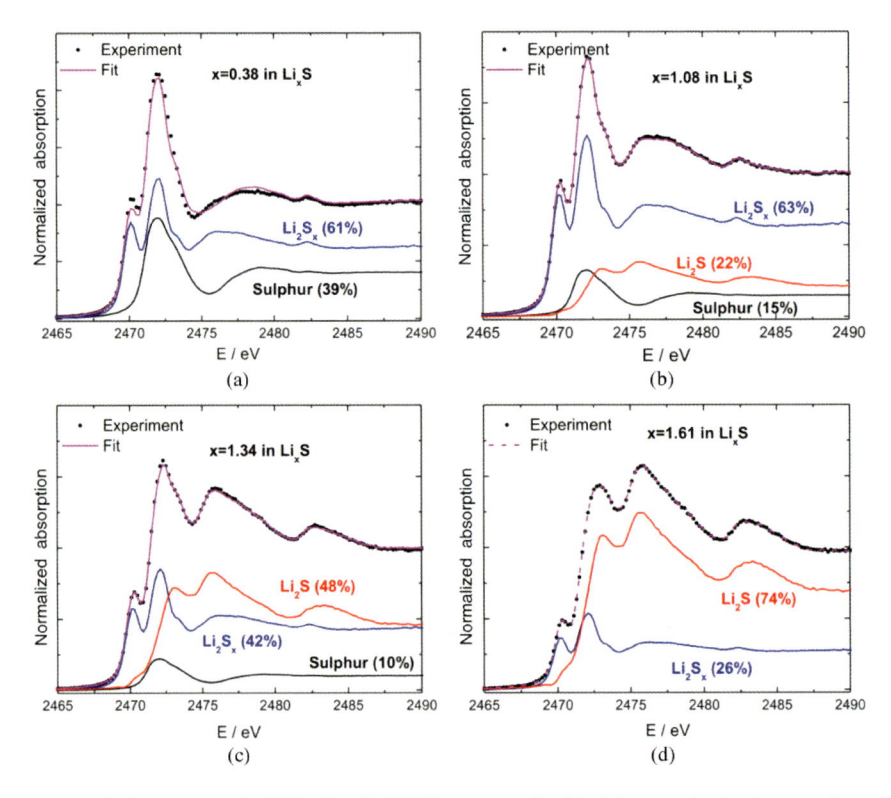

302 *Li–S Batteries: The Challenges, Chemistry, Materials and Future Perspectives*

Figure 6.19 *Operando* S K-edge XANES spectra of a Li–S battery in the intermediate states during discharge at nominal compositions: (a) Li$_{0.38}$S, (b) Li$_{1.08}$S, (c) Li$_{1.34}$S, and (d) Li$_{1.61}$S. Solid squares: experiment; solid line: best fit with linear combination of the three reference XANES profiles (sulfur, Li$_2$S$_x$, and Li$_2$S), plotted below.

Source: Reproduced with permission from Ref. 16.

the battery. The low voltage plateau corresponds to the precipitation of Li$_2$S. This observation is in perfect agreement with some previous reports showing that Li$_2$S starts forming at the beginning of the low voltage plateau.[17,35] Since this process is linear, most probably a direct conversion of polysulfides to Li$_2$S occurs. Interestingly, LCF of spectra obtained during the low voltage plateau show a complete reduction of polysulfides. The final composition contains about 75% of Li$_2$S, and the remaining sulfur components are Li polysulfides.

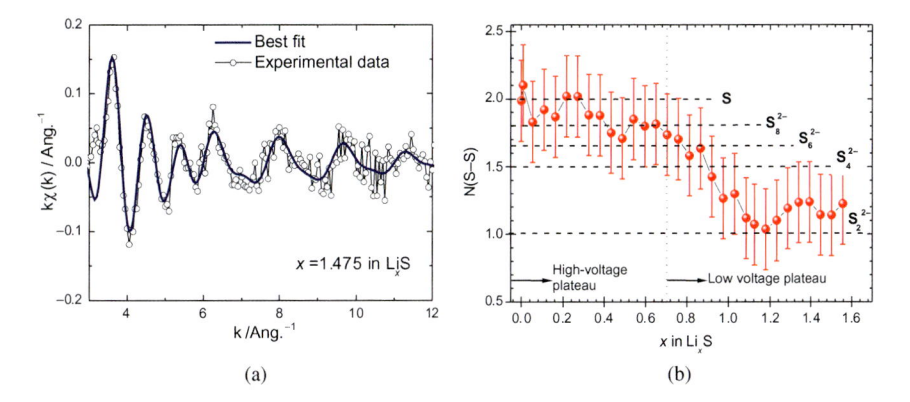

Figure 6.20 (a) Comparison between the experimental EXAFS spectrum at nominal composition $Li_{1.47}S$ (dots) and the best-fit calculation (solid curve). (b) Variation of the average S coordination number during the first discharge. The average coordination of the most important polysulfides is reported for comparison. The vertical line represents the end of the high-voltage plateau.

Source: Reproduced with permission from Ref. 16.

In the end, it should be emphasized that the content of sulfur in the XAS experiments is much lower in comparison to the requirements for the successful commercialization and all measurements are done in the reflection mode, in which the penetration of the beam is about 5–10 μm on the back side of the cathode; this does not necessarily correspond to the overall bulk performance.

The electrolyte used in this study was sulfate-free, which enabled a more detailed EXAFS data analysis and to examine possible interactions between host matrix and polysulfides. However, it is possible to determine the average coordination number for sulfur. A complete fit of the series of spectra collected through the first discharge with the two S–S components, the first one at 2.04 Å, representing sulfur and polysulfides, and the other one at 4.04 Å, representing Li_2S, was also attempted (Figure 6.20(a)). The variation of the average coordination number of the S–S component throughout the entire process is shown in Figure 6.20(b). A gradual decrease in the coordination number is shown in the first part of the discharge, varying from 2 to about 1.6(2) at the end of the high-voltage plateau. The coordination number corresponding to the number of S–S bonds and number of terminating and bonding sulfur atoms can be calculated from the coordination number. At the end of the high voltage plateau, the coordination number corresponds to an average composition between S_6^{2-} and S_4^{2-}. A steeper decrease is observed at the beginning of the low-voltage plateau, where the average coordination number varies rapidly, down to values close to those typical for S_4^{2-} and S_2^{2-}. The average coordination number of the component representing polysulfides does not vary during this part of the discharge, even though the relative amount decreases. At the same time, the amount of Li_2S increases rapidly to about 70% of the total amount of sulfur at the end of the discharge, which is in good agreement with XANES analysis. The most important output from EXAFS study is that apart from these two components, no other signal was clearly detected in the EXAFS spectra throughout the whole series. Taking into account the sensitivity of EXAFS, which is usually above 5%, no specific interaction of the polysulfide species with the matrix or with other species in the electrolyte could be detected.

6.5. Conclusions

In this chapter, a combination of analytical tools is presented as efficient instrumentation, which aids in understanding the mechanisms of Li–S batteries in different environments. Three *in operando* techniques can describe most of the processes occurring within the Li–S battery. Use of *in operando* UV–Vis spectroscopy confirms the diffusion of polysulfides with different chain lengths to the separator. This analytical tool can be used for quantitative and qualitative determination of polysulfides in the separator and in the back side of the sulfur electrode. The 4-electrode Swagelok cell analytical tool is a complimentary method to UV–Vis spectrometry, and the quantitative determination of polysulfides in the separator is feasible. The third technique presented in this chapter is the most powerful since it enables determination of different S components in the battery in a quantitative and qualitative way; however, its limitation is synchrotron source of energy, while the former two techniques can be used on the lab scale.

Bibliography

1. Y. V. Mikhaylik and J. R. Akridge, *Journal of the Electrochemical Society*, **151**(11) (2004) A1969–A1976.
2. M.-K. Song, E. J. Cairns and Y. Zhang, *Nanoscale*, **5** (2013) 2186–2204.
3. S. Eversand and L. Nazar, *Accounts of Chemical Research*, **46**(5) (2013) 1135–1143.
4. A. Manthiram, Y. Fu and Y-S. Su, *Accounts of Chemical Research*, **46**(5) (2013) 1125–1134.
5. V. S. Kolosnitsynz and E. V. Karaseva, *Russian Journal of Electrochemistry*, **44**(5) (2008) 506–509.
6. Y. Zhang, Y. Zhao, K. E. Sun and P. Chen, *The Open Materials Science Journal*, **5**(Suppl 1: M3) (2011) 215–221.
7. Y. Jung and S. Kim, *Electrochemical Communications*, **9** (2007) 249–254.
8. B. S. Kim and S. M. Park, *Journal of the Electrochemical Society*, **140** (1993) 115–122.
9. M. U. M. Patel, R. Demir-Cakan, M. Morcrette, J-M. Tarascon, M. Gaberscek and R. Dominko, *ChemSusChem*, **6** (2013) 1177–1181.
10. M. U. M. Patel and R. Dominko, *ChemSusChem*, **7** (2014) 2167–2175.
11. C. Barchasz, F. Molton, C. Duboc, J.-C. Lepretre, S. Patoux and F. Alloin, *Analytical Chemistry*, **84** (2012) 3973–3980.
12. M. U. M. Patel, I. Arčon, G. Aquilanti, L. Stievano, G. Mali and R. Dominko, *ChemPhysChem*, **15** (2014) 894–904.
13. M. Cuisinier, P.-E. Cabelguen, S. Evers, G. He, M. Kolbeck, A. Garsuch, T. Bolin, M. Balasubramanian and L. F. Nazar, *Journal of Physical and Chemical Letters*, **4** (2013) 3227–3232.
14. Y. Gorlin, A. Siebel, M. Piana, T. Huthwelker, H. Jha, G. Monsch, F. Kraus, H. A. Gasteiger and M. Tromp, *Journal of the Electrochemical Society*, **162** (2015) A1146–A1155.
15. J. Gao, M. A. Lowe, Y. Kiya and H. D. Abruña, *The Journal of Physical Chemistry C*, **115** (2011) 25132–25137.
16. R. Dominko, M. U. M. Patel, V. Lapornik, A. Vizintin, M. Kozelj, N. N. Tusar, I. Arcon, L. Stievano and G. Aquilanti, *The Journal of Physical Chemistry C*, **119** (2015) 19001–19010.
17. S. Walus, C. Barchasz, J-F. Colin, J-F. Martin, E. Elkaïm, J.-C. Leprêtre and F. Alloin, *Chemical Communications*, **49** (2013) 7899–7901.
18. J. Nelson, S. Misra, Y. Yang, A. Jackson, Y. Liu, H. Wang, H. Dai, J. C. Andrews, Y. Cui and M. F. Toney, *Journal of American Chemical Society*, **134** (2012) 6337–6343.

19. J. Hannauer, J. Scheers, J. Fullenwarth, B. Fraisse, L. Stievano and P. Johansson, *ChemPhysChem.*, **16**(13) (2015) 2755–2759.
20. R. Demir-Cakan, M. Morcrette, Gangulibabu, A. Gueguen, R. Dedryvere and J.-M. Tarascon, *Energy and Environmental Science*, **6** (2013) 176–182.
21. Y.-C. Lu, Q. He and H. A. Gasteiger, *The Journal of Physical Chemistry C*, **118** (2014) 5733–5741.
22. R. Dominko, R. Demir-Cakan, M. Morcrette and J.-M. Tarascon, *Electrochemical Communications*, **13** (2011) 117–120.
23. K. A. See, M. Leskes, J. M. Griffin, S. Britto, P. D. Matthews, A. Emly, A. V. Ven, D. S. Wright, A. J. Morris, C. P. Grey and R. Seshadri, *Journal of the American Chemical Society*, **136** (46), (2014) 16368–16377.
24. T. Owen, Principles and applications of UV–visible spectroscopy, In: *Fundamentals of Modern UV–Visible Spectroscopy.* Publication Number 5980–1398E, Agilent Technologies, (2000).
25. R. D. Rauh, F. S. Shuker, J. M. Marston and S. B. Brummer, *Journal of Inorganic and Nuclear Chemistry*, **39** (1977) 1761–1766.
26. S-I. Tobishima, H. Yamamoto and M. Matsuda, *Electrochimica Acta*, **42** (6) (1997) 1019–1029.
27. P. Dubois, J. P. Lelieur and G. Lepoutre, *Inorganic Chemistry*, **27** (1988) 1883–1890.
28. C. Barchasz, J. C. Leprêtre, F. Alloin and S. Patoux, *International Battery Association Meeting*, Waikoloa, Hawaii (2010).
29. L. Suo, Y-S. Hu, H. Li, M. Armand and L. Chen, *Nature Communications*, **4** (2013) 1481.
30. A. Vizintin, M. U. M. Patel, B. Genorio and R. Dominko, *ChemElectroChem*, **1** (2014) 1040–1045.
31. Introduction to XANES and EXAFS analysis, Iztok Arčon, University of Nova Gorica (2008).
32. Fundamentals of XAFS, Matthew Newville, Consortium for Advanced Radiation Sources University of Chicago, Chicago, IL, (2004).
33. A. Di Cicco, G. Aquilanti, M. Minicucci, E. Principi, N. Novello, A. Cognigni and L. Olivi, *Journal of Physics: Conference Series*, **190** (2009) 012043.
34. S. D. Kelly, D. Hesterberg and B. Ravela, *Soil Science Society of America*, (2008) 387–463.

35. M. Cuisinier, C. Hart, M. Balasubramanian, A. Garsuch and L. F. Nazar, *Advanced Energy Materials*, **5** (2015) 1401801.
36. T. A. Pascal, K. H. Wujcik, J. Velasco-Velez, C. Wu, A. A. Teran, M. Kapilashrami, J. Cabana, J. Guo, M. Salmeron, N. Balsara *et al. Journal of Physical Chemistry Letters*, **5** (2014) 1547–1551.

Chapter 7

Other Sulfur Related Rechargeable Batteries: Recent Progress in Li–Se and Na–Se Batteries

Rui Xu,[*,‡] Tianpin Wu,[†,§] Jun Lu[*,¶] and Khalil Amine[*,∥]

[*]*Chemical Sciences and Engineering Division, Argonne National Laboratory, 9700 South Cass Avenue Argonne, IL 60439, USA*
[†]*Advanced Photon Sources, X-ray Science Division, Argonne National Laboratory, Lemont, IL 60439, USA*
[‡]*xur@anl.gov*
[§]*twu@aps.anl.gov*
[¶]*junlu@anl.gov*
[∥]*amine@anl.gov*

7.1. Introduction to Selenium-based Electrode Materials

Efforts are underway worldwide to realize safe and efficient energy storage systems with high-energy density, long cycle life, and low cost. State-of-the-art rechargeable lithium-ion (Li-ion) batteries, mainly based on transition metal oxides and intercalation chemistry, are approaching their energy density boundaries yet are still short of meeting tomorrow's energy storage requirements for advanced transportation, portable, and residential applications. New research directions that differ from conventional intercalation chemistry are needed to meet performance requirements for advanced energy storage systems, and discovery of new cathode materials with high-energy density is key to this research.

Among the candidates for beyond systems, lithium–sulfur (Li–S) and lithium–air (Li–O$_2$) batteries have the potential to provide 2–5 times the energy density of current commercial systems, but these batteries still face tremendous challenges.[1–8] Elemental selenium (Se), a member in the same elemental group as oxygen and sulfur in the periodic table, contains a d-electron and high electrical conductivity, and has recently been proposed as a cathode material for rechargeable batteries because of its potential for low cost and high specific energy. For example, selenium has a theoretical gravimetric capacity (678 mAh g^{-1}) and volumetric capacity (3268 mAh cm^{-3}) that are comparable with those of sulfur. Moreover, unlike the widely studied Li/S system, Se-based cathodes can be cycled to high voltages (up to 4.6 V) without failure. Their high densities (4.82 g cm^{-3}, ~2.4 times that of sulfur) and voltage output offer greater volumetric energy density than that of sulfur-based batteries. In addition, whereas the Li–S system requires an ether-based electrolyte, the Se-based system works with the conventional carbonate-based electrolytes, which are less costly.[9–11] Se is an attractive cathode material for not only rechargeable lithium batteries but also room-temperature sodium batteries.[12–16]

Se has been studied for a long time in medicine because it is an essential trace element in the human body. Also, due to the unique photoconductive and photovoltaic effects, Se-based materials have diverse commercial applications such as glassmaking, pigments, photoelectrochemical and photovoltaic cells, and a few types of electronics that have not been supplanted by silicon semiconductors, such as infrared detectors, solid state lasers, and fluorescent quantum dots.[11,17–19]

Until recently, Se-based materials have not been explored as an electrode material for rechargeable batteries, and thus the electrochemistry of Se is not well understood yet. Fundamental issues in Se-based systems that need to be clarified include the similarity and unique nature of Se in comparison with sulfur-based cathodes, the compatibility and behavior of Se in current organic electrolytes, detailed redox reactions of Se during charge/discharge, and the coulombic efficiency and cycle stability of Se-based cathodes.

This review mainly focuses on the recent research progress on the fundamental Li–Se and Na–Se electrochemistry, which helps gain insight into the battery redox mechanism and direct future cell design. Because Li–Se and Na–Se battery technology is still in its infancy, many hurdles remain to overcome the technical challenges and meet the performance requirements for commercial application. This goal can be achieved via advanced electrode design guided by an improved understanding of basic Li–Se electrochemistry derived from advanced characterization techniques. We begin with a brief overview of Li–Se and Na–Se chemistry and the battery operating principles. We then discuss recent research discoveries in better understanding mechanistic details based on advanced characterization techniques. Next, progress in Se-based cathode design in both reachargeable sodium and lithium batteries is discussed in detail. Finally, concluding remarks and perspectives for future research development of Li–Se and Na–Se batteries are presented.

7.2. Fundamental Selenium Electrochemistry as Compared to Sulfur

Both sulfur (atomic number 16, [Ne]$3s^23p^4$) and Se (atomic number 34, [Ar]$4s^23d^{10}4p^4$) are members of the VIA group, and they share similar chemical and electrochemical properties. While the most common and stable form of sulfur in nature is S_8 rings (cyclic α-octasulfur) with an orthorhombic structure,[20] the most dense and thermodynamically stable form of Se is Se_n chains with a hexagonal structure (gray Se, trigonal Se). Hexagonal Se is different in several respects from the other allotropes of Se (including Se_n rings in monoclinic, rhombohedral, orthorhombic, amorphous, and other metastable forms). For example, the ordered arrangement of Se chains facilitates electronic conduction in hexagonal Se and makes it a p-type semiconductor displaying appreciable photoconductivity, whereas other Se forms are insulators. Also, unlike other forms of Se, hexagonal Se does not dissolve in CS_2.[21]

Sulfur rings undergo unusual viscosity changes when gradually heated because sulfur melts at ≈120°C but exhibits a minimum viscosity when the temperature rises to ≈160°C as the rings in liquid cyclic α-S$_8$ start to open and form linear sulfenyl diradicals.[7] By contrast, selenium chains exhibit the usual viscosity changes.[22] It is also believed that the Se chain structures with good conductivity are more electrochemically active and more electrochemically advantageous than ring structures, since each Se chain has two active terminal atoms.[10] Analogous to the formation of polysulfides as a result of the reduction of elemental sulfur, the reduction of Se gives polyselenides.

Similar to sulfur, Se has several oxidation states, including +6 (selenate), +4 (selenite), 0 (elemental Se), and −2 (selenide), of which the stability greatly depends on the pH and redox potentials of the electrochemical environment. A typical Li–Se or Na–Se cell consists of selenium as the cathode, lithium metal as the anode, and an organic Li/Se solution as the electrolyte. In such a cell, the theoretical energy density is determined by the overall cell reaction, [Se + 2M ↔ M$_2$Se] (M = Li, Na), which is analogous to that of sulfur reactions with Li or Na, [S + 2M ↔ M$_2$S] (M = Li, Na). The redox reaction potentials, specific capacities and energies, and other electrochemical and physical properties of Li–Se and Na–Se cells are summarized in Table 7.1. Note that Se demonstrates approximately 20 orders higher magnitude of electronic conductivity than that of sulfur, which facilitates promising electrochemical performance in reactions with both Li and Na. In addition, although the redox potential and theoretical gravimetric capacity of the Se cathode are lower than those of sulfur, its capacity density and energy density are compensated by the density of Se being 2.4 times higher, 4.82 vs. 2.07 g cm^{-3}. High volumetric capacity and energy densities are greatly favored in energy storage systems for portable devices and electric vehicles, as their packing space is usually limited.

7.3. Research Progress in Li–Se Batteries

7.3.1. Mechanistic investigation on Li–Se batteries

The pioneering studies on Se-based materials as electrode materials for rechargeable batteries were not conducted until recently, initiated by Amine's group in 2012.[9,12,23] Their initial work demonstrated that selenium

Table 7.1: Electrochemical and physical properties of sulfur and selenium.

	Sulfur	Sulfur	Selenium	Selenium
Density (g/cm^3)	2.07 (α-S)	2.07 (α-S)	4.8 (h-Se)	4.8 (h-Se)
Electronic conductivity (S cm^{-1})	5×10^{-30} (25 °C)	5×10^{-30} (25 °C)	10^{-5} (25 °C)	10^{-5} (25 °C)
Melting point (°C)	114	114	217–222	217–222
	Li–S	Na–S	Li–Se	Na–Se
Redox potential (V)	2.2 (versus Li/Li$^+$)	2.0 (versus Na/Na$^+$)	2.0 (versus Li/Li$^+$)	1.5 (versus Na/Na$^+$)
Theoretical gravimetric capacity of cathode (mAh g^{-1})	1675	1675	678	678
Theoretical capacity density of cathode (Ah L^{-1})	3467	3467	3268	3268
Theoretical specific energy of cathode and anode (Wh kg^{-1})	2567	400	1155	644

did show electrochemical reactivity and redox reactions between Se and Li in a coin cell consisting of Se/multiwalled carbon nanotubes (MWCNs) as the cathode, Li-metal as the anode, and a carbonate-based electrolyte. Although Se and MWCNs were ball milled and heated when the composite cathode was being prepared, Se was not completely impregnated into the nanotubes, and thus large crystals of Se were still present in the composite cathode. A Li–Se cell was tested in which the electrolyte (Gen II) was 1.2 M LiPF$_6$ dissolved in a 3:7 volume ratio of ethylene carbonate (EC) and ethyl methyl carbonate (EMC). This cell was charged and discharged in the potential window ranging from 0.8 to 4.0 V under different current densities.

As shown in the voltage profiles of Figure 7.1(a), the typical initial discharge (Li insertion) involves a single well-defined plateau (at 2.0 V) and the initial charge involves two plateaus (at 2.3 and 3.6 V). This behavior indicates a single-phase transition during the initial discharge process from Se to Li$_2$Se as Li is inserted into the cathode. It was observed that the

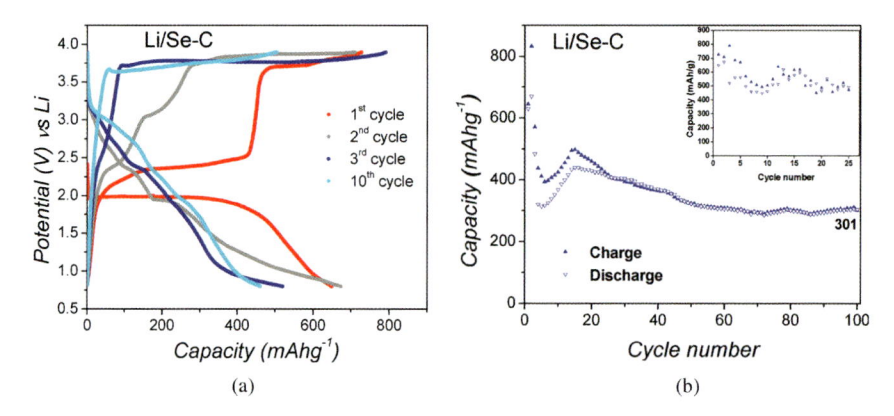

Figure 7.1 Li/Se–C cathode in the presence of carbonate-based electrolyte: (a) voltage profiles during selected cycles and (b) the capacity retention cycled at 50 and 10 mA g^{-1} (inset) current density.[9]

voltage profiles evolved after the first cycle. The discharge plateau and the charge plateau at 2.3 V were less clearly defined starting from the second cycle. In the first couple of cycles, the charge and discharge potentials of the Li/Se–C voltage profile increased. After five cycles, the charge potential stabilized at ~3.7 V, while the average discharge potential increased to ~2.5 V. These voltage profiles were investigated by pair distribution function (PDF) analysis later and were believed to be associated with the change of oxidation potential of Li$_2$Se to Se.

The Se composite cathode has excellent cycling stability, maintaining most of its initial capacity for 100 cycles (Figure 7.1(b)). A reversible capacity of ~500 mAh g^{-1} at low current density (10 mA g^{-1}, ~C/60) and ~300 mAh g^{-1} at higher current density (50 mA g^{-1}, ~C/12) was achieved.

The above experimental evidence indicated that Se-based cathodes are compatible with carbonate-based electrolytes, whereas Li–O$_2$ and most Li–S batteries have been found to be reactive with carbonate-based electrolytes due to nucleophilic reactions.[24,25] The remarkable performance of Se cathodes in carbonated-based electrolyte demonstrates that Se is a promising alternative to sulfur for large-scale applications of beyond Li-ion batteries (LIBs) that still are compatible with commercial electrolytes.

Figure 7.2 shows PDF plots at various stages of charge and discharge superimposed on a voltage profile for Li insertion into Se–C electrodes.[9] Comparison of the PDF data against data obtained for pristine Se confirmed the trigonal structure (Figure 7.2), in which chains of Se atoms are linked by Se–Se bonds of 2.36 Å length. The PDF data for the fully discharged Se–C electrode indicated the formation of Li$_2$Se (Figure 7.2). Both the Li$_2$Se and Se are present at the long discharge plateau (2.0 V, ~240 mA g^{-1}), in agreement with a first-order phase transition. The increase of the initial charge plateau at 2.3 V starting from the second cycle was attributed to Li$_2$Se persisting to higher potentials than during the initial charge, and thus the oxidation of Li$_2$Se to Se taking place at higher potentials.

Besides the carbonate-based electrolyte (Gen II), Amine's group investigated the electrochemical behavior of a Se cathode in ether-based

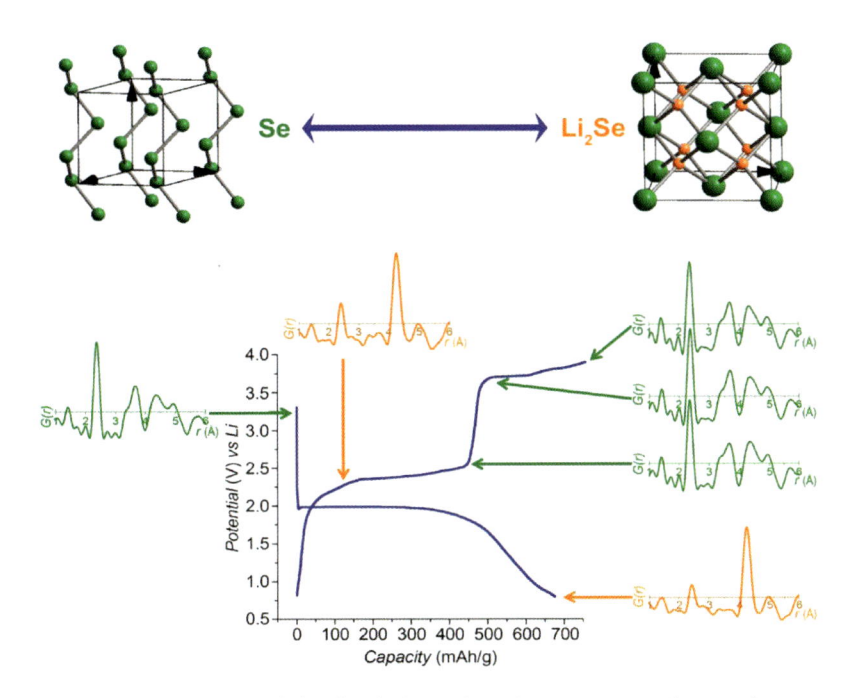

Figure 7.2 PDFs for the pristine Se–C electrode and upon recovery from various states of discharge/charge. Structural representations of the Se and Li$_2$Se (antifluorite) phases are shown.[9]

electrolyte and compared the differences in the lithiation mechanisms in the two electrolytes.[12,23] The ether-based electrolyte (D2) consists of 1 M Li bis(trifluoromethanesulfonyl)imide in 1:1 volume ratio of 1,3-dioxolane (DOL) and 1,2-dimethoxyethane (DME). The voltage profiles of Li–Se cells cycled in the two electrolytes are given in Figure 7.3(a). A most evident difference in the voltage profiles is that the cell with Gen II electrolyte has only one plateau (which is below 2.0 V), while the cell with D2 electrolyte has two plateaus (at around 2.2 and 2.0 V in the first discharge). Also, visual inspection showed neither Se, nor Li_2Se, nor the combination of the two dissolved in the carbonate-based solvent, as shown in Figure 7.3(b). By contrast, both Li_2Se and Li_2Se_n dissolved in the ether-based solvent (Figure 7.3(c)). The mechanism underneath was investigated by various characterization techniques, such as *in situ* X-ray diffraction (XRD) and X-ray absorption near edge structure (XANES) spectroscopy.

The *in situ* XRD results reveal that with discharging of the cell in both electrolytes, the intensity of the Se peak declines, and Li_2Se peaks start to

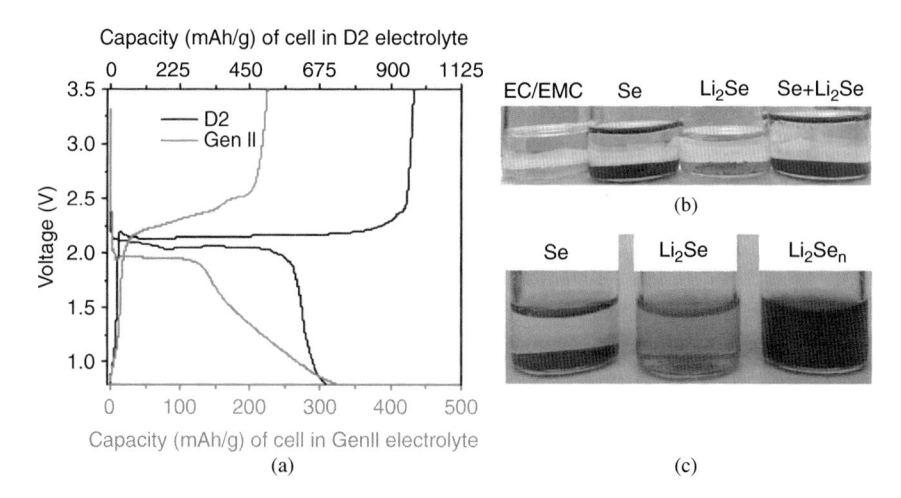

Figure 7.3 (a) Electrochemical performance of Li–Se cells in ether-based (D2) and carbonate-based (Gen II) electrolyte; (b) photograph of Gen II electrolyte solvent (EC-EMC) alone and with insoluble Se, with Li_2Se, and with a combination of the two; and (c) photograph of D2 electrolyte solvent alone and with insoluble Se, with soluble Li_2Se, and with soluble Li_2Se_n.[12,23]

appear and keep increasing in intensity to the end of discharge (as shown in Figures 7.4(a) and 7.4(b)). By the end of discharge, a small amount of Se remains in the cell with Gen II electrolyte, and the amount of unreacted Se (as indicated by discharge capacity) highly depends on the discharge rate. In the cell with D2 electrolyte, however, Se completely reduces into Li_2Se during discharge (as shown in Figures 7.4(c–e)). During the charge of the cells in the two electrolytes, the Se structure starts to grow steadily with accompanying disappearance of Li_2Se without any other phases detected.

In situ XANES spectroscopy, which can provide the average oxidation state of the elements in all phases, was employed to monitor the oxidation state changes during cell cycling with Gen II and D2 electrolytes. For the cell using Gen II electrolyte, the Se-edge positions in the XANES spectrum did not undergo obvious shifts during the cell cycling

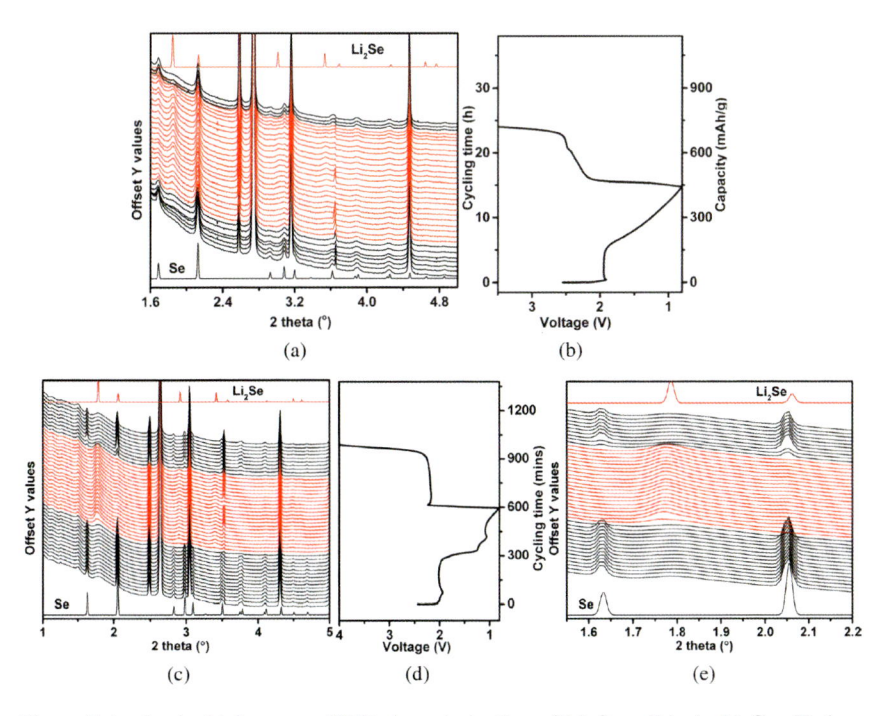

Figure 7.4 *In situ* high-energy XRD characterization of Li–Se cell in (a, b) Gen II electrolyte, and (c, d, and e) D2 electrolyte.[12,23]

318 *Li–S Batteries: The Challenges, Chemistry, Materials and Future Perspectives*

even at the fully discharged state (0.8 V), although the absorption of the Se-edge was weakened when the cell was discharged to 0.8 V (as shown in Figure 7.5). In the Li–Se cell with ether-based electrolyte, there were significant energy shifts with (de)lithiation. As shown in Figure 7.6, at the beginning of the discharge, Se reacts with Li$^+$ ions, and meanwhile, the reduced Se shifts the K-edge to a lower energy. This energy shift was believed to be associated with (de)lithiation. With increasing lithiation, the edge positions do not change much and are accompanied by a voltage plateau around 2 V. Toward the end of discharge as Se is further reduced, the K-edge jumps to a higher energy, suggesting the formation of Li$_2$Se. Here, the inconsistency with the commonly observed phenomenon that

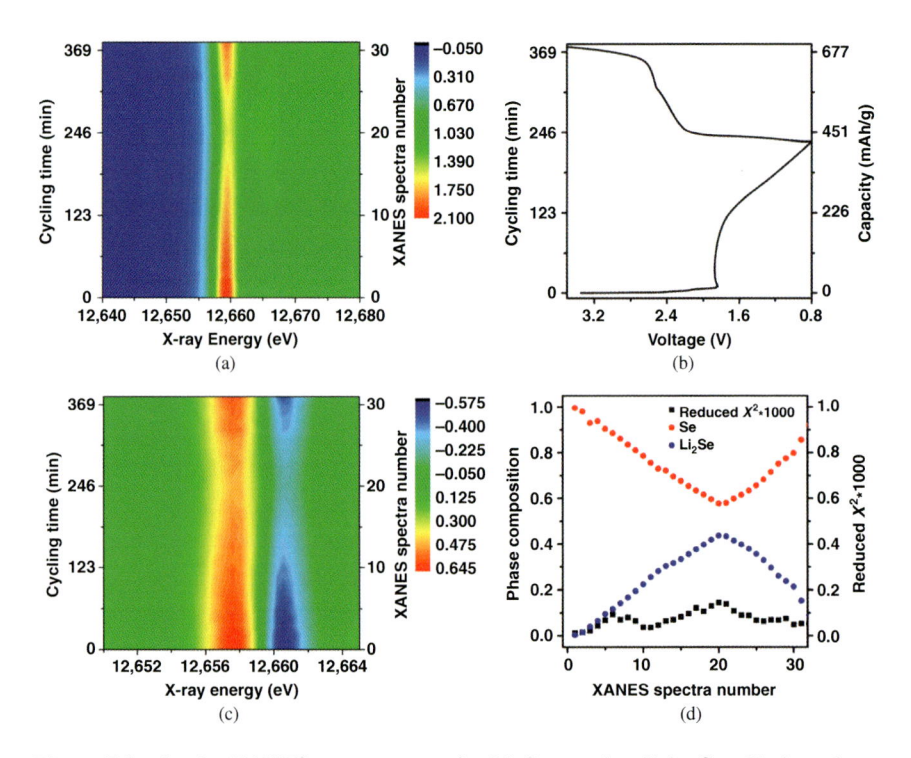

Figure 7.5 *In situ* XANES measurement for Li–Se pouch cell in Gen II electrolyte: (a) normalized XANES spectra of the cycled cell, (b) voltage profile, (c) derivative of normalized XANES spectra, and (d) linear combination fitting of residue values and corresponding phase compositions in different states of charge–discharge.[23]

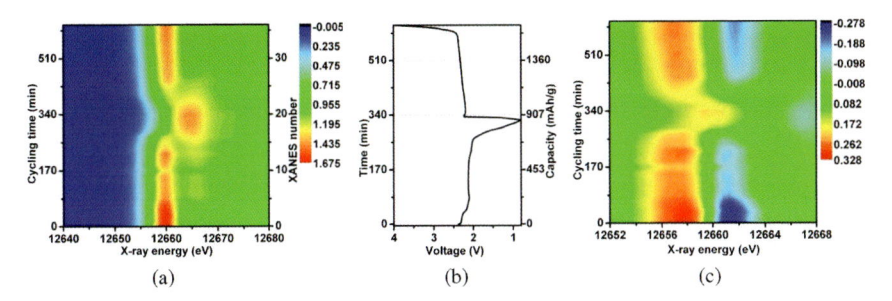

Other Sulfur Related Rechargeable Batteries 319

Figure 7.6 (a) Normalized XANES spectra, (b) Voltage profile, and (c) Derivative of normalized XANES spectra of Li/SeS$_2$ cell during cycling.[12]

the edge shifts to a higher energy with the increasing oxidation state was explained by the reduced screening effect due to the strong coulombic interaction between the Se^{2+} ions and the eight surrounding Li$^+$ ions in crystallized Li$_2$Se formed at the last stage of discharge.[24,26,27] During charge, the Se K-edges slowly shift back to lower energy. The above observations that the Se K-edge first shifts to a lower energy and then to a higher energy during discharge indicate that intermediate species might have been formed in the cell using the D2 electrolyte, very likely, lithium polyselenides. Since they are not detected by XRD, they are probably in a non-crystalline state.

Linear combination fitting of XANES spectra of the Se electrode at different states of charge–discharge also indicated that an intermediate phase was formed in the cell with D2 electrolyte, but not in Gen II electrolyte. Using a Se and Li$_2$Se two-phase fitting, the spectra measured in the latter cell during cycling were well fitted, illustrated by extremely small residues. By contrast, the two phase fitting for the former cell leads to huge residues, but a good fit was obtained by using a multiphase fitting with the combination of Se, Li$_2$Se, and Li$_2$Se$_x$.

The above results from Amine's group show that the Se cathode material exhibits electrochemical reactivity in both carbonate- and ether-based electrolytes. The former system exhibits a single discharge plateau and no polyselenide formation in the electrochemical process, while the latter system exhibits two discharge plateaus and possible Li polyselenide intermediate species.

Although most reports present evidence that the reduction of Se to Li$_2$Se is likely a one-step reaction without producing polyselenides in carbonate-based electrolytes, which is consistent with the conclusion of Amine's group,[8,13–16,28–35] some experiments have indicated the presence of possible intermediate polyselenides during discharge.[36] Liu et al. dismantled a cycled cell comprising a cathode made of commercial selenium particles with the presence of carbonate-based electrolyte, and found that the Celgard membrane had turned from white to red. They believed the color change was caused by the dissolution of the Li polyselenide species. As no such color change was observed in the Se–C composite cathode in their work, and the XANES result by Amine's group confirmed no intermediate species formation in the Se-MWNCs composite cathode, we have reasons to hypothesize that the formation of Li polyselenides prior to Li$_2$Se may occur only when pristine Se crystals are used as the cathode material.[14] Once the polyselenides are produced, they are likely to show high reactivity with carbonyl groups in the carbonate-based electrolytes, since they are more nucleophilic than the sulfur anions.[37] For this reason, the color change observed in Liu et al.'s work may not be caused by polyselenide dissolution, but rather a reaction with the electrolyte solvent.[13,14] However, although Se anions demonstrated even higher reactivity with the carbonate-based solvents than sulfur anions, their formation can be better prevented by preparing Se/porous-carbon composite electrodes due to the strong interaction of Se with the substrate.[15,31,35]

Although multistep reductions for Se/porous-carbon composite cathodes with carbonate-based electrolyte were observed in several reports, they were believed to be caused for reasons other than the production of intermediate polyselenide species.[30,33] A possible reason is the Se molecule transformation (mainly from ring-like Se$_8$ to chain-like Se$_n$) in carbon matrices during the charge/discharge process.[30] As we discussed in Section 2, the most dense and thermodynamically stable form of Se is Se$_n$ chains instead of cyclic Se$_n$. However, in the process of integrating Se into a porous carbon matrix by heat treatment, crystalline Se could partially convert into cyclic Se$_8$ molecules inside the micro/meso pores of the carbon matrix.[13,32] Yang et al. reported a Se composite cathode material in which Se is confined as cyclic Se$_8$ molecules (Raman shift of 267 cm^{-1}) in the mesopores of an ordered mesoporous carbon (CMK-3) matrix.[32] Upon charging, when the Se reduction product, Li$_2$Se, was oxidized back to Se, chain-like Se was formed instead of cyclic Se$_8$ (Raman shift of 256 cm^{-1}). Li et al. also demonstrated this molecular transformation using the X-ray photoelectron spectroscopy (XPS) method.[30] The extra small discharge plateau in their result is believed to correspond to the molecular transformation from cyclic Se$_8$ to the Se$_n$ chains.

Another possible reason for the multistep reductions in carbonate-based electrolyte is the hierarchical structure of carbon matrices in which different Se molecules (chain-like Se$_n$ molecules in micropores and cyclic Se$_8$ molecules in mesopores) exhibited different interaction strengths with nanopores and demonstrated different lithiation peaks.[33]

The chain-like structure of Se is advantageous and could be the reason for its better compatibility in the carbonate-based electrolyte compared to sulfur, which has a cyclic structure. As each chain has two active terminal atoms and can interact strongly with carbon matrices, Se molecules are well confined in carbon nanopores and are reduced directly into Li$_2$Se during discharge. The avoidance of the formation of polyselenides prevents the attack of the nucleophilic selenium anions towards carbonyl groups in the carbonate-based electrolytes, and results in the superior capacity of the Se/C composite cathode.

7.3.2. Carbon–Se composites for Li–Se batteries

As described in Section 3.1, Amine's group demonstrated that selenium, similar to sulfur, can generate soluble intermediate species when working in ether-based electrolytes, resulting in the problem of fast capacity fading and low coulombic efficiency in Li–Se batteries. Therefore, confining selenium locally inside conductive porous carbon matrices avoids its direct contact with the electrolyte, prevents polyselenide dissolution, and enables chemical bonding between Se and carbon. With regard to the carbonate-based system, because nucleophilic Se anions possess high reactivity with carbonyl groups, it is crucial to fabricate Se–C composite materials to prevent the formation of polyselenide species during the redox reactions. To this end, extensive efforts have been devoted to impregnating Se into various porous carbon matrices.[8,13–16,28–36,38–48] The integration and dispersion of Se in conductive matrices can confine Se locally and promote the electronic conductivity and maintain the structural integrity of the cathode, leading to improved electrochemical performance.

To date, published research on Se cathodes can be mainly categorized into three types: (1) adjusting the morphologies and pore structures of the carbon matrices (for example, mesoporous carbon nanofibers, micro-mesoporous carbon spheres, ordered mesoporous carbon (MPC), and metal-organic frameworks (MOFs)[13,30–33,35,39,41]; (2) optimizing the intrinsic textures of the carbon material itself (typically N-doped porous carbon)[28,31,34,45,49]; and (3) designing novel Se structures such as nanoporous or nanofibrous selenium structures.[29,50]

As an example of the first type, Han et al. designed and fabricated a free-standing Se composite cathode material with 3D mesoporous carbon nanoparticles (MCNs) and reduced graphene oxide (RGO) hierarchical architecture to form a binder-free cathode material (Se/MCN–RGO) for high-energy and long-life Li–Se batteries.[39] As shown in Figures 7.7(a–c), the free-standing electrode could be readily bent without breaking. Transmission electron microscopy (TEM) images revealed the confinement of Se within the MCN mesopores and the distribution of Se/MCN particle clusters in graphene sheets, which prevents the restacking of graphene sheets during the reduction process. The as-prepared composite cathode, in which the architecture provides 3D interconnected open channels to enhance Li-ion transportation during cycling, was tested with an ether-based electrolyte and was proved to efficiently suppress polyselenide shuttling and accommodate Se volume change during charge/discharge. As shown in Figure 7.7(d), the typical CV of the Se/MCN–RGO cathode demonstrated three reduction peaks in the first cathodic scan, which represented multistep reductions of Se to polyselenides and finally to Li_2Se. This is similar to the electrochemical process of the sulfur analogues. The oxidation peak during the anodic scan corresponds to the reversible conversion of Li_2Se to polyselenides and elemental selenium. At a selenium content of 62% in the entire cathode, the Se/MCN–RGO electrode exhibits a high discharge capacity of 655 mAh g^{-1} at the 0.1 C rate (97% of theoretical capacity), and as the current rate increases from 0.1 to 0.2, 0.5, 1, 2, and 3 C, the electrode displays good capacity retention, varying from 650 to 593, 525, 462, 377, and 274 mAh g^{-1},

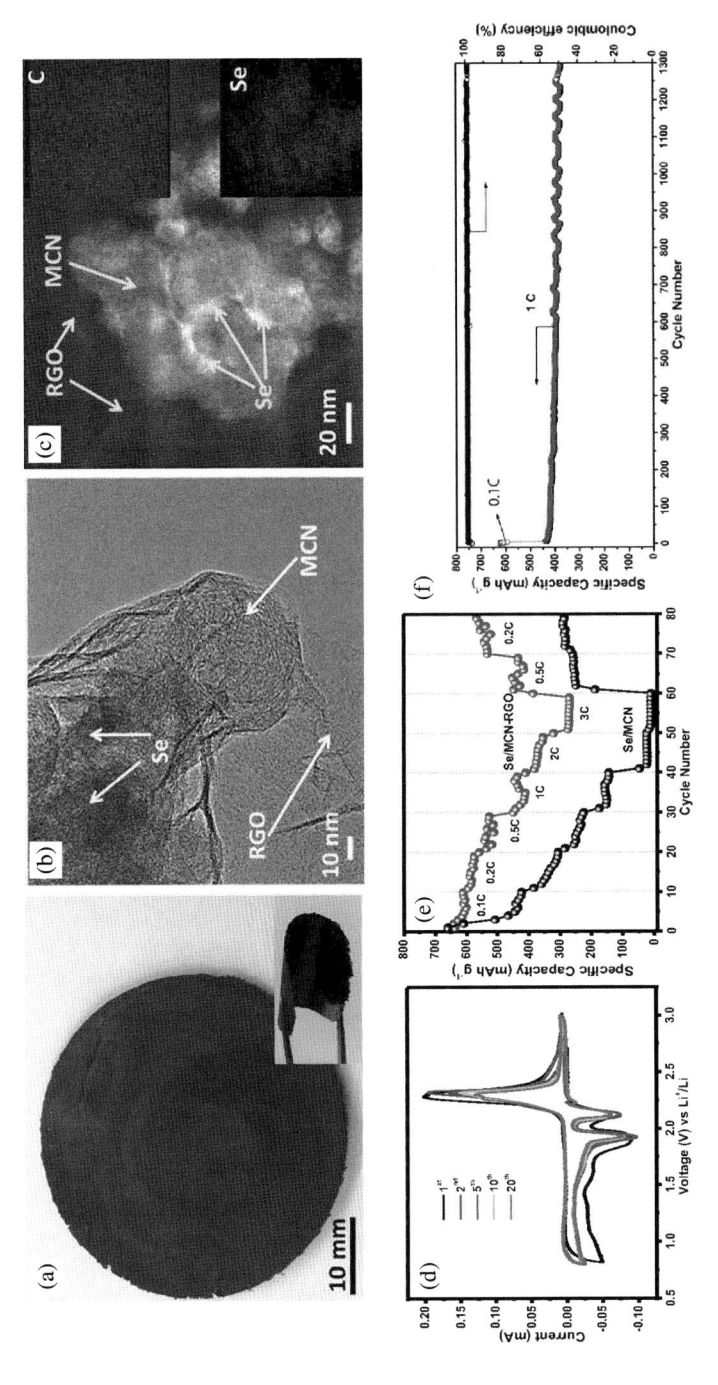

Figure 7.7 (a) Digital photo of the free-standing Se/MCN–RGO paper electrode and electrode bended with tweezer (inset). (b) TEM image of Se/MCN–RGO. (c) Z-contrast scanning transmission electron microscopy (STEM) image of Se/MCN–RGO and corresponding energy dispersive X-ray spectroscopy (EDX) mapping of Se and carbon. (d) CV curves of the 1st, 2nd, 5th, 10th, and 20th cycles for Se/MCN–RGO electrode at a scan rate of 0.1 mV s^{-1}. (e) Rate capability of the Se/MCN–RGO paper electrode and Se/MCN powder electrode at varied current rates of 0.1–3 C. (f) Long-term cycling stability of the Se/MCN–RGO electrode (0.1 C for the first five cycles and 1 C for the next 1300 cycles).[39] (Copyright 2014, John Wiley and Sons).

respectively (Figure 7.7(e)). The free-standing electrode also exhibits long cycling stability with a very small capacity decay of 0.008% per cycle over 1300 cycles at 1 C (Figure 7.7(f)). In contrast, a Se/MCN electrode showed significant capacity fading due to the lack of graphene sheets that could help adsorb the escaped polyselenides from the mesopores of the MCN.

Yi *et al.* confined selenium in N-doped microporous carbon fabricated from carbonized polypyrrole (CP) and investigated the heteroatom-doping effect on the electrochemical performance of the Se/C composites.[34] As shown in Figure 7.8, the as-prepared CP possesses a 3D hierarchical porous structure with macropores on the surface and interconnected micropores. It has a specific surface area as high as 3366 m^2 g^{-1} and a pore volume of up to 1.75 cm^3 g^{-1} (with an average pore diameter of around 0.5 nm). The XPS results confirmed the heteroatom-doping and the chemical bonding state of functional CP. The results show that the N content in CP is 13.28 wt.%, and that the N and O atoms are well introduced into the carbon framework and chemically combine with C. The heteroatom-doping leads to the asymmetrical charge and spin density of the carbon layer, and this charge delocalization effect in CP provides high conductivity. Also, nitrogen has a higher electronegativity (3.5) than that of carbon (3.0) while a smaller atom size, and therefore, N-doping can facilitate Li insertion and enhance the interaction between Se reduction products and the carbon matrices. As a result, heteroatom doping of Se–CP composite material leads to extraordinary electrochemical performance in both ether- and carbonate-based electrolyte.

For the ether electrolyte system, the Se–CP composite exhibits a very high initial discharge capacity of 1504 mAh g^{-1} and a reversible capacity of 480 mAh g^{-1}. The irreversible capacity in the first cycle is believed to be caused by the reactions between the porous carbon and the electrolyte. The composite cathode demonstrated high stability and rate capability in commercial carbonate-based electrolyte, which is relatively inexpensive. A high reversible capacity of 644, 506, and 303 mAh g^{-1} can be stably obtained at 0.5, 1, and 20 C, respectively.

Pure Se crystals with advanced structures have also been explored as cathode materials. Kundu *et al.* synthesized Se nanofibers through a template-free solution route.[29] The as-prepared nanofibers consist of the

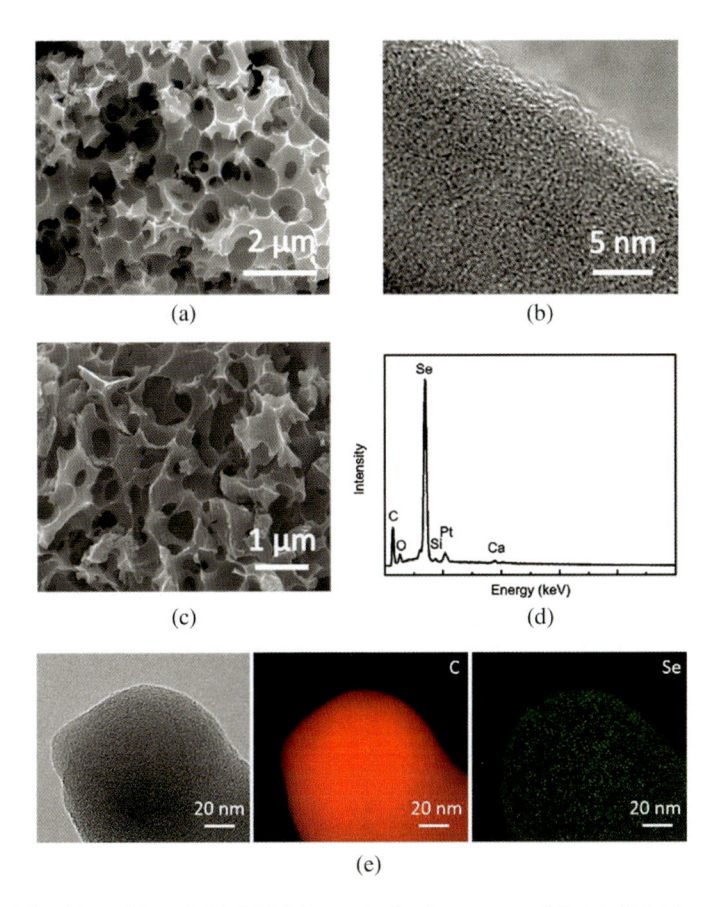

Figure 7.8 (a) SEM and (b) TEM images of microporous CP, (c) SEM image and (d) EDX pattern of the Se–CP composite, and (e) TEM image of Se–CP composite and the EDX elemental mapping images of carbon and Se in the Se–CP composite.[34] (Copyright 2014, Royal Society of Chemistry).

trigonal phase of selenium (t-Se) and have a fiber-like morphology. The nanofibers are several micrometers in length, with diameters varying from 50 to 200 nm. The t-Se nanofibers were made into electrodes with a high loading of 7–10 mg cm^{-2} and tested in carbonate-based electrolyte. At a very low current rate of C/120, the theoretical specific capacity is achieved (~678 mAh g^{-1}). At an increased current rate of C/60, the discharge capacity drops to half of the theoretical value (~340 mAh g^{-1}), and the plateau region around 2.5 V becomes shorter. The electrochemical performance of these nanofibers was improved by composite forming of t-Se/polypyrrole and graphene wrapping of t–Se. Although the cycling performance and rate capabilities of bulk Se crystals are comparatively lower than those of the Se/porous carbon composite cathodes, their unique structure provides the capability of future mechanistic investigations of the redox reactions in Se-based systems by using novel characterization techniques, such as *in situ* XRD and *in situ* TEM.

In addition to the studies of various Se/C composites, Amine's group has investigated mixed Se$_x$S$_y$ systems.[9] These systems combine the high capacities of S-rich systems with the high electrical conductivity of the d-electron containing Se, and can offer higher theoretical capacities than the Se alone with improved performance and conductivity compared to S. For the Li–Se$_x$S$_y$ systems, the theoretical capacities are 678–1550 mAh g^{-1}, compared to 678 mAh g^{-1} for the Li–Se system. Amine's group used the carbonate-based electrolyte and achieved a stable discharge capacity for a Li/SeS$_2$–C cell of 512 mAh g^{-1} at current density of 50 mA g^{-1} in the voltage range from 0.8 to 4.6 V. This capacity is 30% greater than that of the Li/Se–C cell, which is 394 mAh g^{-1}.

7.3.3. Impact of electrolytes in Li–Se batteries

While impregnating Se in porous carbon improves the electrochemical performance of Se cathodes in both ether- and carbonate-based electrolytes, further improvements may be achieved by electrolyte optimization. Jiang *et al.* tested carbon/Se composite material in basic ether-based electrolyte, LiNO$_3$-modified ether-based electrolyte, and an ether-based electrolyte modified with the ionic liquid N-methyl-(n-butyl) pyrrolidinium bis(trifluoromethanesulfonyl) imide (PYR$_{14}$TFSI). They found that the latter two improved the coulombic efficiency of the cells.[40] Lee *et al.* applied ether-based electrolytes with high Li salt concentration in nanocomposites of Se and ordered mesoporous silicon carbide-derived carbon (OM–SiC–CDC) and greatly enhanced capacity utilization and reduced Se dissolution in the nanocomposite cathodes.[31] The origin of this

7.4. Research Progress in Na–Se Batteries

In addition to the growing research interest in Li–Se batteries, Na–Se batteries have also gained much attention recently due to their abundance, low cost, and low environmental impact of Na. Unlike existing Na–S batteries that mostly operate at high temperatures, the Se cathodes are capable of cycling at room temperature against Na. Among the first researchers studying the Na–Se system, Abouimrane *et al.* in Amine's group reported the electrochemical performance and structural mechanism during Na insertion in a Na–Se cell using a Se/MWCN cathode and a carbonate-based electrolyte.[9] The initial discharge of the Na–Se cell involves two plateaus: a short one at 1.9 V and a longer one at 1.5 V (as shown in Figure 7.9). This finding suggests that the full discharge product of Na_2Se is formed from an intermediate phase with the carbonate-based electrolyte. During the charge process in which Se is oxidized, similar to the

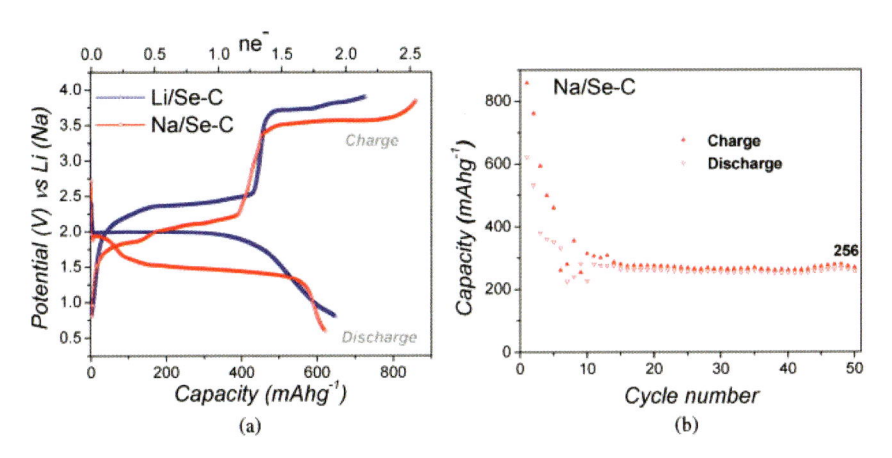

Figure 7.9 (a) Voltage profiles for Li/Se–C and Na/Se–C cells during the first discharge/charge and (b) Capacity retention of Na/Se–C cell cycled at 50 mA g^{-1} current density.[9]

Li–Se cells in Abouimrane's report, the Na–Se cell also shows two plateaus (at 2.1 and 3.5 V, compared to 2.4 and 3.75 V for Li–Se cells in Abouimrane's research). In the PDF analysis, Na insertion into the Se cathode with the carbonate-based electrolyte induced a dramatic broadening of the peaks (at 1.4 V, ~350 mAh g^{-1}, ~1 Na per Se), suggesting the formation of a poorly ordered intermediate phase before the Se transformed to a well-defined antifluorite-type Na$_2$Se phase upon complete discharge.[9,52] This behavior is different from that of the Li–Se cell in the carbonate-based electrolyte in Abouimrane's studies, which shows only one well-defined plateau in the initial discharge.

The advanced Se/C composite materials prepared by many groups, in which Se was dispersed and confined locally, were also tested in Na–Se cells. Interestingly, all the reported tests were done in carbonate–based electrolyte. Yan Yu's group prepared a Se/carbon composite material in which Se was chemically bonded with and physically impregnated in electrospun polyacrylonitrile and carbon nanotubes. The composite cathode maintained a reversible capacity of 410 mAh g^{-1} after 240 cycles at 0.5 A g^{-1}.[16] Yu *et al.* also prepared a flexible and free-standing Se/porous carbon nanofiber composite electrode, which is capable of a capacity of 520 mAh g^{-1} after 80 cycles at 0.05 A g^{-1} and a rate performance of 230 mAh g^{-1} at 1 A g^{-1}.[14] Wang's group synthesized a Se/C composite via an *in situ* formation process, which maintains a discharge capacity of 280 mAh g^{-1} after 50 cycles at a current density of 0.1 A g^{-1}.[15] The encapsulation and dispersion of Se in carbon matrices promote the electronic conductivity and maintain the structural integrity of the cathode, leading to the above significant electrochemical performance of Na–Se cells.

7.5 Summary and Future Design Strategies for Se-based Energy Storage Systems

Advanced Li–Se and Na–Se batteries show great promise in attaining high volumetric capacity density but challenges remain. In the past couple of years, considerable progress has been achieved in understanding the fundamental reaction mechanism and in designing Se-based batteries, including electrodes and electrolyte. To summarize, the knowledge gained to date on Se-based batteries includes:

(1) Se-based cathode materials exhibit electrochemical reactivity in both carbonate- and ether-based electrolytes. The overall cell reaction for Li/Na–Se batteries is analogous to that of sulfur reactions with Li or Na, [S + 2M ↔ M$_2$S] (M = Li, Na).
(2) In the ether-based electrolyte system, polyselenides are formed before the full reduction of Se to Li$_2$Se in the discharge process.
(3) In the Li–Se cell with carbonate-based electrolyte system, the reduction of Se in the Se/C composite cathode to Li$_2$Se is believed to be a single-phase transition without producing polyselenides. In the Na–Se cell, by contrast, the reduction of Se to Na$_2$Se is a two-step reaction, with the formation of a poorly ordered intermediate before Se is transformed to a well-defined antifluorite-type Na$_2$Se phase.
(4) It is necessary to confine Se locally inside conductive porous carbon matrices to avoid its direct contact with the electrolyte, prevent polyselenide dissolution, promote the electronic conductivity, and maintain the structural integrity of the Se cathode. Studies to date on preparing advanced Se/C composite materials have led to Li–Se and Na–Se batteries with improved electrochemical performance.

The Se-based battery technology is still in its infancy. Many hurdles remain to overcome the technical challenges and meet the performance requirements for future application. Based on the studies reviewed in this chapter, suggestions for the rational design of future practical Se-based batteries include: (1) further investigation of the reaction mechanism in Li/Na–Se batteries; (2) further development of advanced Se/C composite cathode that has high electric conductivity, less volumetric change during charge/discharge, and higher utilization of the active material; (3) development of organic electrolyte that is favorable for Se redox reactions and compatible with the possible presence of polyselenides; and (4) novel design of cell configurations that can improve the specific capacity and cycle life of Se-based batteries, which may benefit from the tremendous achievements in Li–S cell design. Although the development of a Se-based electrochemical storage system is still far from practical utilization, the future is bright with the growing knowledge of its reaction mechanism and improvement of cell design.

Acknowledgments

This work was supported by the U.S. Department of Energy under Contract DE-AC0206CH11357 with the main support provided by the Vehicle Technologies Office, Department of Energy (DOE) Office of Energy Efficiency and Renewable Energy (EERE). Use of the Advanced Photon Source, an Office of Science User Facility operated for the U.S. Department of Energy (DOE) Office of Science by Argonne National Laboratory, was supported by the U.S. DOE under Contract No. DE-AC02-06CH11357.

Bibliography

1. R. Xu, I. Belharouak, J. C. M. Li, X. F. Zhang, I. Bloom and J. Bareno, *Advanced Energy Materials*, **3** (2013) 833.
2. J. Lu, L. Li, J. B. Park, Y. K. Sun, F. Wu and K. Amine, *Chemical Reviewing*, **114** (2014) 5611.
3. A. Manthiram, S. H. Chung and C. X. Zu, *Advanced Materials*, **27** (2015) 1980.
4. L. F. Nazar, M. Cuisinier and Q. Pang, *MRS Bulletin*, **39** (2014) 436.
5. P. G. Bruce, S. A. Freunberger, L. J. Hardwick and J. M. Tarascon, *Nature Materials*, **11** (2012) 19.
6. R. Xu, I. Belharouak, X. F. Zhang, R. Chamoun, C. Yu, Y. Ren, A. M. Nie, R. Shahbazian-Yassar, J. Lu, J. C. M. Li and K. Amine, *ACS Applied Materials & Interfaces*, **6** (2014) 21938.
7. R. Xu, J. Lu and K. Amine, *Advanced Energy Materials*, **5** (2015) 1500408.
8. K. Han, Z. Liu, H. Q. Ye and F. Dai, *Journal of Power Sources*, **263** (2014) 85.
9. A. Abouimrane, D. Dambournet, K. W. Chapman, P. J. Chupas, W. Weng and K. Amine, *Journal of the American Chemical Society*, **134** (2012) 4505.
10. C.-P. Yang, Y.-X. Yin and Y.-G. Guo, *Journal of Physical Chemistry Letters*, **6** (2015) 256.

11. V. S. Saji and C. W. Lee, *RSC Advances*, **3** (2013) 10058.
12. Y. J. Cui, A. Abouimrane, J. Lu, T. Bolin, Y. Ren, W. Weng, C. J. Sun, V. A. Maroni, S. M. Heald and K. Amine, *Journal of the American Chemical Society*, **135** (2013) 8047.
13. C. Luo, Y. H. Xu, Y. J. Zhu, Y. H. Liu, S. Y. Zheng, Y. Liu, A. Langrock and C. S. Wang, *ACS Nano*, **7** (2013) 8003.
14. L. C. Zeng, W. C. Zeng, Y. Jiang, X. Wei, W. H. Li, C. L. Yang, Y. W. Zhu and Y. Yu, *Advanced Energy Materials*, **5** (2015).
15. C. Luo, J. J. Wang, L. M. Suo, J. F. Mao, X. L. Fan and C. S. Wang, *Journal of Materials Chemistry A*, **3** (2015) 555.
16. L. C. Zeng, X. Wei, J. Q. Wang, Y. Jiang, W. H. Li and Y. Yu, *Journal of Power Sources*, **281** (2015) 461.
17. K. B. Subila, G. K. Kumar, S. M. Shivaprasad and K. G. Thomas, *Journal of Physical Chemical Letters*, **4** (2013) 2774.
18. V. S. Saji, I. H. Choi and C. W. Lee, *Sol Energy*, **85** (2011) 2666.
19. M. Harati, J. Jia, K. Giffard, K. Pellarin, C. Hewson, D. A. Love, W. M. Lau and Z. F. Ding, *Physical Chemistry Chemical Physics,* **12** (2010) 15282.
20. B. Meyer, *Chemical Reviewing*, **76** (1976) 367.
21. N. N. Greenwood and A. Earnshaw, *Chemistry of the Elements*, Oxford, Boston: Butterworth-Heinemann (1997).
22. J. E. House, *Inorganic Chemistry*, Academic Press/Elsevier, Amsterdam, Boston (2008).
23. Y. J. Cui, A. Abouimrane, C. J. Sun, Y. Ren and K. Amine, *Chemical Communications*, **50** (2014) 5576.
24. J. Gao, M. A. Lowe, Y. Kiya and H. D. Abruna, *The Journal of Physical Chemistry C*, **115** (2011) 25132.
25. S. A. Freunberger, Y. H. Chen, Z. Q. Peng, J. M. Griffin, L. J. Hardwick, F. Barde, P. Novak and P. G. Bruce, *Journal of the American Chemical Society*, **133** (2011) 8040.
26. S. C. B. Myneni, T. K. Tokunaga and G. E. Brown, *Science*, **278** (1997) 1106.
27. R. A. Mori, E. Paris, G. Giuli, S. G. Eeckhout, M. Kavcic, M. Zitnik, K. Bucar, L. G. M. Pettersson and P. Glatzel, *Analytical Chemistry*, **81** (2009) 6516.
28. Y. Jiang, X. J. Ma, J. K. Feng and S. L. Xiong, *Journal of Materials Chemistry A*, **3** (2015) 4539.
29. D. Kundu, F. Krumeich and R. Nesper, *Journal of Power Sources*, **236** (2013) 112.

30. Z. Li, L. X. Yuan, Z. Q. Yil, Y. Liu and Y. H. Huang, *Nano Energy*, **9** (2014) 229.
31. Z. Q. Li and L. W. Yin, *Nanoscale*, **7** (2015) 9597.
32. C. P. Yang, S. Xin, Y. X. Yin, H. Ye, J. Zhang and Y. G. Guo, *Angewandte Chemie International Edition*, **52** (2013) 8363.
33. H. Ye, Y. X. Yin, S. F. Zhang and Y. G. Guo, *Journal of Materials Chemistry A*, **2** (2014) 13293.
34. Z. Q. Yi, L. X. Yuan, D. Sun, Z. Li, C. Wu, W. J. Yang, Y. W. Wen, B. Shan and Y. H. Huang, *Journal of Materials Chemistry A*, **3** (2015) 3059.
35. J. J. Zhang, L. Fan, Y. C. Zhu, Y. H. Xu, J. W. Liang, D. H. Wei and Y. T. Qian, *Nanoscale*, **6** (2014) 12952.
36. L. L. Liu, Y. Y. Hou, Y. Q. Yang, M. X. Li, X. W. Wang and Y. P. Wu, *RSC Advances*, **4** (2014) 9086.
37. D. Steinmann, T. Nauser and W. H. Koppenol, *Journal of Organic Chemistry*, **75** (2010) 6696.
38. R. P. Fang, G. M. Zhou, S. F. Pei, F. Li and H. M. Cheng, *Chemical Communications*, **51** (2015) 3667.
39. K. Han, Z. Liu, J. M. Shen, Y. Y. Lin, F. Dai and H. Q. Ye, *Advanced Functional Materials*, **25** (2015) 455.
40. S. F. Jiang, Z. Zhang, Y. Q. Lai, Y. H. Qu, X. W. Wang and J. Li, *Journal of Power Sources*, **267** (2014) 394.
41. J. T. Lee, H. Kim, M. Oschatz, D. C. Lee, F. X. Wu, H. T. Lin, B. Zdyrko, W. I. Cho, S. Kaskel and G. Yushin, *Advanced Energy Materials*, **27** (2015) 101–108.
42. J. Li, X. X. Zhao, Z. Zhang and Y. Q. Lai, *Journal of Alloys and Compounds*, **619** (2015) 794.
43. L. Liu, Y. J. Wei, C. F. Zhang, C. Zhang, X. Li, J. T. Wang, L. C. Ling, W. M. Qiao and D. H. Long, *Electrochimica Acta*, **153** (2015) 140.
44. X. Peng, L. Wang, X. M. Zhang, B. Gao, J. J. Fu, S. Xiao, K. F. Huo and P. K. Chu, *Journal of Power Sources*, **288** (2015) 214.
45. Y. H. Qu, Z. A. Zhang, S. F. Jiang, X. W. Wang, Y. Q. Lai, Y. X. Liu and J. Li, *Journal of Materials Chemistry A*, **2** (2014) 12255.
46. Y. H. Qu, Z. A. Zhang, Y. Q. Lai, Y. X. Liu and J. Li, *Solid State Ionics*, **274** (2015) 71.
47. Z. A. Zhang, X. Yang, Z. P. Guo, Y. H. Qu, J. Li and Y. Q. Lai, *Journal of Power Sources*, **279** (2015) 88.
48. Z. A. Zhang, Z. Y. Zhang, K. Zhang, X. Yang and Q. Li, *RSC Advances*, **4** (2014) 15489.

49. J. Zhang, Z. A. Zhang, Q. Li, Y. H. Qu and S. F. Jiang, *Journal of the Electrochemical Society*, **161** (2014) A2093.
50. L. L. Liu, Y. Y. Hou, X. W. Wu, S. Y. Xiao, Z. Chang, Y. Q. Yang and Y. P. Wu, *Chemical Communications*, **49** (2013) 11515.
51. L. M. Suo, Y. S. Hu, H. Li, M. Armand and L. Q. Chen, *Nature Communications*, **4**, (2013) 1481.
52. J. Sangster and A. D. Pelton, *Journal Phase Equilibria*, **18** (1997) 185.

Chapter 8

Computational Modeling of Lithium–Sulfur Batteries: Myths, Facts, and Controversies

Alejandro A. Franco

*Laboratoire de Réactivité et Chimie des Solides (LRCS),
CNRS UMR 7314, Université de Picardie Jules Verne,
33 rue Saint Leu, 80039 Amiens Cedex, France
Réseau sur le Stockage Electrochimique de l'Energie (RS2E),
FR CNRS 3459, France
ALISTORE-European Research Institute,
Fédération de Recherche CNRS 3104,
33 rue Saint Leu, 80039 Amiens cedex, France
Institut Universitaire de France, 75005 Paris, France
alejandro.franco@u-picardie.fr*

8.1. Introduction

From a practical point of view regarding automotive applications, it is crucial to accurately predict the performance, state-of-health and remaining lifetime of lithium–sulfur batteries (LSBs). For that purpose, it is necessary to develop diagnostic schemes that can evaluate the electrochemical cells performance and its state-of-health adequately.[1] In order to achieve this, several steps are required:

- A better understanding of the several individual processes in the cell components;
- The understanding of the interplay between electrochemical/transport mechanisms and (thermo-) mechanical behaviors at multiple spatiotemporal scales with their possible competitions and synergies;
- The identification of the contribution of each mechanism into the global cell response under operating conditions;
- The design of controllers for the online control of the LSB cells behavior to enhance its durability under specific operation conditions (e.g. by controlling the dynamics of the cycling, the temperature, etc.).

In view of the complexity of the LSB operation principles, achieving these goals is not a trivial task, and then mathematical modeling (hereafter referred to as *modeling* or *theory*) is called to play a key role to understand and correlate the different observed behaviors, and ultimately lead to the optimization of the cell design and operation conditions.[2]

Despite the strong need of theory, it is quite surprising to see that there are only very few modeling efforts reported so far in the literature. These models fall mainly in two categories:

- **Continuum approaches:** Within these approaches, transport and electrochemical reactions are described through a set of coupled partial and ordinary differential equations; the equations are written as functions of effective electrode structural parameters (e.g. porosity) and reactivity parameters (e.g. activation energy).
- **Atomistic/molecular approaches:** Within these approaches, the electronic structure and inter-molecular forces are calculated by solving Schrödinger equation and/or N-body Newton equations; equations are more or less "independent" on empirical parameters (e.g. case of Density Functional Theory, DFT).

In the following text, the progresses achieved on these different modeling approaches are reviewed, and their limitations are discussed, the goal being to identify opportunities for further model developments.

8.2. Continuum Modeling

Regarding the continuum modeling approach, Mikhaylik and Akridge from the Sion Power Corporation[3] reported in 2004 the first cell model, devoted mainly to the study of the polysulfide shuttle reactions.[4] This model is engineering-oriented and is designed to calculate reduction potentials in the high and low voltage-plateaus by using the Nernst equation. It neglects the transport along the separator and S/C composite electrode, and cannot capture the activation overpotentials, the electrolyte resistance change with the polysulfides concentrations and dissolution/precipitation reactions.

Four years later, White *et al.*[5] from University of South Carolina, in collaboration with Sion Power Corporation, reported a more sophisticated cell model based on the seminal electrode porous theory proposed by Newman almost 50 years ago.[6,7] This model managed to capture several potential vs. capacity characteristics experimentally observed in discharging LSBs with sulfur/carbon cathodes, such as the two plateaus and the presence of a tip between them. The model accounts for a one-dimensional description of the lithium (Li) and polysulfides diffusion along the separator/cathode thicknesses, as well as the sulfur dissolution and subsequent polysulfides reduction and precipitation reactions along the cathode thickness (Figure 8.1). The cathode microstructure is taken into account through an averaged porosity.

White *et al.* model assumes the cathode active surface area to be, mathematically, directly related to the cathode porosity (evolving with the calculated amount of precipitates), and thus it does not describe the impact on the electrochemistry of the carbon particles surface passivation by the polysulfides precipitation. Furthermore, their model does not take into account the initial spatial distribution of sulfur within the cathode.

Later, Bessler *et al.* (working at DLR Sttugart, and later at Offenburg University),[8,9] reported a similar model to White *et al.* one, by adding some new features such as the consideration of an electrochemical double layer capacitance. The model capabilities are illustrated through the simulation of a full discharge/charge cycle and electrochemical impedance (EIS) experiments (Figure 8.2). The authors find that the calculated overall electrolyte conductivity is lower during charging vs. discharging, because of the calculated lower values of polysulfides concentrations.

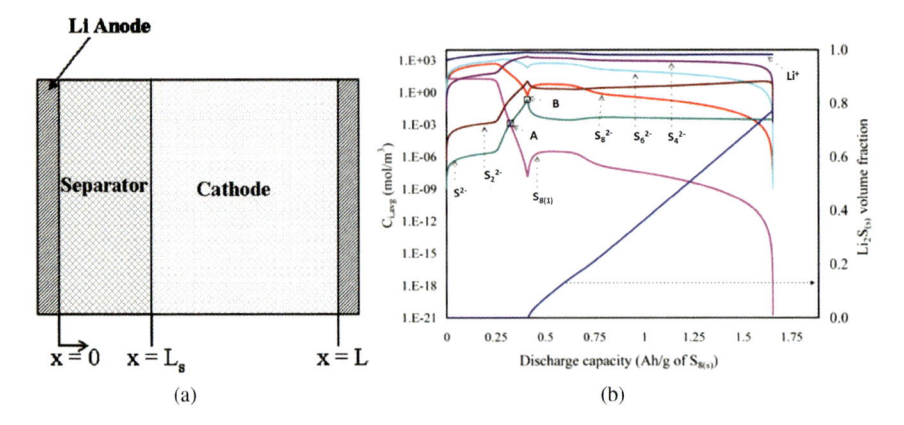

338 *Li–S Batteries: The Challenges, Chemistry, Materials and Future Perspectives*

Figure 8.1 (a) Domains simulated in White *et al.* model; (b) Calculated sulfur and polysulfides concentrations along the battery discharge.
Source: Reproduced with the permission from Ref. 5.

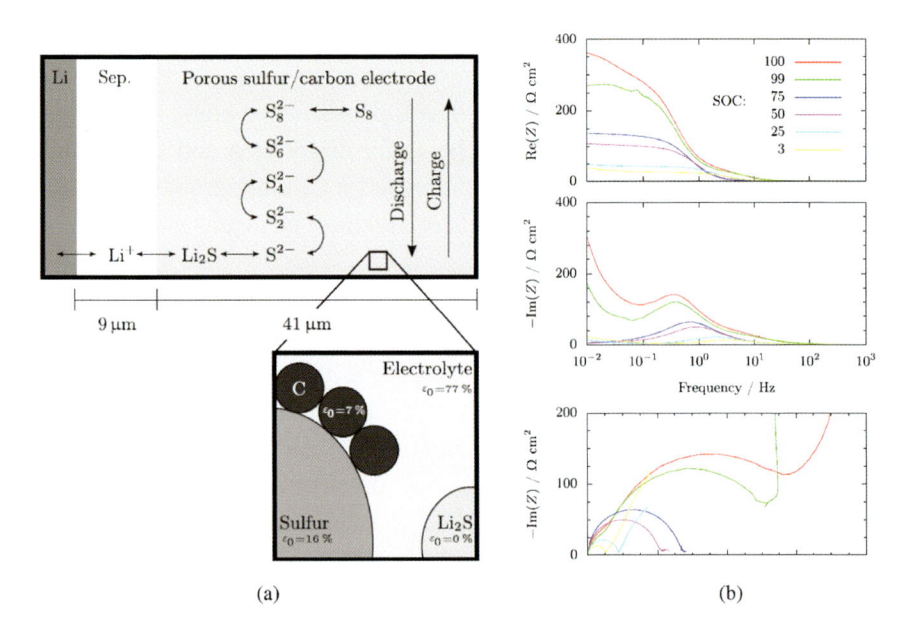

Figure 8.2 (a) Scheme of the Bessler *et al.* continuum modeling approach; (b) Calculated EIS signals.
Source: Reproduced with the permission from Ref. 9.

Such a conductivity is implicitly derived through the utilization of the Nernst–Planck equations for the polysulfides and Li-ion transport (as in the White *et al.* model).

The authors stress about the difficulty of the determination of the used kinetic parameters for the polysulfides reactions and neglect the shuttle reactions, implemented later by the same authors in a paper published in 2014.[10] In the latter, the model for the polysulfides reduction reactions is simplified to four steps capturing the reduction of S_8 to S_4^{2-}, of S_4^{2-} to S^{2-}, and the Li_2S formation and precipitation. The authors justify their simplification supported on the assumption that the ionization of the dissolved S_8 can only happen at the triple phase boundary between solid sulfur, carbon and electrolyte. This triple phase boundary is believed by them to be very small and unstable in Li/S cells. The concept of triple point is an oversimplification which applies here because the authors represent the S/C electrode as a homogeneous medium: note that if the real microstructure would be considered, ionization of dissolved S_8 is possible when it will enter in direct contact with the carbon (assuming that it is not fully covered by sulfur and/or precipitates) and/or because of the electronic tunneling effect through the precipitates deposited on the carbon surface. Last but not least, in their work, Bessler *et al.* report fitted kinetic parameter values which seem unrealistic (order of 10^{-101}),[10] perhaps implicating that some of the assumptions made are not physically consistent.

In 2014, Ghazvani and Chen (University of Waterloo) reported, in a series of three papers, a comprehensive parameter sensitivity study of the White *et al.* model aiming to identify its limits.[11–13] The calculated discharge and concentration profiles are analyzed as a function of the values adopted for the precipitation rate constants and sulfur content. The authors report that the values for the kinetic parameters of the polysulfide reduction reactions are critical for an appropriate adjustment of the prediction with the experimental trends. They find also that this model predicts the existence of an upper limit on the sulfur content of the cathode to ensure an optimal performance. Furthermore, a minimum cathode solid material conductivity to activate the cell is identified, beyond which no significant improvement of the capacity performance occurs. The authors also report the existence of an optimal cathode thickness for charging but highlight the need for more refined models for the reduction reactions as well as the consideration of the insulating nature of sulfur to improve the predictive capabilities of the White et al. model.

More recently, Zhang et al. from Imperial College of London (UK), and in collaboration with Oxis Energy,[14] reported a model predicting the variation of the electrolyte conductivity along the discharge.[15] The model was also presented in the 229th Meeting of the ECS (San Diego, May 29–June 2, 2016). Their model is actually a lumped-version of White et al. model, thus zero-dimensional, neglecting the transport of Li, dissolved sulfur and polysulfides along the electrode and separator thicknesses. The authors justify their assumption on the basis of Ghaznavi and Chen findings, indicating that mass transport does not have a significant impact on the White et al. model predictions, except if the ionic diffusion coefficients are reduced by more than one order of magnitude. Zhang et al. utilize the same mathematical expressions as White et al. for the calculation of the variation of the active area as a function of the electrode porosity, and thus do not use an explicit (or physical-based) formulation for the surface passivation of the carbon particles and/or active sulfur by the precipitates.

Zhang et al. claim that White et al. and Bessler et al. models fail on capturing the impact of the polysulfides concentrations on the electrolyte conductivity, attributing the dip between the two plateaus to the interplay between the Li-ion concentration variation and the precipitation reactions leading to the formation of Li_2S (Figure 8.3). As mentioned earlier, White and Bessler describe ionic species transport by using the Nernst–Planck formalism. In the associated equation, the conductivity can be derived from the calculated ionic concentrations by using the Einstein relation. White et al. and Bessler et al. assumed the validity of diluted concentrations, a hypothesis inherent to the Nernst–Planck formalism. In contrast to this, Zhang et al. use an empirical formula to capture the impact of ionic concentration on the electrolyte conductivity. The authors do not solve the contribution of each dissolved polysulfide onto the overall conductivity value, because the individual conductivities are unknown. It is then unclear why Zhang et al. would be more correct than White or Bessler models, as the experimental relevance of the values adopted for the diffusion coefficients and kinetic parameters is not discussed.

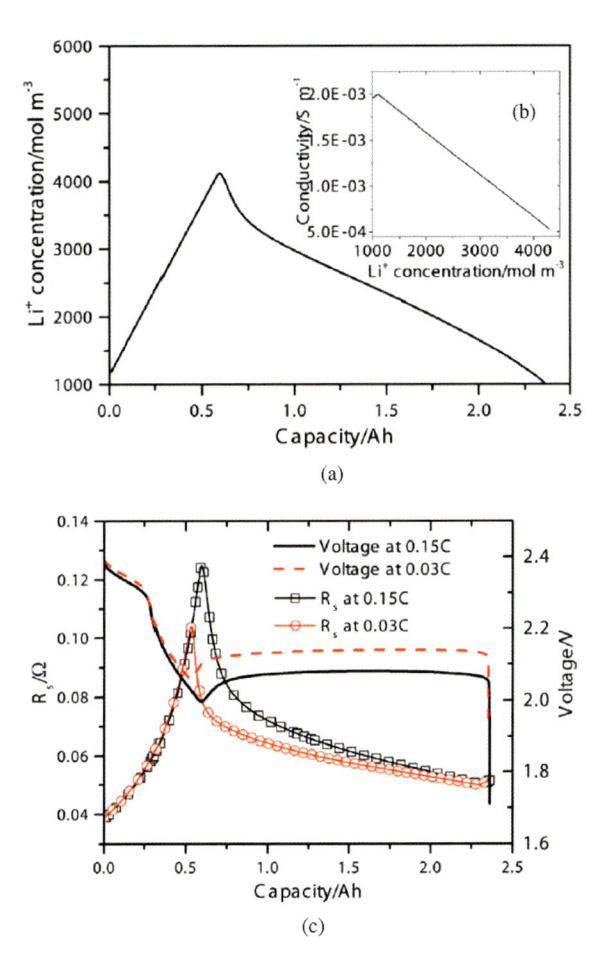

Figure 8.3 (a) Calculated concentration of Li$^+$ during 0.15 C discharge; inset (b) shows the electrolyte conductivity as a function of Li$^+$ concentration; (c) Calculated discharge voltages (symbols) and electrolyte resistances (lines) at 0.15 and 0.03 C.[15] Not subject to copyright.

While the models described here can be more or less successful in describing the discharge curves and/or degradation during cycling, they cannot provide indications for the optimization of the cathode microstructure (e.g. in regard of the spatial distribution of sulfur between different scales of carbon porosity) because the electrode is considered as an effective medium described through a single overall porosity.

A first attempt to describe the electrode microstructure in more detail was reported by Lutz et al. (DLR, Stuttgart). The authors' model is a full cycle model for the cell using a mesoporous cathode containing microporous carbon particles.[16] Essentially, their model is very similar to White et al. but it considers two scales of porosities: (1) mesopores between carbon particles, where only the Li-ion transport is assumed to occur; (2) microporosity inside the carbon particles containing the sulfur. This model assumes that the sulfur-based species are electrochemically active as long as they remain confined in the microporous within the carbon particles. This model can only be utilized to study the performance of the Li–S batteries which is based on the strategy of confining sulfur-based species inside the microporosity in the carbon particles, since it ignores the transport of these species along the cell and their electrochemistry over the external surface of the carbon particles.

Very recently, Mukherjee's group at Texas A&M University (TAMU) reported a multiscale analysis of Li–S batteries, also presented in a series of presentations at the 229th Meeting of the ECS (Figure 8.4).[17] In their study, the authors used a stochastic model to reconstruct the microstructure of a carbon host matrix with impregnated sulfur. The authors calculate the associated effective transport properties (as an overall effective conductivity) by using the commercial software GeoDict,[18] but lack on predicting how the microstructural properties evolve along the discharge.

Their model is a lumped model, as Zhang et al. one, i.e. neglecting the description of the transport processes along the electrode thickness. Furthermore, the calculated evolutions of the polysulfides concentrations along the discharge are not reported. Their simulations reveal that the sulfur loading and the C-rate impact the activation overpotentials but not the maximal capacity.

Continuum models have so far focused the attention to the simulation of the electrochemistry and polysulfides and ionic transport. Only very recently one pioneering work, again by Mukherjee et al. at TAMU, has been reported on the modeling of the mechanical stresses due to the solid sulfur dissolution and precipitation (Figure 8.5).[19] In their model, the authors assume pores which can shrink or swell to accommodate the changes in the pore volume resulting from the electrolyte-induced hydrostatic pressure. The results are then incorporated in their lumped

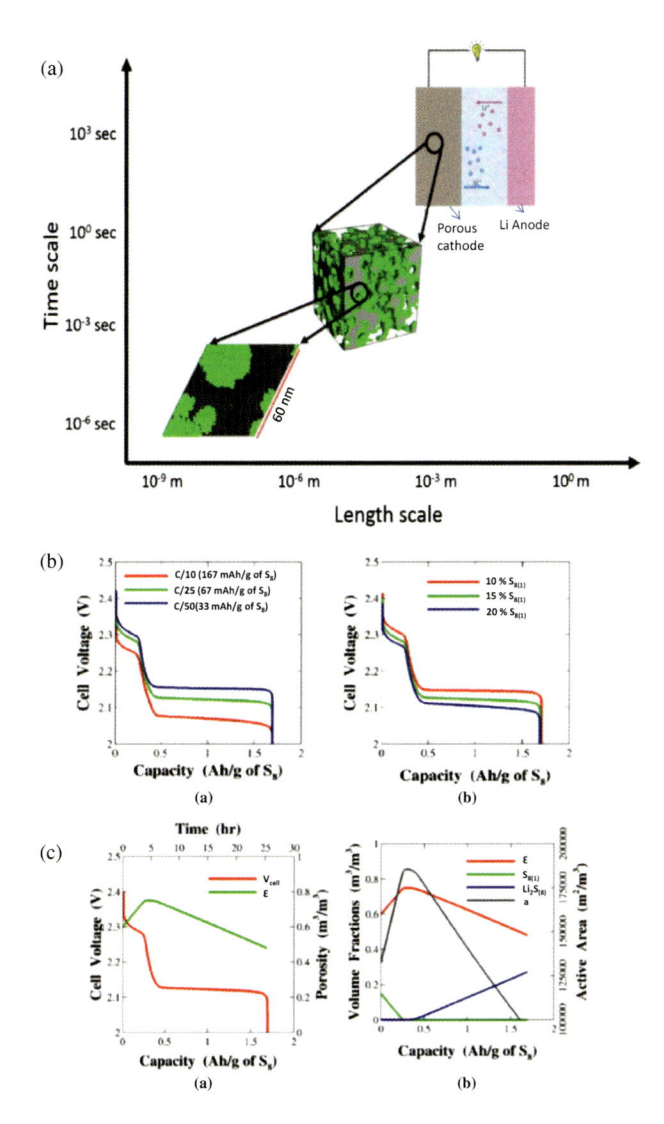

Figure 8.4 (a) Mukherjee's multiscale modeling approach (presented at the 229th ECS Meeting, San Diego, May 29–June 2, 2016); (b) Calculated discharge curves at different C-rates and sulfur loadings; (c) Calculated solid volume, porosity and active surface area along the discharge.

Source: Reproduced with permission from Ref. 17.

Figure 8.5 (a) The mesh used in the computational analysis of the single pore microstructure by Mukherjee et al. During dissolution, compressive stress is assumed to occur, whereas, at the time of precipitation tensile hydrostatic stress acts on the pore walls; (b) Variation in microstructure porosity during discharge of a Li–S cell at C/4 rate. Cell porosity with and without mechanical expansion is indicated using the solid and dashed line, respectively. The shape of the pore during medium compression, large compression, medium tension, and large tension is shown by A, B, C, and D, respectively.

Source: Reproduced with the permission from Ref. 19.

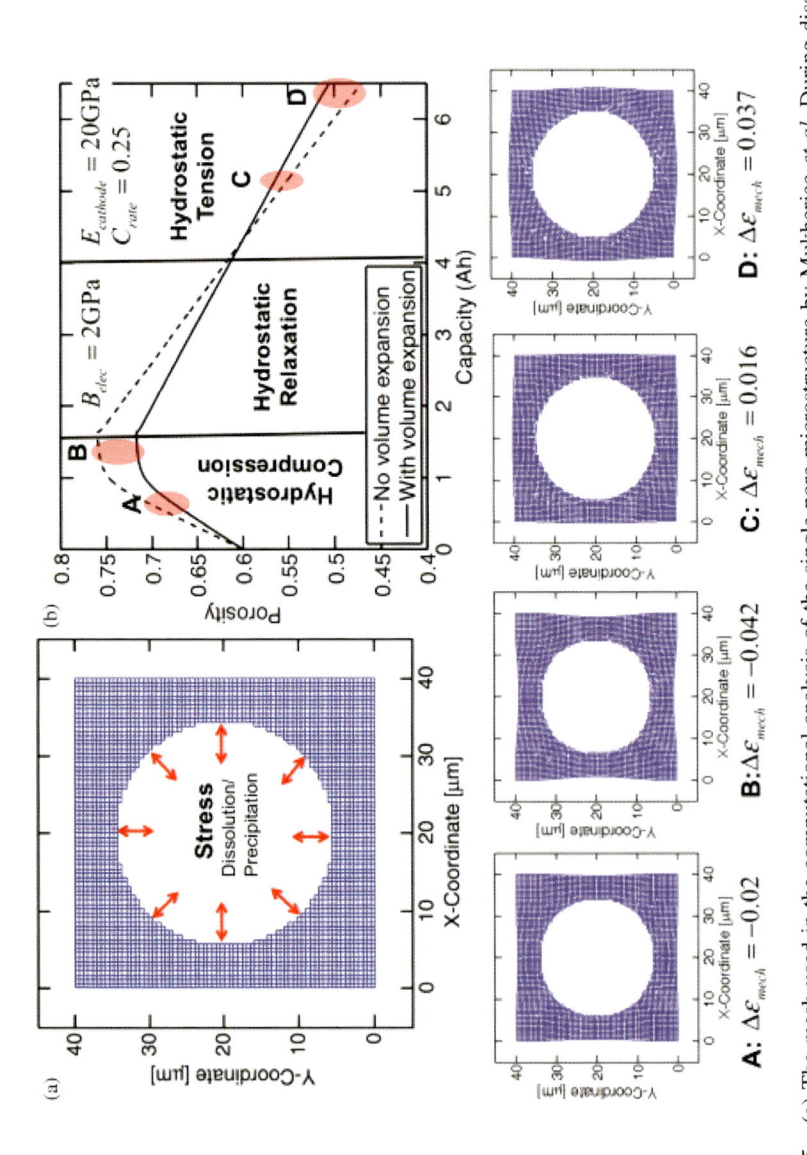

performance model to analyze the impact of volume expansion on the discharge voltage. The arising performance model captures the impact of both precipitation/reduction reactions and mechanical stresses on the macroscopic overall cathode porosity evolution. As a consequence, the authors found that the mechanical stresses impact mainly the second plateau characteristics. They also report that non-uniform precipitation may cause microcrack formation in the pore walls and lead to the creation of new paths for the Li and sulfur-based species transport.

We underline here that the model describes stress–strain in a single plane. However, out-of-plane elastic strain induced by the component of pressure along the thickness direction will change the in-plane stress distribution. The authors make this approximation for the sake of simplicity and plan to extend their model to three-dimensions in the future.

8.3. Atomistic Modeling

Going down into the scales, only very few work has been reported so far on the understanding of the polysulfides reduction chemistry at the atomistic scale. Regarding the interface between the electrolyte and the solid substrate, pioneering work has been reported by Balbuena *et al.* They reported an atomistic study of the Li_2S precipitation on graphite and silicene materials.[20] This work extends the previous work by the authors, showing by DFT that direct Li_2S deposition on Li_2S is energetically more favorable than the Li_2S_2 deposition and reduction process.[21] The authors motivate their new work by showing that the silicene may act as a polysulfide "trapping" cathode material: indeed, their DFT calculations show that Li_2S_x (x = 1, 2, 4) molecules interact more strongly with silicene (formation of chemical bonds) than with graphite. The effect of dopants on the arising adsorptions is also studied, finding that N-doped silicene can further facilitate the adsorption and reduction of Li_2S_4 and Li_2S_2. Based on further DFT calculations of the Li_2S layer structural properties as a function of the number of LiS molecules on both silicene and graphene, a Kinetic Monte Carlo model is proposed to predict the Li_2S precipitation and resulting coverage on both types of materials. This model coarse-grains the geometry of the Li_2S molecules as a bead, and it is based on a simple first-order law of deposition and Arrhenius-type description

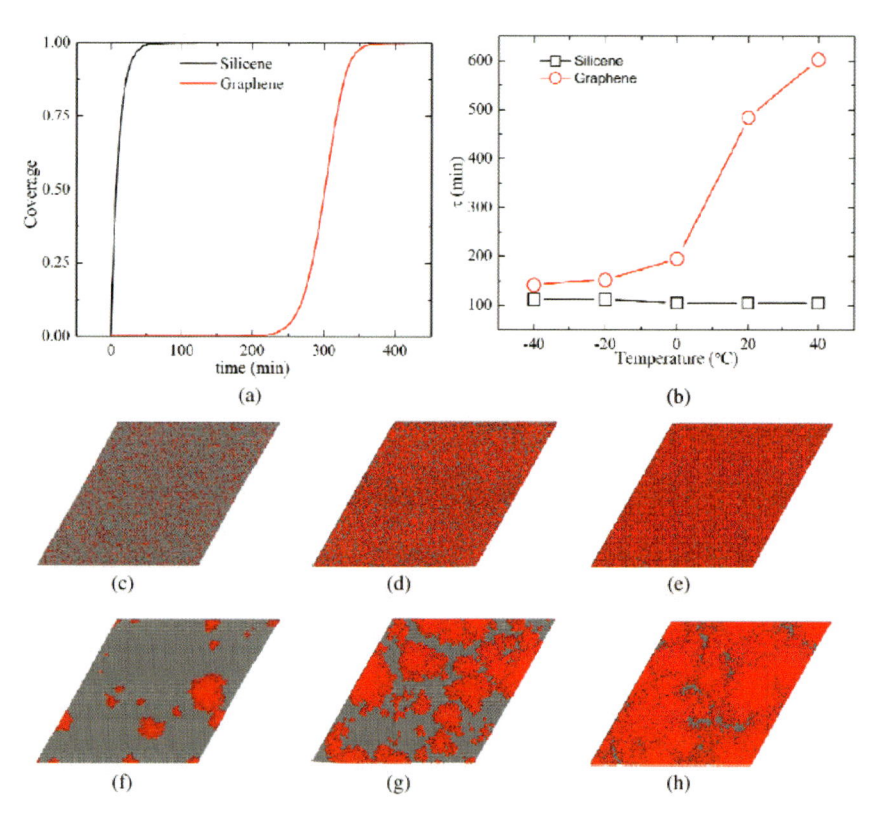

Figure 8.6 (a) Li$_2$S coverage on substrate as a function of time at 20°C. (b) The time required for the substrate getting fully covered at different temperatures; (c)–(e) calculated substrates configurations at 20°C for Li$_2$S coverages on silicene of 0.1, 0.5, and 0.9; (f)–(h) calculated substrates configurations at 20°C for Li$_2$S coverages on graphene of 0.1, 0.5, and 0.9.

Source: Reproduced with the permission from Ref. 20.

of the desorption rates as a function of the temperature and activation barrier. The authors find that Li$_2$S coverage increases faster on silicene than graphene (Figure 8.6). They conclude that silicene-based cathode will suffer greater surface passivation than graphene-based cathode, and plan further modeling studies to avoid this passivation by maintaining reasonable levels of Li$_2$S coverage. From the methodological point of view, their Monte Carlo model does not take into account the electrochemical conditions of the interface (electric field is neglected, and assumed Li and

S^{2-} concentrations are set to constant values). The parameterization of the model is not fully first-principles (the value for the deposition rate parameter is actually assumed).

Within a similar scope, but by employing a combination of DFT and *ab initio* Molecular Dynamics (AIMD) calculations, thus remaining within the atomistic view, Balbuena *et al.* reported an interesting theoretical study of the Li_2S formation on Li metal.[22] This investigation, with relevance for developing a better understanding of the shuttle reactions, leads to conclude that the formation of Li_2S film formation on Li metal is thermodynamically favorable. Other materials were also studied in another publication, namely nanopores of graphene, MoS_2 and Mo-doped graphene, MnO_2 and Fe_2O_3.[23] It is found that Mo-containing materials and oxides are more reactive towards the polysulfides adsorption than graphene nanopores.

Furthermore, van Duin *et al.* reported a Reactive Force Field potential of the Li–S interactions, and performed molecular dynamics simulations for various structural, mechanical, and diffusion properties in Li_xS compounds.[24] However, it is technically challenging to transfer their approach for simulations of full Li–S batteries.

8.4. Conclusions and Open Challenges

In conclusion, there is a very little amount of theoretical studies of the Li–S batteries reported in literature from both the continuum and the atomistic viewpoints. This may be justified by the challenging physico-chemistry which these types of batteries are offering to us, underlying the need for highly multidisciplinary modeling approaches combining aspects of chemistry, electrochemistry, transport phenomena and even thermo-mechanics.

In particular, the models reported at the continuum level are still not able to predict important features from an engineering perspective, such as the impact of the initial sulfur spatial distribution within different scales of porosity onto the overall cell performance. Other aspects, such as the impact of the carbon particle size distribution and the cathode microstructure evolution along the discharge (and charge) have not been investigated in depth.

Generally, the published continuum models show large discrepancy when fitted to experimental discharge curves. Indeed, parameterization of the continuum models on these systems is not a trivial task, and may generate some controversies. For instance, Ghazvani, in his PhD manuscript writes, when referring to White *et al.* model (Ref. 25, page 11):

"Nevertheless, their results corresponded well with a class of experimental results. However, implementing their model with the given parameters, one realizes that the reported parameters cause the simulation result to diverge."

Therefore, there is a strong need of model experiments to feed the models developments. More efforts should also be put on the description of the shuttle reactions[4,10,26] and other degradation mechanisms.

Thanks to the development of the modern computational science over the past few decades, multiscale modeling and numerical simulation are emerging as powerful tools for *in silico* studies of mechanisms and processes in other types of electrochemical devices, such as fuel cells, Li-ion and even, since more recently, Li–O_2 batteries.[27,28] These innovative approaches, for which we have a strong expertise, allow linking the chemical/microstructural properties of materials and components with their macroscopic efficiency. In combination with dedicated experiments, they can tremendously support the progress in designing and optimizing the next generation of Li–S cells through the development of more predictive and more reliable physics-based models: efforts within this sense involving us have already started.[29] For instance, we have recently reported a multiscale model resolving the impact of the cathode microstructure onto the discharge performance of Li–S cells.[30] Among other novel aspects, this model allows investigating the impact on the electrochemical response of the sulfur repartition within the cathode. Polysulfides transport within different scales of the carbon porosity is taken into account explicitly. It is found that inhomogeneous Li_2S formation upon discharge can occur along the cathode thickness. We believe that this model and results contribute at paving the way towards the computational-based optimization of Li–S electrodes at multiple scales.

Bibliography

1. A. Fotouhi, D. J. Auger, K. Propp, S. Longo and M. Wild, *Renewable and Sustainable Energy Reviews*, **56** (2016) 1008.
2. A. A. Franco, Fuel cells and batteries in silico experimentation through integrative multiscale modeling. In: *Physical Multiscale Modeling and Numerical Simulation of Electrochemical Devices for Energy Conversion and Storage*, Springer, London, (2016) pp. 191–233.
3. Retrieved from http://www.sionpower.com/.
4. Y. V. Mikhaylik and J. R. Akridge, *Journal of the Electrochemical Society*, **154**(11) (2004) A1969.
5. K. Kumaresan, Y. Mikhaylik and R. E. White, *Journal of the Electrochemical Society*, **155**(8) (2008) A576.
6. J. S. Newman and C. W. Tobias, *Journal of the Electrochemical Society*, **109**(12) (1962) 1183.
7. J. S. Newman and W. Tiedemann, *AIChE Journal*, **21**(1) (1975) 25.
8. J. P. Neidhardt et al., *Journal of the Electrochemical Society*, **159**(9) (2012) A1528.
9. D. N. Fronczek and W. G. Bessler, *Journal of Power Sources*, **244** (2013) 183.
10. A. F. Hofmann, D. N. Fronczek and W. G. Bessler, *Journal of Power Sources*, **259** (2014) 300.
11. M. Ghaznavi and P. Chen, *Journal of Power Sources*, **257** (2014) 394.
12. M. Ghaznavi and P. Chen, *Journal of Power Sources*, **257** (2014) 402.
13. M. Ghaznavi and P. Chen, *Electrochimica Acta*, **137** (2014) 575.
14. Retrieved from http://www.oxisenergy.com/.
15. T. Zhang, M. Marinescu, L. O'Neill and M. Wild, Gregory offer, *Physical Chemistry Chemical Physics*, **17** (2015) 22581.
16. T. Danner, G. Zhu, A. F. Hofmann and A. Latz, *Electrochimica Acta*, **184** (2015) 124.
17. A. D. Dysart, J. C. Burgos, A. Mistry, C. Chen, Z. Liu, C. N. P. Balbuena, P. Mukherjee and V. G. Pol, *Journal of the Electrochemical Society*, **163**(5) (2016) A730.
18. Retrieved from http://www.geodict.com/.
19. P. Barai, A. Mistry and P. P. Mukherjee, *Extreme Mechanics Letters*, **9**(3) (2016) 359.
20. Z. Liu, P. B. Balbuena and P. P. Mukherjee, *Journal of Coordination Chemistry*, **69**(11–13) (2016) 2090.

21. Z. Liu, D. Hubble, P. B. Balbuena and P. P. Mukherjee, *Physical Chemistry Chemical Physics*, **17**(14) (2015) 9032.
22. Z. Liu, S. Bertolini, P. B. Balbuena and P. P. Mukherjee, *ACS Applied Materials & Interfaces*, **8**(7) (2016) 4700.
23. E. P. Kamphaus and P. B. Balbuena, *The Journal of Physical Chemistry C*, **120**(8) (2016) 4296.
24. M. Islam, A. Ostadhossein, O. Borodin, A. T. Yeates, W. W. Tipton, R. Hennig, N. Kumar and A. C. T. van Duin, *Physical Chemistry Chemical Physics*, **17**(5) (2015) 3383.
25. M. Ghazvani, Continuum modeling of two battery systems: Lithium-Sulfur, and Rechargeable Hybrid Aqueous Cells, PhD manuscript, University of Waterloo (2016).
26. S. M. Al-Mahmoud Saddam, J. W. Dibden, J. R. Owen, G. Denuault and N. Garcia-Araez, *Journal of Power Sources*, **306** (2016) 323.
27. A. A. Franco, M. Alfredsson, D. Brandell, C. Frayret, M. Gaberscek, S. Islam and P. Johansson, Boosting rechargeable batteries R&D by multiscale modeling: Myth or reality? *Chemical Reviews*, submitted (2016).
28. A. A. Franco, *RSC Advances*, **3**(32) (2013) 13027.
29. Retrieved from www.helis-project.eu.
30. V. Thangavel, K. H. Xue, Y. Mammeri, M. Quiroga, A. Mastouri, C. Guéry, P. Johansson, M. Morcrette and A. A. Franco, *Journal of the Electrochemical Society*, **163**(13) (2016) A2817–A2829.

Chapter 9

Conclusion: Challenges and Future Directions

Elton J. Cairns

*Chemical and Biomolecular Engineering Department,
University of California, Berkeley, CA 94720, USA
Lawrence Berkeley National Laboratory, Berkeley,
CA 94720, USA
ejcairns@lbl.gov*

This book provides the reader with an excellent survey and review of the state of knowledge of the lithium–sulfur (Li–S) cell at ambient temperature. This system is still in the early stages of development and commercialization, so is not easily compared to the mature Li-ion technology. Li–S offers the opportunity to develop a commercial system that provides a specific energy much higher than that of the best Li-ion system, and much safer, at a significantly lower cost.

Chapter 2 reviews the many approaches that have been explored for the structure and composition of the sulfur electrode. This topic is still under active investigation, with several approaches showing various degrees of promise. In general, sulfur electrode structures with very short Li diffusion distances (less than one micrometer), well-distributed current collectors, and accommodation of the expansion and contraction accompanying the S–Li$_2$S conversion are showing the best cycling stability. Within the last few years, some investigators have reported cycle lives in excess of 1000 cycles, but with low sulfur loadings (~1 mg S/cm^2), and low sulfur contents (<70% S) in the electrode mixture (active material, conductive additive, and binder). Progress towards a commercial Li–S cell will require much higher S loadings (4–15 mg S/cm^2) and high sulfur contents (>70% S), while maintaining good S utilization (at least 1000 mAh/g S), long cycle life (~1000 cycles), and good rate capability (C/3 to C, depending on application).

The important topic of the most appropriate electrolyte is still unsettled. It is clear that the electrolytes that work best for Li-ion cells are not suitable for Li–S cells. This topic is not very deeply explored yet, and it can be expected that there will be new developments that provide longer cycle lives and wider operating temperature ranges than have been reported to date. Important issues here include the solubility of polysulfides, the conductivity (Li$^+$ transport rate), and the stability of the electrolyte in contact with Li metal and the sulfur electrode.

The promise of very high specific energy for Li–S cells depends on the use of Li metal electrodes, which form dendrites when used in many non-aqueous electrolytes. Various approaches for avoiding the formation of Li dendrites during the charging process have been reported with only a few being promising for real-world applications. The formation and maintenance of a protective SEI layer is a common theme, and is important for good cycling stability. It is clear that the electrolyte plays an important role in the formation and maintenance of a stable, renewable, and protective SEI layer on the Li metal.

The general strategy for producing a high specific energy cell must include the minimization of the weights of all electrochemically inactive components of the cell, including current collectors, tabs, separators, electrolyte, and cell container. A commonly overlooked but important weight is that of the electrolyte. For Li–S cells, the electrolyte: sulfur weight ratio is a key variable, which should be minimized. A reasonable target is a ratio of about 3; the current literature reveals much higher ratios, often as high as 50 or more. The cell container should not weigh more than about 15% of the total cell weight.

The progress represented by the work reviewed in this book is very encouraging. Long cycle lives and excellent rate capability have been achieved with low-loading electrodes. These excellent results now should be extended to more practical sulfur loadings in cells that approach commercial requirements of specific energy and cost. Some formulations such as those using ionic liquid-containing electrolytes offer significant safety advantages. The future of the Li–S cell looks very promising as the next generation of rechargeable cell beyond Li-ion cells.

Index

A

ab initio molecular dynamics (AIMD), 347
absorbance, 278
absorbing materials to trap, 71
absorption spectra, 49
absorptivity coefficient, 277
activation process, 126
adhesion, 177
adsorption agents, 158
Al_2O_3 coatings, 245
allotropes, 311
alloy anodes, 138
all-solid-state Li–S batteries, 171, 248, 250
anionic block copolymer electrolyte, 259
aprotic solvents, 14
aqueous electrolyte, 165, 167
argyrodite-type, 175
artificial SEI, 240
asymmetric cells, 214
atomic layer deposition (ALD), 245
atomistic modeling, 345
atomistic/molecular approaches, 336

B

battery pack, v
battery pack price, 12
beam damages, 208
beam-induced complications, 213
Beer–Lambert law, 277
binders, 81

C

carbide-derived carbon, 326
carbonate-based electrolytes, 119
carbonate based solvents, 153–154
carbon dioxide, 1
carbon nanotubes, 13
ceramic electrolytes, 250
ceramic glass membranes, 252
ceramic ionic conductors, 249
characterization tools, vi
charge transfer resistance, 46, 69, 85, 218
chemical protection, 234
coin-cells, 159
combustion engine, v, 2
common ion effect, 327
conducting polymer, 68

355

confinement strategies, 4
continuum approaches, 336
continuum modeling, 337
control experiments, 216
coordination number, 295, 303
coulometer, 215
coulometry technique, 214
cumulative charge, 293
current collector, 33–35, 58, 76, 81–87, 90, 231
cyclic voltammetry, 7
cyclovoltammogram, 292

D
1D structured carbonaceous materials
  CNFs, 53
  CNTs, 53
dendrite, 6, 20, 196, 198–200, 209–213, 352
dendrite growth, 230, 243–244
density functional theory (DFT), 336, 345
DFT simulations, 164
diagnostic schemes, 335
dielectric constant, 119, 202
dissociation ability, 203
dissolution of polysulfides, 5
double electrolyte, 168
driving range, 2, 10, 24
dual-layer carbon electrode, 114

E
Einstein relation, 340
Einstein's theory of viscosity, 160
electric traction, 2
electrochemical impedance spectroscopy (EIS), 46, 215
electrochemical stability window (ESW), 151

electrode interfaces, 4
  SEI, 6
  solid electrolyte interphase, 6
electrode swelling, 82
electrolyte additive, 4, 122, 223
  $LiNO_3$, 6
electrolyte formulations, 152
electrolyte-to-sulfur ratio, 12, 257
electron affinity, 121
electronegativity, 220
electron pair donor solvents, 119
energy dispersive X-ray spectroscopy, 237
excited state, 278
ex situ, 159, 163, 240, 279

F
Fermi energy, 199, 201
fluorescence/phosphorescence, 277
fluorinated ethers, 220
fluoroethylene carbonate (FEC), 136
fossil fuels, 1
fourier transform infrared (FTIR), 207
free-standing electrode, 324
functional group, 45–46, 49–53, 62, 68, 71

G
gel polymer electrolytes, 247
Gibbs energy, 10
Gibbs energy change, 177
glass-ceramic electrolyte, 180, 210
glass electrolytes, 173, 177
global warming, 1
grain boundary, 175, 177
ground state, 278

H

hard and soft acids and bases (HSAB), 177
heteroatom-doping, 324
high energy ball milling, 184
high temperature ball milling, 184
history of Li–S chemistry, 3
hydrofluoroether, 119
hydrothermal method, 37

I

identification of the electrolyte formulation, 4
impedance, 206, 225
infrared spectroscopy, 278
in operando, 279, 288
in operando techniques, 304
inorganic solid electrolytes, 171–172
in situ, 163, 207, 209, 279
in situ XANES, 317
insulating nature of sulfur, 13
    carbon skeletons, 13
    doped carbons, 13
    graphene, 13
    graphene oxides, 13
    hollow carbon nanofiber, 13
interlayer concept, 87
ionically-conductive organic films, 242
ionization potential, 121
ion-selective polymer membrane, 158, 169
    Nafion membrane, 169

K

key players in the field, 22

L

large-sized cells, 114
Li conducting polymer interface, 242
Li deposition/stripping, 226
Li-ion conductive coatings, 235
Li-ion sulfur batteries, 131
Li metal plating/stripping, 213
linear combination fits (LCF), 301
Li phosphorous oxynitride, 236
Li–S dioxide, 107
lithiated graphite, 132
lithium–thionyl chloride, 107

M

magnetic field-controlled cell, 112
magnetic resonance spectroscopy, 277
mechanochemically, 184
mechanochemistry, 129
metal-organic frameworks (MOFs), 36, 75
methods to fabricate sulfur carbon nanocomposites, 37
    high energy mechanical ball-milling, 37
    infiltration method, 37
    precipitation of sulfur, 37
    wrapping sulfur with 2D carbonaceous materials, 37
modeling of Li–S batteries, vi
modeling of the Li–S cell, 8
modeling or theory, 336
molecular dynamics calculations, 222
Monte Carlo, 345
Monte Carlo model, 346
Monte Carlo simulations, 228

N

Nafion, 241–242
NASICON, 175

near-edge X-ray absorption fine structure, 50
Nernst equation, 230, 337
Nernst–Planck equations, 339
Nernst–Planck formalism, 340
Newman, 337
niche applications, 12
    military purposes, 13
    space batteries, 13
    unmanned aerial vehicles, 13
nuclear magnetic resonance (NMR), 211
    chemical shift, 211
non-polar solvents, 220
nucleation of the polysulfide species, 126
nucleophilic attacks, 154

O
Ohara, 252–253, 257
Ohara Corporation, 112
operando, 207, 209, 215
organosulfur compound, 77–78
other sulfur based nanocomposites
    organosulfur compounds, 71
    sulfur metal oxides, 71
    sulfur MOF, 71
over potential of Li_2S, 125–126
Oxis Energy, 12, 340

P
patent landscape of Li–S batteries, 3
phase transition, 315
physical protection, 240
plating, 200
polarization, 213–214, 220, 223, 241
polymer coatings, 240
polysulfide binding effect, 73
polysulfide dissolution, 4
    absorption, 4
    adsorption, 4
    redox flow, 4
polysulfide reservoir(s), 35, 39, 41, 72–73, 75, 80, 87
    adsorption, 36
    porous carbon, 36
polysulfides dissolution, 7
pouch cells, 159
protected lithium electrode, 253
protection layer, 35, 61, 67, 72, 80
protection of Li metal anodes, 124
protective coating, 64, 76
protective layer, 112, 128, 131, 157, 225, 235, 239

Q
qualitative, 288
quantitative determination, 288

R
Raman shift, 320
Raman spectroscopy, 163, 277
redox mediators, 121–122, 126–127
reflection curve, 280
ring-opening polymerization, 204
rotating ring disc electrode (RRDE), 277

S
SEI-forming electrolyte, 157
Selenium electrochemistry, 311
self-discharge, 17–19
self-healing, 4
semiconductor, 310–311

semiliquid battery, 113
semiliquid redox flow batteries, 115–116
shuttle constant, 15, 18
shuttle factor, 15
silane-based coatings, 237–238
silicon anodes, 134
    columnar amorphous silicon structures, 135
silicon electrodes, 260
Sion Power, 12, 22, 337
solid electrolyte interphase (SEI), 14, 19, 198, 352
solid electrolytes, 247
solid–solid interfaces, 171
solid-solutions, 183
solvation properties, 203
spectroscopy techniques, 163
stabilized Li metal powder (SLMP), 129, 132
static condition, 109–110, 120
steric hindrance, 220
stripping, 200
stripping and plating, 228
sulfide glass, 176
sulfur fraction, 12
sulfur impregnation, 4, 16
sulfur loading, 12, 20–21
sulfur–polymer nanocomposites
    conducting polymers, 62
    poly(3,4-ethylenedioxythiophene), 62
    polyaniline, 62
    polypyrrole, 62
    polythiophene, 62
superconcentrated solutions, 221
surface chemistry, 109, 136, 140
surface growth mechanism, 6

sustainable road transport, 13
Swagelok cell, 159, 291, 293, 295
symmetrical cells, 213–215, 225, 232, 236, 241, 247
synchrotron, 297

T
thermodynamically stable, 320
thermodynamic stability window, 166
thermo-mechanics, 347
thick electrodes, 12
thick sulfur electrodes, 20, 114
thick sulfur positive electrodes, 228
Tin anodes, 136
transference number, 160
transmission electron microscopy (TEM), 322
transport phenomena, 347

U
unique ultrahigh vacuum (UHV), 246
UV–visible (UV–Vis) spectroscopy, 7, 277, 279, 283, 304

V
vinylene carbonate (VC), 225
viscosity, 153, 159–160, 184

X
X-ray absorption near edge spectroscopy (XANES), 295, 298, 319
X-ray absorption spectroscopy (XAS), 119, 164, 277, 295, 299
X-ray diffraction (XRD), 277
X-ray emission spectroscopy, 50
X-ray photoelectron spectroscopy (XPS), 50, 237, 277